Ammonia emissions in agriculture

Ammonia

emissions in agriculture

edited by:
Gert-Jan Monteny
Eberhard Hartung

Wageningen Academic
Publishers

ISBN: 978-90-8686-029-6

First published, 2007

Wageningen Academic Publishers
The Netherlands, 2007

Preface

On 19, 20 and 21 March 2007, Wageningen UR (The Netherlands) and the University of Kiel (Germany) organised the *International Conference on Ammonia in Agriculture: Policy, Science, Control and Implementation*. It addresses ammonia in agriculture in a broad sense; from the policy on ammonia inventory and abatement in different world regions to measuring methods and mitigation strategies.
The conference was sponsored by the Commission International du Genie Rurale (CIGR), the European Association of Agricultural Engineers (EurAgEng), and the Dutch Association of Agricultural Engineers (NVTL).
During the conference, around 180 representatives from a wide variety of countries gathered to present and discuss the state of the art in ammonia related science, and to debate solutions with representatives from governmental bodies and stakeholders. The following policy related questions were formulated as input to the conference, the workshops, and the discussions:

1. Techniques and abatement strategies:
   a. Main categories of abatement options; latest developments and costs.
   b. How to implement on farm level (and verify, control)?
   c. Most important pollution swapping issues (farm level), integrated measures (animal welfare, NO_3, etc).
   d. Contribution of animal feeding and animal efficiency.
   e. Relevance for application in other parts of the world (see also 3).
2. Emission measurement and inventories:
   a. How to improve.
   b. Evolution of NH_3 sources in the past and in the future.
   c. Progress in measurement technology.
   d. Sources other than agriculture (e.g. cars, natural sources).
3. Communication and exchange of information:
   a. Efforts (policies) needed to achieve global level playing field.
   b. Adaptation of inventory methods on global level.
   c. How to reach society, communities, practice, consumers with clear and objective messages concerning reactive N (e.g. address the impact of meat consumption and alternatives).
4. (Political) instruments:
   a. Assessment of current policies and instruments on reactive N.
   b. Alternatives for critical loads (e.g. air quality limits).
   c. Impact of further liberalisation of agriculture, biofuel scenario's.
   d. How to co-ordinate various instruments.

This proceedings book contains an overview of all papers presented, the open space workshops, and the discussions. Furthermore, a final conference statement is included, summarising the major outcome of the conference.

A sincere word of 'thanks' is addressed to all of the attendees to the conference. Especially the Scientific Committees' work is highly appreciated, since they contributed to the scope and structure of the conference, and to the reviewing of the technical papers.

The Organising Committee put major effort in the process and content during the Conference, and has successfully formulated the conference statement. In this perspective, the contributions of Aad Jongebreur, Dennis Schulte, Eberhard Hartung, and Marry van den Top – assisted by Corien Fopma and Hedy Wessels – is greatly appreciated, and has made the conference an unforgettable and important event.

On behalf of the Organising and Scientific Committee, I sincerely hope and trust that the outcome of the conference contributes to the process of interaction between science, policy makers, and other stakeholders, to contribute to sustainable agriculture.

Gert-Jan Monteny
Wageningen University

Table of contents

Invited papers

Ammonia emission: history and future of research, the effects of research and the relation between policy development and research

A.A. Jongebreur
President of the Dutch Association of Agricultural Engineers (NVTL) and president of EurAgEng

The first signals that ammonia is a threat for the environment were already found in the 19th century, when in the direct vicinity of livestock houses damage to the vegetation was observed. It was also wellknown that ammonia in the soil through the process of nitrification contributes to the acidification of the soil. Another fact was the observation of the presence of ammonia in rain, by scientists in those period considered as fertiliser for free (of charge) from the atmosphere. However the recognition of the ammonia emission problem as a real threat to nature, but also to sustainable production methods can be marked by the wellknown article in Nature 25 years ago: Soil acidification from atmospheric ammonium sulphate published by Van Breemen *et al.* (1982). High concentrations of ammonium were measured in the water of the troughfall of the canopy of trees. This discovery was not only of importance from the view of plant nutrition and ecology but also for the sources of ammonia emissions, the livestock production units. On the occasion of the 25th anniversary of a weekly magazine, Science and Education the first author of this important article states that all inventions which are really innovative in science are discovered by coincidence (Anonymous, 2007).
From that time more emphasis on ammonia in research and policy making was a fact. Untill 1982 the urgent environmental topics related to livestock production were odour emission, water pollution through eutrophication and leakage of nitrate and phosphate due to the excessive amounts of manure applicated. In the next period many research projects of universities and institutions were started with respect to the ammonia emission. The objectives of the projects were the quantification and the qualitative and quantitative effects of the emissions.
In 1990 one of the first international meetings on ammonia was organised by the Commission of the European Communities. Its objective: to review the present state of knowledge concerning the measurement and control of ammonia and odours emissions from livestock production (Nielsen *et al.*, 1991). You can conclude that research in Europe had got priority in quite a number of countries of the European Community. It is interesting to notice that in the conclusions and recommendations of this conference the reliable measurement methods and protocols of ammonia concentrations and air volumes from naturally ventilated houses were key problems. Still questions of current interest, I should think.
From the nineties in the last century on research could hardly meet the high demand for facts and figures from policy makers. On a national basis there are good examples of co-operation between policy development and research e.g. in the area of slurry incorporation, the coverage of manure silos. Improving the accuracy of measurement methods of ammonia emissions played a crucial role in that period!

I already stated that a large number of projects started in the areas of animal nutrition, application of manure, housing systems, on-farm handling of manure and economic evaluation. An example of a co-operative EU-project was 'Emissons of Aerial Pollutants in Livestock Buildings in

Northern Europe' (Wathes *et al.*, 1998) and was by far the most comprehensive study of its kind with the help of 329 livestock producers, carried out in the years 1992-1996. Not only ammonia concentrations were determined but also airborne dust concentrations and microorganisms concentrations. Moreover, a cleverly designed measurement set up resulted in a valuable data base about emissions of various gases and substances. Cattle was found to be the largest source of ammonia emissions and this added weight to the concern of this enivironmental pollutant. It striked me that the greatest uncertainty in the estimates of the emission rates was the calculation of the ventilation rates, especially in naturally ventilated buildings. This study also stressed the importance of the concentration of aerial pollutants in pig and poultry houses which may be harmful to human and animal health.
In the period after 1995 the priority in R&D remained on volatile emissions from livestock business but the greenhouse gas emissions and emission of fine dust particles became more actual, and research on ammonia emission, prevention and control continued on a somewhat lower capacity. It may be noticed that ammonia emission and control research in the different states of the USA became urgent in the period after 1995 with the major reason the increase and scaling–up of the livestock business. Whereas in the seventies environmental research on manure and emissions had high priority in USA before this was the case in Europe.

The latest knowledge and developments on ammonia emission, prevention and control come available on this international conference and the relation with the EU, UNECE and more or less national policies will be discussed.
National policies, EU regulations and UNECE protocols influence the future of livestock production.The treaty of Rome, which started the development towards the European Union as we know today celibrates its 50th anniversary next week. The emphasis on the environmental problems did not start untill the beginning of the 70's. Important to mention is the EU directive from April 22th 1999 (Anonymous, 1999), where the National Emission Ceilings(NECs) are mentioned which originate from the Gorthenburg protocol (UNECE, 1999). It regulates the maximum emission on ammonia, other nitrogen oxides (NO_x) and volatile organic compounds (VOCs) and SO_2 for 2010. Under this heading the EU Directive on integrated pollution prevention and control (IPPC), which has important consequences for the ammonia emission on the large units with intensive pigs and poultry production must be included. For the future the 'Clean Air for Europe' programme is the scientific basis for the strategy in the EU policy. In this programme a further reduction of the targets of the emission ceilings on ammonia, nitrous oxide, nitrogen oxides, volatile organic compounds, SO_2 and particulate matter is foreseen for 2020 and 2030.
Also, the UNECE has in 2004 in celebrated the 25th anniversary of the Convention on Long Range Transboundary Air Pollution (CLRTAP), which includes protocols on the abatement of – amongst others – ammonia (UNECE, 1999). A growing interaction between various international air pollution abatement bodies across the globe is of major importance to successfully abate volatile emissions.

The public concern of the quality of our living environment has increased and the demand for sustainable production and consumption is growing. The expected effects of the climate change – caused mainly by human activities as increased mobility and consumption – is bothering many people in the world of today.

To achieve the high ambitions of ammonia emission, prevention and control (lower than the National Emission Ceilings for 2010) on the longer term the attitude of the livestock producer is crucial. I am convinced that livestock producers are ready to make steps for more sustainable production methods. In this frame we must be aware that not only the targets for ammonia emission reduction or even voaltile emissions must be met but also adaptations in the production methods from the view of animal welfare and health must be made. Therefore for the future an integral approach along the trple P concept: Planet, Profit and People (Slingerland *et al.*, 2003) may be helpful. In simple words smart combinations of prevention and abatement technology must be found. It would be a great achievement if, lets say 15 years from now, the technologies that meet those demands could be regarded as elements of Good Agricultural Practice, with a clear view on the benefits, instead of additional measures that put a burden on agriculture.

In the beginning of my speech I have mentioned the breakthrough /inovation in the thinking of plant nutrition science. We need more breakthroughs and innovations to achieve the goals of air quality levels which meet the standards of the triple P concept 'Planet, Profit, and People', representing ecological, economic and socio-cultural values.

On behalf of the Dutch Association of Agricultural Engineers and the European Society of Agricultural Engineers I wish all participants a fruitful conference.

References

Anonymous, 1999. Council Directive 1999/30/ECof 22 April 1999 relating to limit values for sulfur dioxide, nitrogen dioxide, and oxides of nitrogen, particulate matter and lead in ambient air. Official Journal of the European Communities L163, 29/06/1999, 41-60.

Anonymous, 2007. Vieze Stikstof. NRC Handelsblad 24 Februari 2007, Wetenschap & Onderwijs 25 jaar, p.4.

Nielsen,V.C., J.H. Voorburg and P. L'Hermite (Eds.), 1991. Odour and Ammonia Emissions from Livestock Farming. Elsevier Science Publishers LTD, ECSC, EEC, EAEC, Brussels and Luxembourg. 221 pp.

Slingerland, M.A., J.A. Klijn, R.H.G. Jongman en J.W. Van der Schans, 2003. The unifying power of sustainable development. Towards balanced options between People, Planet and Profit in agricultural production chains and rural land use: the role of science. Wageningen Universiteit: WUR Report Sustainable Development. p. 29-33.

UNECE, 1999. Protocol to the 1979 Convention on Longe Range Transboundary Air Pollution to abate acidification, eutrophication and ground level ozone, Geneva.

Van Breemen, N., P.A. Burrough, E.J. Velthorst, H.F. Van Dobben, T. De Wit, T.B. Ridder and H.F.R. Reinders, 1982. Soil acidification from atmospheric ammonium sulphate in forest canopy throughfall. Nature 299: 548-550.

Wathes, C.M., V.R. Philips, M.R. Holden, R.W. Sneath, J.L. Short, R.P. White, J. Hartung, J. Seedorf, M. Schröder, K.H. Linkert, S. Pedersen, H. Takai, J.O. Johnsen, P.W. Groot Koerkamp, G.H. Uenk, J.H.M. Metz, T. Hinz, V. Caspary and S. Linke., 1998. Emissions of Aerial Pollutants in Livestock Buildings in Northern Europe. Overview of a Multinational Project. J. Agric. Eng. Res. 70: 3-9.

Towards sustainable ammonia emission levels

K. Plug
Ministry of Environment, The Netherlands

Ladies and gentlemen,

As the representative of the Ministry of Housing, Spatial Planning and the Environment, I welcome you all to Ede. It is good to see that so many people have come to talk about this important subject. I hope the Conference will help bring policy and science closer to each other.

At current levels, ammonia emissions are harmful for nature. In The Netherlands and other parts of the world, nitrogen depositions exceed the critical threshold of what native vegetation can bear. The result is loss of biodiversity. In 2010, only twenty per cent of the total Dutch nature area will not be burdened by excessive nitrogen. Only there will our biodiversity be adequately protected. Ammonia is just one of the many forms of reactive nitrogen that threaten our natural environment. Nitrogen oxide emitted by industry and traffic also contribute to nitrogen deposition. That is why we need an integrated approach to the nitrogen problem. We know that reactive nitrogen can be converted from one form into another. It spreads easily through water, air and soil. Loss of biodiversity is only one of the problems it can cause. It also has harmful impacts on human health and aggravates climate change. This is known as the cascade effect. The negative impacts continue as long as the nitrogen is active and stop only when the reactive nitrogen is stored for a very long time or is converted back to non-reactive nitrogen.
This is why we need an integrated approach to nitrogen. Without it, ammonia policy could lead to pollutant swapping. Ideally, we should aim for policy that prevents the excess production of reactive nitrogen. Feed measures are a good example. I will talk more about those shortly.

I am happy to say that the European Commission and UNECE understand the need for an integrated approach. The Commission is looking at how the various nitrogen directives can be harmonised. And UNECE is investigating how it may apply the new scientific findings. Other speakers will also address the need for integrated nitrogen later on today.
We cannot talk about ammonia without addressing livestock production. In the Netherlands, our 13 million pigs, 105 million poultry and 1.5 million dairy cows are responsible for more than ninety per cent of ammonia emissions. In 2004, ammonia emissions totalled 134 million kg. Granted, that is a lot less than the record 250 million kg produced in 1985. We owe the reduction mainly to low-emission housing and manure application methods and the covering of manure storage. Our manure policy has also cut nitrogen surpluses at farm level and resulted in fewer livestock.
The outlook for 2010 is that we will manage – just barely – to stay under the national emission ceiling of 128 million kg agreed with UNECE and the EU. But forecasts are always accompanied by a measure of uncertainty, so we cannot sit back and relax. That is especially true for ammonia, because we measure more ammonia in the air then we calculate with our models. This probably means we will have to increase our national emission total. There is also a strong chance that the emission ceiling for 2020 will be lowered again. And The Netherlands wants to do more than the bare minimum. In the long term, we want to achieve sustainable levels of ammonia emission.

Scientists have calculated that an emission of 30 to 50 million kg prevents excessive nitrogen deposition and protects biodiversity. This means that, even if we adopt less ambitious goals for nature, we still have a long way to go. Conferences like this one can be important in finding new ways of achieving our long-term aim.

Until now, most technical measures have focused on emissions from intensive poultry and pig production. The latest technology is an advanced air filter system that reduces the emission of particulates, odour and ammonia from animal housing. The government has a grants scheme to promote these air scrubbers. The scheme has two aims: to help develop cheaper and more effective technology and to tackle air quality hot spots. Hundreds of scrubbers will be fitted over the next few years. This will help lower the total national emission of particulates and ammonia.
So far, policy for the dairy sector has not targeted ammonia emissions explicitly. There has been a reduction due to the smaller dairy herd and improved nitrogen efficiency. But the dairy sector's share in national ammonia emissions has increased to one third. This is the only major livestock sector where there has been little innovation and next to no application of low-emission housing. I am keen to know whether some of you have interesting ideas on this.
Besides looking at low emission technologies, policy should also take account of the fact that ammonia is deposited, for a large part, close to the source. The burden on nature is greatest in areas with high concentrations of intensive livestock farms. We make an extra effort to limit emissions close to nature areas. Perhaps we can improve our score on this point, although at the same time, we must tackle the ammonia that blankets all of The Netherlands.

As I said, more ammonia cuts could be achieved by making changes to animal feed. A trial is being carried out in the north of the Netherlands in which dairy farmers use less chemical fertiliser to grow roughage in order to reduce the animals' protein intake. The animals produce manure with a lower mineral nitrogen content, which in turn changes the protein content of the sward. Farmers are also trying different methods of surface manure application. We will have to wait a few years before we know just how much all these measures contribute to ammonia reduction.
A new development that ties in well with the integrated approach is manure digestion. In theory, this method could reduce dependence on chemical fertilisers and help to close the nutrient cycle. Another advantage of the method is that it would prevent the emission of methane from stored manure, while at the same time producing extra methane as a source of renewable energy. Manure digestion also changes the composition of the manure. After separation, the liquid fraction has a better nitrogen efficiency while nutrients in the solid fraction are biologically fixed. At the moment, we are investigating whether animal manure processed in this way could be applied under the EU Nitrates Directive.

In conclusion, ammonia is an important type of reactive nitrogen. By tackling ammonia emissions we can improve our nitrogen efficiency and protect biodiversity. But there are still many questions to be answered and options to be explored. That is why I support this conference. I hope and expect that the exchange of knowledge here will accelerate our progress towards sustainable agricultural production. Come Wednesday afternoon, I hope there will be tangible results for policymakers to start working on right away. I wish you all a productive conference.

Thank you.

Towards a strong, forward-thinking agricultural sector in harmony with its surroundings

R. Bergkamp
Ministry of Agriculture, Nature and Food Quality, The Netherlands

Ladies and gentlemen,

On behalf of the Ministry of Agriculture, Nature and Food Quality, I would like to welcome you to the 'First International Ammonia Conference'.
The international character of this conference is its most valuable asset. It's important that researchers and policymakers from around the world can meet and share their thoughts and experiences, and inspire one another. After all, research and expertise transcend national boundaries.

In the Netherlands, we're accustomed to the phenomenon of policy-driven research, and there is a close relationship between research and policy.
Let me take you back for a moment to the late 70s and early 80s, when agriculture and industry were booming sectors in Europe. Agricultural policies focused on the P of Profit. This resulted in an intensification of agriculture, the side effects of which quickly started to become visible. Agriculture and environmental groups soon became pitted against one other. Environmental groups predicted, for example, large-scale forest degradation due to acid rain, while agriculture denied there was a problem.
This conflict served as the impetus for the Ministry of Agriculture, Nature and Food Quality to fund research into the acid rain issue. After all, an insight was required into the relationship between ammonia emissions and the factors that influence these emissions. This insight was to be used as a basis for resolving the problem. The resulting research led of course to a wealth of information on manure and ammonia. But this information was rather fundamental and technical in nature and therefore could not always be applied in policy and in practice.

As the focus of policy shifted to the P of Planet, the 'Polluter Pays' principle was introduced, and technical measures were applied at the source. This shift led to requirements such as keeping manure stores covered, using low-emission application methods for animal manure, and employing low-emission housing. Research was also refocused on ammonia and on assessing the different animal housing systems. Low-emission housing – in intensive livestock production in particular – was then designated as a 'Green label system' and awarded funding. This resulted in a multitude of detailed regulations that were primarily prescriptive in nature.
Everything in this period was aimed at solving this specific problem with technological fixes. The limitations of this single-issue approach have since become clear, as the solutions brought new problems of their own. Low-emission application of animal manure, for example, impacts the soil and meadowbird populations adversely. And some low-emission housing has a negative impact on animal welfare.

Entrepreneurs, too, need more room to manoeuvre. Solutions which used to be provided in the form of specific means and methods, have now expanded to be couched in terms of targets and goals instead. And in addition to technological fixes, we now look to farming practices and operational measures for solutions. Instead of the single issue of ammonia, we now understand the importance of also considering other issues such as odour pollution, particulate matter and greenhouse gases.

We have entered a new phase in which a balance must be sought between the three P's: People, Profit, and Planet. We must consider the economic prospects for forms of agriculture that meet ecological and other requirements demanded by society. Of course, this shift also has consequences for research.

So, where do we go from here? How do we achieve a strong, forward-looking agricultural sector?

Firstly, strong, future-oriented agriculture must operate within a given environmental bandwidth. In addition to limitations relating to ammonia, the area in which agriculture can operate is bounded by other gaseous emission standards governing odour pollution, particulate matter, and greenhouse gases.

Secondly, activities must be performed within a certain ecological bandwidth. By operating within these limits, a symbiosis can again be formed between agriculture and nature.

Lastly, agriculture must operate within a certain societal bandwidth. Society's limits on agriculture can include issues of animal welfare and animal health, public health, climate concerns, maintaining tranquillity, space for recreation, and scenic rural areas. In short, consideration for the surroundings.

And what is the role of policy in this? Policy's role is to formulate a clear vision for the future of agriculture. The Ministry's publication 'The Choice for Agriculture' is one such example. The essence of this vision is to give entrepreneurs more freedom (After all, it's the entrepreneurs who ultimately make the choices in agriculture). In doing so, this vision indicates a particular course for policy to take: namely a less pronounced role than it had in the past – more of a guiding and supportive role than a prescriptive one. Expressions of policy's new role include giving entrepreneurs more freedom where possible, stimulating ideas and initiatives, and facilitating any major changes.

Policy also plays a role here, by defining the environment's carrying capacity. The ambitious agreements on cutting greenhouse gas emissions and energy saving are recent examples in this regard. These agreements are not only reflected in the new coalition agreement, but also on a European level.

So what is the role of knowledge and expertise here? A strong, forward-thinking agriculture sector not only requires expertise on the environment, ecology, and social concerns, but also needs applied knowledge that is of direct, practical use. In short, strong agriculture requires new, integral business systems designed for the real world.

This conference offers you the opportunity to exchange knowledge and experiences on familiar issues, such as assessing and calculating emissions, reducing emissions from housing and manure applications, and management measures. But in addition, this conference will also give you the chance to think about the future – a future which I believe lies in finding and strengthening connections: connections between ammonia and other gaseous emissions such as odour pollution, particulate matter and greenhouse gases, as well as the connections between society and the agricultural sector, and policy and research.

Ammonia and the UNECE convention on long-range transboundary air pollution

J. Sliggers[1] and K. Bull[2]
[1]Vice-chair of Working Group on Strategies and Policies of the UNECE Convention on Long-range Transboundary Air Pollution. Ministry of Environment, The Hague, The Netherlands
[2]Head of the secretariat for the UNECE Convention on Long-range Transboundary Air Pollution. Geneva, Switzerland

Introduction

The Convention on Long-range Transboundary Air Pollution has a long history of developing international emission controls but has only relatively recently sought cuts of ammonia emissions through its protocols. This paper provides the background to the work of the Convention and looks in detail at its current work on nitrogen. It considers the important issues still to be addressed and looks to future action at the international level.

The Convention and its protocols

The Convention on Long-range Transboundary Air Pollution, signed in Geneva in 1979, is a landmark international agreement. For more than 25 years it has been instrumental in reducing emissions contributing to transboundary air pollution in the UNECE region through coordinated efforts on research, monitoring and the development of emission reduction strategies on regional air pollution and its effects (Sliggers and Kakebeeke, 2004).

With the accession of Albania to the Convention in December 2005, the Convention's Parties now number 51. Virtually the entire area of the UNECE region in Europe and North America is now covered by the Convention. While only two countries from Central Asia are Party to the Convention (Kazakhstan and Kyrgyzstan), the remaining three (Tajikistan, Turkmenistan, and Uzbekistan) are involved in work that aims at accession.

The Convention has successfully negotiated and adopted eight legally binding protocols to control specific pollutants[1]. With the entry into force of the 1998 Aarhus Protocol on Persistent Organic Pollutants, the 1998 Aarhus Protocol on Heavy Metals in 2003, and the 1999 Gothenburg Protocol to Abate Acidification, Eutrophication and Ground-level Ozone in 2005, all eight protocols to the Convention are now in force.

Reflecting the Convention's science-based approach to emission control strategies, the Executive Body has two scientific subsidiary bodies, the Working Group on Effects and the EMEP Steering Body (Figure 1). The Working Group on Strategies and Review is the main negotiating body for the Convention and is responsible for reviewing protocols, identifying any need for amendment or revision and making recommendations for such changes. Task Forces and Expert Groups provide the expert scientific and technical input to these Working Groups.

[1] On the website of the Convention there are texts of the treaty and its protocols, http://www.unece.org/env/lrtap/status/lrtap_s.htm

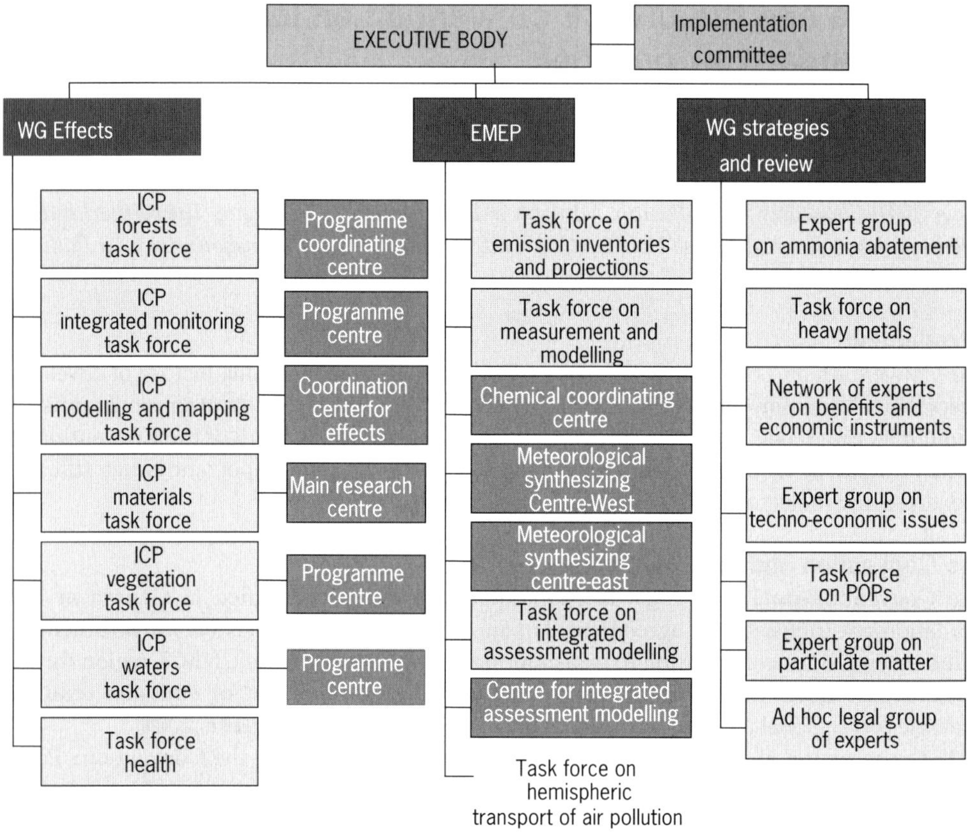

Figure 1. Organogram of working groups, task forces, expert groups and centres under the Convention on Long-range Transboundary Air Pollution. The executive body takes the decisions in its yearly meeting in December, the 'Conference of Parties.'

Critical loads and integrated assessment modelling

Through the 1990s the Convention developed an 'effects-based' approach for negotiating some of its protocols. The 1994 Oslo Protocol on Further Reduction of Sulphur Emissions was the first instrument to be developed in this way and that was followed in 1999 by the adoption of the Gothenburg Protocol on Acidification, Eutrophication and Ground-level Ozone (see more details below).

The effects-based approach relies upon an understanding of the damage done by different levels of pollution; with this information strategies can be devised to avoid harmful effects by decreasing pollution below certain levels. In the Convention critical levels and critical loads have been used to define the thresholds for effects cause by concentrations of gases (levels) and deposition of pollution (loads)[2]. A critical load is defined as a deposition level under which

[2] Under the Working Group on Effects of the Convention there is information on the Convention's mapping programme and critical loads work, http://www.unece.org/env/wge/mapping.htm.

no harmful effects occur, based on the current scientific knowledge. In other words, it is a sustainable deposition level.

Using European maps of critical loads it has been possible to define the sensitive European areas that merit protection and, through the use of models, determine what pollution sources should be controlled to achieve this. In practice, 'integrated assessment models' bring together the information on critical loads, the emissions from sources, the transport and deposition from these sources and the costs of taking pollution abatement measures for each source category. Such models can provide optimised solutions where maximum protection is achieved for minimum cost. The first time the approach was used several models provided input to the negotiation of the 1994 Oslo Protocol. For the Gothenburg Protocol, the RAINS model[3] for the International Institute of Applied Systems Analysis played a leading role.

The Gothenburg Protocol

The most recent protocol to the Convention, adopted in Gothenburg in December 1999, is the Protocol to Abate Acidification, Eutrophication and Ground-level Ozone. This multi-pollutant, multi-effect protocol marks a new era for pollution control and recognition of the need to consider the overall pollution climate if we are to address environmental effects effectively. It deals not just with the three effects in its title but does this through emission control of four pollutants - sulphur dioxide, oxides of nitrogen, volatile organic compounds and ammonia (see Figure 2).

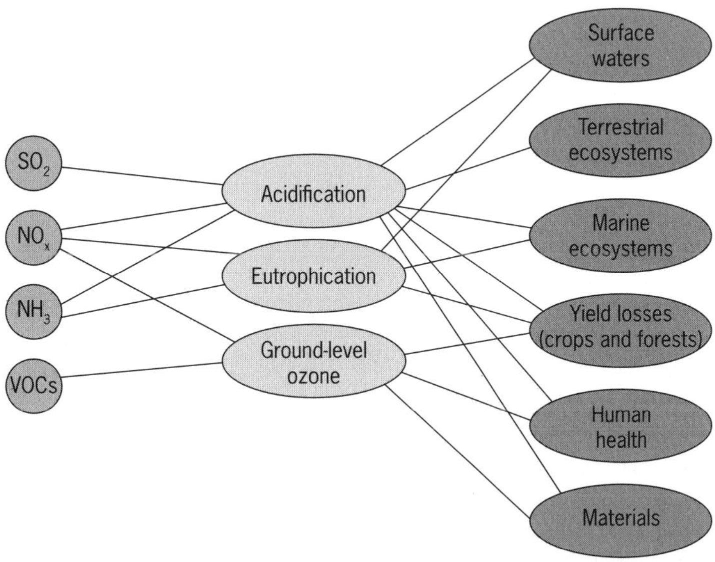

Figure 2. The Gothenburg Protocol sets emission ceilings for 4 pollutants to reduce 3 effects and decreases the pollution load of many receptors.

[3] IIASA maintains a website on the RAINS model. Here there are many publications on calculated scenarios for Europe. Scenarios can even be calculated online. The website also provides information on the GAINS model, http://www.iiasa.ac.at/rains/ciam.html.

It is the first protocol to the Convention to deal with ammonia. Parties had, for many years, been concerned with agriculture as sector that suffered from air pollution. Now they were focusing on the polluting aspects of this activity. Furthermore, the preamble to the Protocol indicates, 'Measures taken to reduce the emissions of nitrogen oxides and ammonia should involve consideration of the full biogeochemical nitrogen cycle'.
But we know ammonia is difficult to control. It is a diffuse source and it is difficult to quantify since fluxes move both into and out of the soil. Control measures are also difficult to apply and are not always effective. And, although the Protocol indicates the need to take into account the full biogeochemical cycle of nitrogen, this cycle is a very complex one, with great uncertainties in certain processes.
The Protocol operates through two mechanisms to achieve its objectives. First, it sets targets for national emission ceilings (the total emission of a pollutant by a country each year). Since 1980, NOx and NH_3 emissions in Europe have been reduced by 25% and 16% respectively (see Table 1). The agreed European reductions in the Gothenburg Protocol to be achieved by 2010 are 53% for NOx and 15% for NH_3, also compared to 1980. Second, the Protocol also sets limit values on certain types fuels or emissions from sources, e.g. the concentration of pollutants of many specific stationary sources or in exhaust gases from motor vehicles.
For ammonia, there are special provisions to be met and these are spelled out in annex XI to the Protocol[4]. They include the requirement of a Party to publish, within 12 months of entry into force for it, an advisory code of good agricultural practice to control ammonia emissions. There are also requirements for urea and ammonium carbonate fertilisers, manure application, manure storage and animal housing. The Protocol also requires the application of best available techniques as listed in a guidance document on ammonia[5] that was adopted at the same time as the Protocol.

Table 1. Nitrogen emissions 1980-2000.

Countries	NO_x	NH_3
CE = Czech Rep., Hungary, Poland and Slovak Rep.	-42%	-46%
CW = Austria, Switzerland and Germany	-49%	-23%
E = Estonia, Latvia, Lithuania and Russia (European part)	+21%	-48%
N = Denmark Finland Iceland, Norway and Sweden	-21%	-10%
NW = Belgium, Luxemburg, the Netherlands, Ireland and United Kingdom	-36%	-13%
S = France, Greece, Italy, Portugal and Spain	-4%	+1%
SE = Albania, Armenia, Belarus, Bosnia-Herzegovina, Bulgaria, Croatia, Cyprus, Georgia, Kazakhstan, Republic of Moldova, Romania, Slovenia, The FYROM Macedonia, Turkey, Ukraine and Yugoslavia	-26%	-12%
Total Europe (excluding ships)	-25%	-16%

[4] Annex XI of the Protocol contains the mandatory obligations, http://www.unece.org/env/lrtap/multi_h1.htm.

[5] Besides annexes, that form an integral part of the Protocol, BAT documents (to apply BAT is mandatory under the Protocol) have been drafted to assist countries in achieving their ceilings and other emission obligations (guidance document V is on ammonia), http://www.unece.org/env/documents/1999/eb/eb.air.1999.2.e.pdf.

The Expert Group on Ammonia Abatement

After adoption of the Protocol and before it entered into force in May 2005, an Expert Group on Ammonia Abatement[6] was established under the Working Group on Strategies and Review. It was charged with preparing input for review of the Gothenburg Protocol, including promotion of a Framework Advisory Code of Good Agricultural Practice for Reducing Ammonia Emissions. This was aimed at assisting Parties to establish their required national codes by using the Framework Code as a starting point. In addition, the Expert Group has set about quantifying relationships between recommended control options/techniques and the resulting ammonia emissions.

Currently, the Expert Group is devising ways of improving the estimates of current emissions and future projections of ammonia and other nitrogen species from agricultural and non-agricultural sources. It is also revising the text for the Guidance Document on Control Techniques for Preventing and Abating Emissions of Ammonia and updating the Framework Advisory Code, taking into account the European Union Integrated Pollution Prevention and Control (IPPC) Best Available Techniques (BAT) reference document (BREF) for pigs and poultry (IPPC, 2005).

The Convention's agenda for the future

One of the most important issues being dealt with by the Convention at present is the first review of the 1999 Gothenburg Protocol. As required by article 10 of the Protocol, the first review of the Protocol started in December 2005 (within one year of its entry into force). The Parties have set themselves an ambitious target, to complete this work and finalise the review in December 2007.

In the review of the Protocol's obligations and assessment of their adequacy, it will be important to take into account best scientific information since data and models are much improved since 1999. There are other related issues, too, that Parties are keen to include in the review process. Particulate matter has been flagged up by WHO as a serious threat to human health throughout Europe; it is associated with the emissions of the pollutants already in the Protocol (secondary particulates), and this includes ammonia.

Many Parties, too, are keen to include consideration of climate change issues. These are interlinked in several ways with the work of the Convention: carbon dioxide, the most important greenhouse gas is associated with most of the emission sources of sulphur and nitrogen oxides; links with the nitrogen cycle need to be considered both from a pollution and greenhouse gas perspective; and, effects of traditional pollutants will be affected, possibly exacerbated, by a changing climate. The idea of linked climate change and air pollution strategies appears to be an attractive one and calculations have shown that there are financial benefits if we go down this road (see Figure 3).

In addition, there is recognition that more needs to be done to understand and take into account the full nitrogen cycle. There are implications here not just for air pollution and climate change but also for biodiversity, which is another big challenge to the world for this century. However, there is currently no group under the Convention dealing with the nitrogen cycle in its entirety. In April, the UK, now lead country for the Expert Group on Ammonia Abatement, will propose

---

[6] Just as for all groups under the Convention, also the Expert Group on Ammonia Abatement has pages under the website of the Convention for its documents, its meetings etc, http://www.unece.org/env/aa/welcome.htm.

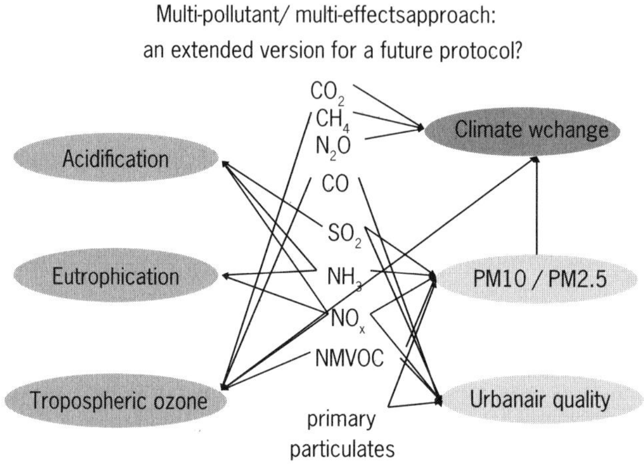

Figure 3. A revised Gothenburg Protocol will address more substances and effects than those of the 1999 'Multi-pollutant, multi-effect Protocol', see Figure 1. The diagram spans the urban to global scale. The diagram covers the gasses and effects now included in the GAINS model (the smaller Kyoto gasses are included in the model but only the three major are shown in the diagram). Compared to Figure 1 the receptors are not included in the diagram. This is done to keep the diagram manageable.

to the Working Group on Strategies and Review to let the Expert Group (from now on with a co-lead by the Netherlands) evolve into a Task Force on Integrated Nitrogen. The Expert Group will consider a draft mandate for such a Task Force, which could start its work in 2008[7]. In November 2007 the Convention's Task Force on Integrated Assessment Modelling together with COST Action 729[8] will hold a workshop on Integrated Assessment Modelling of Nitrogen at IIASA in Laxenburg. Already now NOx, NH_3 and N_2O are included in IIASA's GAINS model, which is an extension of the RAINS model that includes the Kyoto greenhouse gases. Through this extension (commissioned and financed by the Dutch ministry of Environment) it is possible to use GAINS to show the benefits of taking measures that are mutually beneficial to air pollution and climate change (synergies) and to avoid trade-offs. Other calculations can be made with the extended model; for instance, a burden sharing arrangement can be calculated for the EU-27 for a specific climate change target.

In December 2007 it is likely that the Executive Body will decide to embark on the revision of the Gothenburg Protocol, possibly a new, 'replacement' protocol. For this, a great deal of scientific and technical work has already been done but more still needs to be done. The Convention already has many scientific groups to do its work and these cover most of the likely requirements for developing a protocol. However, there might be a need to establish links with other conventions, such as the UN Framework Convention on Climate Change and the Convention on Biodiversity, if we are looking for a further harmonised approach.

---

[7] The proposal was accepted by the Working Group and will be developed further.

[8] COST Action 729 – Assessing and Managing Nitrogen Fluxes in the Atmosphere-Bioshere System in Europe, http://www.cost729.org.

References

Integrated Pollution Prevention and Control (IPPC), 2005. BAT reference document (BREF) for intensive farming pigs and poultry, http://eippcb.jrc.es/pages/FActivities.htm.

Sliggers J. and W. Kakebeeke (eds), 2004. Clearing the Air: 25 years of the Convention on Long-range Transboundary Air Pollution. ECE/EB.AIR/84, United Nations Publication, Sales No. E.04.II.E.20, Geneva (a free copy of the book can be obtained from both authors of this article).

The regulation of ammonia emissions in the United States

S. Shaver
Associate Counselor for Agricultural Policy, Office of Air and Radiation, United States Environmental Protection Agency, USA

Currently, in the United States there is very limited regulation of ammonia emissions as a matter of federal policy. The regulation of air emissions rests primarily with the United States Environmental Protection Agency. The Clean Air Act (CAA), including its various amendments, provides the primary statutory authority for this regulatory agenda. There are some requirements under the Comprehensive Environmental Response, Compensation and Liability Act (CERCLA) and Emergency Planning and Community Right-To-Know (EPCRA) which also apply to some air emissions. This paper will address relevant provisions of these statutes and will relate the status of the implementation of these provisions to ammonia, in particular, to the agricultural sources of those ammonia emissions. Ammonia in discharges to waterways is regulated under the Clean Water Act (CWA). However, agricultural sources are exempt from regulation under this statute, with the exception of concentrated animal feeding operations (CAFO). The paper will further highlight recent and upcoming actions regarding the regulation of CAFO under these statutes and the ongoing discussions to attempt to address nutrient management at CAFO in a holistic, cross-media fashion.

The regulation of pollutants under the CAA is divided into regulation of 'criteria' pollutants and hazardous air pollutants. Criteria pollutants are those which are ubiquitous in the atmosphere and which are determined by the Administrator of the USEPA to need regulation to adequately protect the public health and welfare. The six criteria pollutants are ozone, particulate matter (PM10 and PM2.5), sulfur dioxide, carbon monoxide, nitrogen oxides, and lead. National ambient air quality standards are established for these pollutants, national ambient monitoring networks are established, inventories of pollutants or their precursors are required to be developed, and areas are determined to be either in attainment or nonattainment of these standards. The states, tribes, or local air districts develop implementation plans to reach attainment of these standards within their jurisdictions (For purposes of this paper, future use of the word 'states' will include tribes and local air districts; and state implementation plans (SIP) will refer to implementation plans from any of these entities). These implementation plans identify the requirements necessary for individual sources which are needed to bring the area into attainment. The requirements must address reductions in precursors of the criteria pollutant where appropriate. Depending on the pollutant, the severity of the nonattainment may impact the stringency of the requirements on sources within the nonattainment area. Failure to comply with the implementation plans can lead to sanctions on the state.

For individual stationary sources, the CAA also has provisions which apply nationally. There are New Source Performance Standards which have been established for a significant number of stationary sources. These standards are required to be updated on an ongoing basis and apply only to new sources. However, there is a provision whereby performance standards may be set for existing sources by requiring states to adopt these into their implementation plans. Municipal waste combustor regulations are a good example of this process. Another requirement is the need for an operating permit for all major sources (title V permit). For purposes of title V permitting, a major source is one with 100 tons/year or more of non-fugitive emissions of any criteria pollutant or its precursors. These permits establish no new or additional requirements for these sources, but they must have all of the CAA regulatory requirements for the source included in the permit. There are also requirements for New Source Review permits for new sources or significant modifications to existing sources which are separate permit requirements from the title V permits. New sources or those undergoing significant modifications must put on best available control technology or lowest achievable emission rate technology depending on the area's nonattainment status. Offsets for increased emissions in nonattainment areas are also required as a part of this permitting process. In attainment areas, the permit is called a Prevention of Significant Deterioration (PSD) permit. Major sources are required to obtain these permits, and the definition for major varies by pollutant or precursor and by the area's nonattainment status. Also, the definition for major source is different from that used for title V permitting. For sources in attainment areas, a major source is 100 tons/year for certain listed source categories and 250 tons/year for all other source categories. For nonattainment areas, the range for major source or major modification definition is from 10 tons/year to 100 tons/year. There are other rules which are national in scope as well. Emissions from mobile sources and fuels are regulated on a national basis and reductions achieved from these programs are credited in the state's implementation plans. There have been national rules put into place recently to address emissions from large coal-fired power plants, which will significantly reduce NO_x and SO_x emissions from these sources. There are provisions for a trading program which allows for more cost-effective strategies to achieve these reductions. For very large coal-fired power plants (greater than 250MW) there are provisions for a cap and trade program for NO_x and SO_x to address acid rain.

Hazardous air pollutants are governed by different sections of the CAA. There are 186 hazardous air pollutants (HAP) identified in the statute and the USEPA has the authority to add or delete pollutants based on the weight of the scientific information available. There is also a petition process in place which allows non-governmental entities or individuals to request that a substance be added or deleted. To date, three pollutants have been removed from the list. However, there is a request to add hydrogen sulfide to the list, which has been submitted to the Agency in recent years. No action has been taken on that request.

The Agency is required to identify sources of these hazardous air pollutants, list the appropriate industrial source categories, and then promulgate standards which at a minimum require that all major sources meet the emissions levels of the best performing twelve percent of sources in that source category. There are provisions to sub-categorise within a source category based on the uniqueness of facilities' configuration and processes. For purposes of HAP, a major source is defined as a source which emits 10 tons/year of one HAP or 25 tons/year of a combination of HAPs. These standards are called Maximum Achievable Control Technology Standards (MACT) and have been established for 174 source categories of industries or of processes at industries (For example, there is a category listed for industrial boilers and one for primary smelters). The MACT standards address all of the 186 HAP for each source category as appropriate. However, in reality only a subset of them is actually reduced. Once every eight years thereafter, the USEPA is to evaluate these standards and identify any new technologies which may have emerged and might be feasible to apply to the individual source category. Additionally, there are requirements that the remaining risk after implementation of MACT standards should be evaluated; and if there is a significant health risk remaining, then additional reductions in emissions are required.

The statute also has a program for addressing smaller sources of HAP, and this program is called the Area Source Program. Ammonia is not a HAP under the CAA.
The CAA has enforcement provisions which address the impacts on states which fail to attain the ambient standards and which fail to implement their SIPs. Sources face enforcement actions when they fail to obtain the appropriate permits or when they do not meet the specific requirements of a SIP or underlying state or federal regulation. Under some scenarios, sources may face fines of $37,500 per day. States may face sanctions and loss of federal highway funds. The CAA also has provisions which allow individual citizens to sue.

The lists of hazardous pollutants under the CERCLA and EPCRA statutes are different from the hazardous lists under the CAA. It is also important to note that the definition of a 'facility' under these two statutes is different from the definitions of 'source' under the CAA. Both ammonia and hydrogen sulfide are hazardous substances under these two statutes. Emissions of 100 pounds/day of these hazardous substances meet the threshold for reporting releases to the National Response Center under CERCLA and to the local emergency response centers and to the states likely to be affected by the release under EPCRA. These statutes were enacted to require reporting of releases of hazardous substances which when released into the environment may present substantial danger to the public health or welfare or the environment. Specifically, EPCRA is a statute that provides information to the public based on the principle that the community is entitled to know what extremely hazardous substances may potentially be released in their area and so that local emergency response personnel can be better prepared in the event of an accidental release. Under EPCRA there is an exemption in the definition of a 'hazardous chemical' to the extent that it is used in routine agricultural operations or is a fertiliser held for sale by a retailer to the ultimate customer. Under CERCLA there are a few exemptions to the notification requirements of certain releases. The are other provisions regarding liability for the costs of response, removal or remedial action; as well as damages for injury to, destruction of, or loss of natural resources; and costs of any health assessment or health effects studies carried out under the response authorities of CERCLA. There are also significant monetary penalties for failing to report releases of a hazardous substance (other than a federally permitted release) as soon as the person in charge of a facility has knowledge of the release.
At this time, these three statutes are the only ones which deal with air emissions from sources; and there are no specific exemptions in the statutes for agricultural practices with exception of the one for the use of hazardous chemicals in 'routine agricultural operations' under EPCRA. Fertiliser held for sale by a retailer to the ultimate customer is not considered to be a 'hazardous chemical' under EPCRA and the normal application of fertiliser is by definition, not a release under CERCLA and therefore not required to be reported as a 'release'. The Clean Water Act (CWA) has provisions for establishing Total Maximum Daily Loads for watersheds which includes determining the input from atmospheric deposition. The CAA has provisions for addressing deposition to certain water bodies to protect the quality of those waters; however, to date, we have not sought to address deposition using the CAA authorities.

Under the CAA, there are no specific NSPS which apply to CAFOs. Since ammonia is not a HAP, there is no MACT standard for CAFOs nor have CAFOs been listed as a source category to be regulated because of HAP emissions. There is no statutory authority for the USEPA to regulate odor although a number of states do have state regulations to address odor per se. Although several states have standards for hydrogen sulfide for these sources, and some states have requirements in their state implementation plans regarding emissions from these sources, there are no specific federally mandated regulations in place for CAFOs. Some SIPs have requirements regarding emissions of particulate matter which have the potential to affect some of the CAFOs. The only potential requirements for CAFOs are the requirement to have an operating permit under CAA title V provisions if the source emits more than 100 tons/year of a criteria pollutant or one of its precursors and the requirement to have a PSD/NSR permit if the source emits more than the specified amount for that source category of a criteria pollutant or one of its precursors in an attainment area or a lesser quantity in a nonattainment area depending on the pollutant and the severity of the nonattainment.

Recently, the USEPA enacted modifications to the fine particulate matter standards by changing the daily standard to 35 μg/m3 of PM2.5. The annual standard remains at 15 μg/m3 of PM2.5. There are separate standards for PM10. This strengthening of the PM2.5 standard may potentially cause more areas to be classified as nonattainment areas, some of which may contain agricultural sources. The impact of ammonia on fine particle concentrations and on the secondary formation of PM2.5 is being studied; however, for the current standard the agency has chosen not to consider ammonia as a regulated pollutant under the CAA based on the lack of information and our current understanding of the science. In the state of California, which has some of our most difficult air pollution problems, we do know that the reduction of ammonia sources does not improve the air quality as measured by fine particulate matter. We do acknowledge that in other places the science could yield different results, and we are giving states the ability to pursue regulation of ammonia as a precursor to PM2.5 when appropriately justified by the scientific information.

With the significant increase of large animal feeding operations in the United States and with the corresponding spread of population centers to formerly agricultural areas, there has been a growing concern about the health and environmental implications of these operations. In response to numerous citizens' complaints about these sources, there were some early USEPA enforcement actions taken against a couple of operations based on violations of the CAA and CERCLA and EPCRA. Additionally, there have been several citizens' suits against a number of operations in different states. There is at least one case pending which attempts to declare an entire watershed as a hazardous waste site requiring cleanup. Primarily because of the objectionable odours from these facilities, the air quality surrounding these sources quickly became a matter of concern.

As a result, some time ago the USEPA began to try to determine the impact of the air emissions from these sources on air quality and to determine what regulations might be feasible. Although we were able to identify a number of emissions estimates from literature values, we found very limited information from US sources. Our practice was to base emissions on the available data in the literature using emission factors multiplied by the number of animals at the site. Needless to say, there was significant concern about the use of this type of approach as the basis for any effort which might lead to considerable expenditure of funds to meet any potential CAA permitting requirements. In conjunction with the US Department of Agriculture, the USEPA requested that the National Research Council of the National Academies of Science review our compilation of the available data and our use of that information to generate emission estimates. It was further requested that they make recommendations on the approach and on short and

long term research that is needed. The results of this effort were published in 2003 under the title of 'Air Emissions from Animal Feeding Operations, Current Knowledge, Future Needs.' Further information on this document may be obtained at www.nap.edu.
One of the primary underlying themes from the report was the need for more actual data from these sources, representative data based on various management practices which could then be used to characterise more accurately the emissions from these sources. Over a period of several years, the industry and USEPA negotiated a consent agreement which provides for an industry-funded two-year monitoring effort and a limited covenant not to sue for those participating in the agreement. The USEPA has responsibility for approving the study, oversight of the study, and analysing the data and developing emissions estimating methodologies for CAFOs. Upon completion of the methodologies development and publication, the industry will then have a limited time period to apply these methodologies, determine their emissions, and comply with any applicable permitting and reporting requirements under the CAA, CERCLA, and EPCRA.
Over 2,700 agreements have been signed and approved, and these agreements represent over 13,000 farming operations. Forty-four percent of these operations are in only three states, Iowa, North Carolina, and Arkansas. Forty-two percent of these farms are from the swine sector. Dairies represent four percent of the farms. Broilers represent thirty-five percent of the farms, while layers represent nineteen percent of the farms. The USEPA determined that the numbers and locations of the operations which signed up for the agreement provided a representative sample for that sector. Although a number of turkey operations applied, there were not sufficient numbers to support the study for those operations; thus, no agreement was approved for that sector. It should also be noted that open cattle feedlots are not addressed by this agreement. These operations have primarily fugitive emissions which are not generally regulated by the CAA permitting requirements.

Monitoring plans have been approved for this multi-million dollar study, and the monitoring will be conducted by an independent monitoring contractor. The industry has established a board to oversee the project and monitor the expenditure of the funds for the study since this is a privately funded effort. The monitoring should begin in the spring of 2007 and will continue at the same sites for a period of two years. The data from the study will be made publicly available as they are quality assured. The study focuses on swine, poultry and dairy operations and will be conducted at twenty-five sites in ten states. Considerable effort has been put into quality assuring project plans, identifying sites, developing standard operating procedures, and developing site monitoring plans. Some of the challenges of the study must deal with the changing climatic conditions, partially enclosed and naturally ventilated systems, large open sources, animal movements, and changing feed rations all of which could have impact on the emissions from the source.
The USEPA will concurrently initiate its work on emissions estimating methodologies and will use the data as they become available in this effort. It is not anticipated that this monitoring effort alone will result in sufficient information to convert to a complete process-based approach to estimating emissions as recommended by the National Research Council; however, it should provide for a much more credible and scientific approach and should be a good first step towards a full process-based approach. The USEPA is aware of efforts underway by some sectors to proceed more rapidly towards a process-based approach on their own initiative, an effort which

is strongly encouraged. The development of the emissions estimating methodologies will be a process which engages the industry as well as other interested stakeholders and will provide for public review.

In regulating CAFOs under the CWA, there are requirements for nutrient management plans. The USEPA is in the final rulemaking stages on this issue. These plans will become a part of the NPDES permits under the CWA and we are encouraging states and industry to look at these plans in a cross-media manner. The USEPA has established a cross-media team which is, among other things, investigating the possibility of cross-media voluntary and regulatory approaches to addressing CAFOs under the existing environmental statutes. If successful, this effort could provide an approach which could deal with the management of ammonia from all sources on the operation and its impact on all media. Without a more holistic approach, it is often expedient to simply transfer the pollutant to another medium.

Although agriculture is the largest contributor to the ammonia inventory in the United States, we are continuing to investigate the other contributors as well; and research is underway regarding ammonia emissions from vehicles. Our 2002 emissions inventory estimates that about eighty-eight percent of the ammonia emissions are from agricultural sources, both crop and livestock production. The CAFO monitoring study which is just beginning will clearly provide much needed information regarding CAFOs, which represent a significant portion of the inventory as it is currently understood. Much research is already underway at various agencies and universities regarding how to reduce emissions from these sources, so the next few years should yield new insights into these sources and what response actions are possible. There are also pilot projects underway in the private sector to analyse and test ammonia reduction techniques. Last summer the US Department of Agriculture, North Carolina State University, and others sponsored the first 'Workshop on Agricultural Air Quality: State of the Science' in Potomac, Maryland. The USEPA's Scientific Advisory Board has established an Integrated Nitrogen Committee to convene over the next two year period and advise the Agency on effects, production and technology.

Clearly there is much work to be done; and clearly, there is a need from an environmental perspective to understand the issues and to devise appropriate reduction and management strategies where deemed necessary. Before we move forward with further regulation, we must better understand the science. We must be able to measure the emissions, determine their impact in the environment and be prepared to take appropriate action.

Acknowledgements

I thank Mr. Bill Harnett, Ms. Lynn Beasley, Mr. Bill Schrock, Mr. Larry Elmore, Mr. Doug Solomon, and Ms. Teresa Clemons of the US Environmental Protection Agency for their assistance in reviewing this paper. I also thank Dr. Viney Aneja of North Carolina State University for his constructive comments.

Ammonia assessment from agriculture: status and national needs

V.P. Aneja[1], J. Blunden[1], K. James[1], W.H. Schlesinger[2], R. Knighton[3], W. Gilliam[4], D. Niyogi[5] and S. Cole[5]
[1]Department of Marine, Earth, and Atmospheric Sciences, North Carolina State University, Raleigh, NC, USA
[2]Duke University, Durham, NC, USA
[3]US Department of Agriculture, Washington, DC, USA
[4]Department of Soil Sciences, North Carolina State University, Raleigh, NC, USA
[5]Purdue University, West Lafayette, IN, USA

Biologically active, photochemically reactive, and radiatively active nitrogen compounds in the atmosphere, hydrosphere, and biosphere are collectively referred to as reactive nitrogen (Galloway *et al.*, 2003). Over the past few decades, human activities leading to the production of reactive nitrogen from diatomic nitrogen (N_2) exceed that of nitrogen fixation in the natural terrestrial ecosystem at the global scale.

Ammonia (NH_3) is the most reduced form of reactive nitrogen. It is also the most abundant alkaline constituent in the atmosphere (Aneja *et al.*, 2006a). Figure 1 illustrates the major processes (emissions, chemical transformation, transport, and removal) that drive the global cycle of NH_3 in the atmosphere (Aneja *et al.*, 2006b). In the past 50 years, emissions and subsequent deposition of NH_3 have increased significantly in parallel with the development of intensive agricultural management and increased livestock numbers (Sutton *et al.*, 1993). Globally, domestic animals are the largest source [32×10^{12} g NH_3-N (ammonia-nitrogen) yr^{-1}] of atmospheric NH_3, comprising approximately 40% of natural and anthropogenic emissions combined. Additionally, synthetic fertilisers and agricultural crops together contribute 9×10^{12}

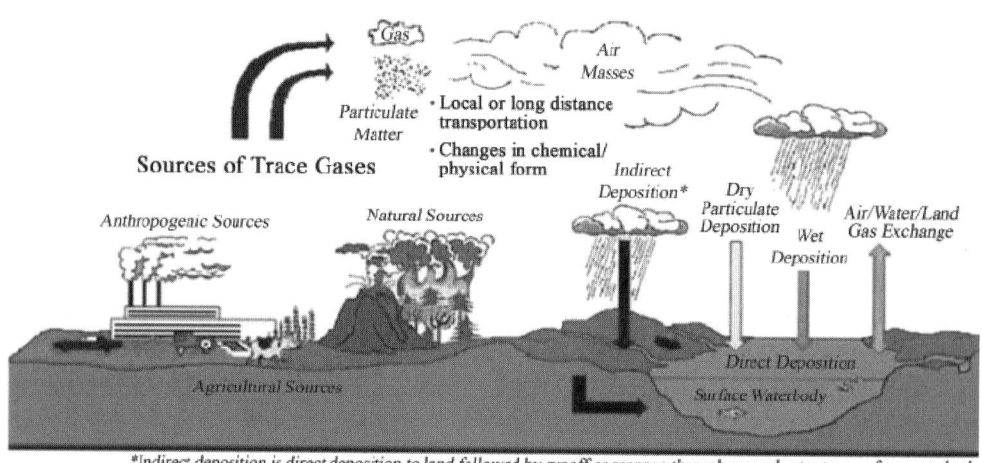

*Indirect deposition is direct deposition to land followed by runoff or seepage through groundwater to a surface waterbody.

Figure 1. Atmospheric emissions, transport, transformation, and deposition of trace gases (Aneja et al., *2003).*

g NH_3-N yr^{-1} (12% of total emissions) (Schlesinger and Hartley, 1992). Table 1 shows the total 2002 estimated NH_3 emissions from various animal husbandry operations in the United States (US EPA, 2005). Several other countries have also developed specific NH_3 emission factors for these types of operations. For example, Table 2 lists emission factors estimated for 2002 in the Czech Republic and in Denmark during the 1990's. The emission factors vary by a factor of two to three between the countries and some estimates for individual species in Denmark are nonexistent. Several countries (e.g. U.S.A.) are in the process of developing national emission factors for both animal and crop agriculture, Table 3 (e.g. Battye *et al.*, 2003 and Aneja *et al.*, 2003). However, emissions factors from emerging agricultural producers in Southeast Asia (e.g. China and India) and other developing countries are limited.

Once released into the atmosphere, NH_3 has a relatively short residence time of ~1-5 days (Warneck, 2000). Once airborne, it is either readily converted to aerosol or it is subjected to dry or wet deposition processes. Ammonia is reactive with a variety of acidic atmospheric species, including nitric acid (HNO_3), hydrochloric acid (HCl) and sulfuric acid (H_2SO_4), which result in the formation of ammonium aerosols, i.e. fine particulate matter (aerodynamic diameter <2.5 μm). Due to the extended lifetime of these aerosols (~1-15 days), nitrogen may be transported to

Table 1. Ammonia emission estimates in the United States for the year 2002 (US EPA, 2005).

Source	Ammonia Emissions (tons/yr)	% of Total Emissions
Animal		
Dairy(dairy cows and dairy heifers)	487,253	11.9
Beef(beef cattle, bulls, and calves)	573,297	14.0
Poultry(chickens and turkeys)	579,924	14.2
Swine(breeding and marketing pigs)	374,954	9.2
Sheep	21,683	0.5
Goats(milking and Angora goats)	12,247	0.3
Horses	62,237	1.5
Total animal	2,111,594	51.7
Fertiliser	1,140,396	27.9
Total agriculture	3,251,990	79.6
Other		
Chemical and applied product MFG	23,123	0.6
Fuel comb. elec. util.	30,256	0.7
Fuel comb. industrial	15,959	0.4
Miscellaneous	282,166	6.9
Fuel comb. other	17,602	0.4
Mobile sources	289,871	7.1
Waste disposal and recycling	25,770	0.6
Other industrial processes	148,288	3.6
Total other	833,035	20.4
Total emissions	4,085,025	100.0

Table 2. Ammonia emission factors for individual animals categories in the Czech Republic for the year 2002 and in Denmark during the 1990s.

Animal Category	Emission factor (kg NH_3 animal^{-1} year^{-1})	
	Czech Republic 2002[1]	Denmark 1990s[2]
Dairy Cows	27.9	13.55
Cows	16.2	
Heifers	16.2	
Calves	16.2	2.91
Bulls	16.2	
Other Cattle	16.2	5.08
Total Sows	17.44	6.39
Sucking-pigs	6.5	3.45
Pigs	8.3	2.79
Broilers	0.21	0.08
Layers	0.27	
Turkey cocks and hens	0.92	0.40
Other poultry	0.21	
Geese and dugs	0.73	
Horses	8	
Sheep	1.34	
Goats	1.34	

[1] Zapletal and Chroust (2006); [2] Pederson (2006).

previously pristine regions far from the pollutant sources. Assuming an atmospheric residence time of six days and a wind velocity of 5 ms^{-1}, Irwin and Williams (1988) estimate that transport of ammonium aerosols might be as large as 2500 km; however, dry and wet deposition may reduce this transport significantly.
Both ammonia and subsequently derived ammonium (NH_4^+) may be removed from the atmosphere through both wet and dry deposition. Dry deposition occurs by diffusion of NH_3 in the atmospheric boundary layer and the surface layer, and by molecular diffusion or Brownian motion of fine particle NH_4^+ within the molecular sublayer. Wet deposition occurs by below cloud scavenging (washout) and rainout (in-cloud processes). Overall, wet deposition is more important in regions with low NH_3 emissions. Conversely, dry deposition is more important in regions of high NH_3 emissions (Krupa, 2003).

Scientific information suggests that reactive nitrogen (e.g. NH_3) is accumulating in the environment, and that nitrogen cycling through biogeochemical pathways has a variety of negative environmental consequences including acidification and eutrophication, photochemical air pollution, reduced visibility, ecosystem fertilisation, global warming, and stratospheric ozone depletion. A number of studies have evaluated the effects of nitrogen deposition. Significant excess nitrogen deposition has occurred is the eastern coastal areas of the United States (Paerl, 1995). A particular area of concern is the coastal plain rivers and their estuaries. Atmospheric deposition of nitrogen compounds may contribute as much as 35-60% of total nitrogen loading

Table 3. Ammonia emission factors from animal and crop agriculture in the US.

Animal	Animal Agriculture Emission factor (kg-NH_3/animal/ yr)	
	Battye *et al.* (2003)	**US EPA (Battye *et al.*, 1994)**
Cattle		
Dairy cow	28	40
Beef cow	10.2	27 (steers)
Pigs		
Sow	16.4	16
Finishing pig	6.4	7
Poultry		
Laying hen	0.31	0.31
Broiler	0.28	0.17
Sheep	1.34	3.4
Horses	8.0	12
Fertiliser	**Crop Agriculture Emission factor (kg NH_3/Mg N)**	
	Aneja *et al.* (2003)	**US EPA (Battye *et al.*, 1994)**
N-P-K	48	48
Nitrogen solutions	30	30
Ammonium phosphates	48	48
Anhydrous NH_3	12	12
Urea	182	182
Ammonium nitrate	25	25
Other straight nitrogen	30	30
Ammonium sulfate	97	97
Aqua NH_3	12	12
Ammonium thiosulfate	30	30

to North Carolina coastal waters (Paerl, 1995). This excess nitrogen can result in toxic and non-toxic phytoplankton blooms, which can lead to fish kills and reductions of 'clean water' species (Paerl, 1995). Soil acidification is a problem experienced in the UK and the Netherlands, due to the high density of animal operations. Van Breeman *et al.* (1982) identified the deposition of ammonium bisulfate (NH_4HSO_4) as the main cause of soil acidification in The Netherlands. Research conducted by Barthelmie and Pryor (1998) in the Lower Fraser Valley, British Columbia, Canada showed that NH_3 and NH_4^+ species and emissions play a particularly critical role in visibility degradation. Fine particulate aerosols have also been linked to human respiratory health problems. Studies suggest that the smaller the particle the greater the potential health effect. For example, Lippmann (1998) found fine particles ($PM_{2.5}$) to be more toxic than coarse particles (PM_{10}-$PM_{2.5}$). Donaldson and MacNee, (1998) examined ultra-fine particles (<100 nm) and found that toxicity increases as particle size decreases. In the eastern part of the United States it is estimated that 47% of $PM_{2.5}$ consists of ammonium sulfate (US EPA, 1995).

On the positive side, reactive nitrogen has been responsible for dramatically improved food production, which becomes more important as the human population is expected to grow to 8-12 billion by the end of the 21st century. Therefore, the challenge for the scientific community is to find ways to maximise beneficial use of reactive nitrogen while simultaneously minimising adverse environmental impacts. One way to approach this challenge is through the deliberate integration of reactive nitrogen research, management, and control strategies. Integrated research and control strategies that consider urban-rural air quality connections and interactions are necessary for optimal ammonia/nitrogen management.

Several best management practices (BMPs) to curtail ammonia emissions from agricultural sources have been tested, mainly dealing with emissions from cattle and swine. A relatively simple solution was undertaken by Lefcourt and Meisinger (2001) who tested the addition of alum and zeolite to cattle slurry in an effort to curb the volatilisation of ammonia. When alum was added at 2.5% and 6.25%, there were reductions of 58 ± 6% and 57 ± 10 % respectively. Slightly lower reductions were seen with the addition of zeolite with additions of 2.5% and 6.25% resulting in reductions of 22 ± 6% and 47 ± 10%. Similar tests were performed by Berg (2006), with attempts to lower ammonia emissions from cattle slurry by acidifying it with lactic and nitric acid. Lactic acid was applied reaching pH levels of 5.73, 5.14 and 4.18, yielding decreased emissions by 65%, 72% and 88%, respectively. Nitric acid reduced emissions by about half of the success seen with lactic acid. When nitric acid was added to reach pH levels of 5.20 and 4.49, emissions were lowered by 29% and 49%, respectively. Another relatively simple solution was the reduction of crude protein in the cattle's diet. Frank and Swensson (2002) found that as the crude protein was reduced, the NH_3 concentrations were reduced in parallel. In addition, when condensed tannin was added to the drinking water of both cattle and sheep, less ammonia was volatilised. The amount of nitrogen in solid and liquid waste was similar to that of regular water, but the nitrogen was nitrified/denitrified rather than volatilised into ammonia (Kronberg, 2006).

More detailed solutions to reduce dairy cattle emission include changes in floor design and ventilation. It was found that if a flat floor in a cattle barn was scraped 96 times per day there was a 5% reduction in the ammonia emitted from the barn. Yet, with a sloped floor, reductions were much greater. With 12 scrapings per day, there was a 21% reduction and, with 96 scrapings per day, there was a 26% reduction (Braam *et al.*, 1997). Using a filtered, custom built, double-polytube ventilation system in their calf barns, Hillman *et al.*, (1992) were able to reduce the ammonia concentration from 10 ppm in the air to 5 ppm and below.

Both simple and more detailed BMPs have been tested for the reduction of ammonia from swine sources. A simple solution was the addition of the manure additive Alliance® to swine manure, resulting in a 24% reduction in ammonia emissions (Heber *et al.*, 2000); while an example of a more complicated solution involves ventilation and indoor air climate control (Hartung *et al.*, 2006). There was a reduction of 10–14% of ammonia when there was a reduction of indoor air temperature and ventilation rate. Yet, to obtain these results an evaporative indoor air cooling system with an 'optimisation of the fogging control with regard to a continuously complete evaporation of water' was needed. Other solutions are more complicated in design and setup, but are easier to use in order to obtain results. Using biotrickling filters for the manure, Hansen and Jensen (2006) were able to reduce both odor and ammonia from manure. Loyon *et al.*, (2006) found that with the use of a storage spreading system with biological treatment of manure, there was a 30–50 % reduction with separated manure and a 68% reduction with unseparated manure.

Promising results have been reported for reducing ammonia from swine manure through the use of an 'engineered system,' i.e. treatment plant with solid-liquid separation. Szögi (2006) reported a 73% reduction in ammonia emissions from the implementation of such a system. Vanotti (2006) found that when manure from such a system was applied there was a 98.8% reduction in greenhouse gases emissions, as well as a potential additional income of $9,100 to $27,500/year (approximately $0.91/finished pig) from implementing cleaner technology through the Supersoil program. In addition, when organic fertilisers with gypsum added are applied, they can reduce ammonia volatilisation by 11% (Model, 2006).

In the U.S., there are currently no regulations or incentive programs that require reductions in ammonia emissions. It is striking when comparing the different pollutants, sulfur dioxide (SO_2), oxides of nitrogen (NO_x), volatile organic compounds (VOCs), and ammonia (NH_3), that the extensive control measures applied to SO_2, NO_x, and anthropogenic VOCs have not been extended to ammonia. However, future policies and control measures are necessary to successfully decrease ammonia emissions and its related problems. Current research issues include the quantification of agricultural point and non-point sources; the atmosphere-biosphere exchange of ammonia; the quantification of landscape processes; the primary and secondary emissions of PM; and the gas-to-particle conversion to PM fine.

In the past, control of criteria pollutants policies mainly focused on single pollutants, while addressing, in general, single-effects. Today, multi-pollutant / multi-effect approaches are being considered, thus offering unique opportunities for the development of ammonia abatement measures offering integrated approach strategies.

Acknowledgments

We acknowledge support from the Cooperative State Research, Education, and Extension Service (CSREES), USDA National Research Initiative Competitive Grants Program, contract 2005-35112-15377. We thank Ms. Sally Shaver, US Environmental Protection Agency; and Mr. Bill Battye of ECR for their constructive comments.

References

Aneja, V.P., D. Nelson, P. Roelle and J. Walker, 2003. Agricultural ammonia emissions and ammonium concentrations associated with aerosols and precipitation in the southeast United States. Journal of Geophysical Research 108: 4152.

Aneja, V.P., W. Schlesinger, R. Knighton, G. Jennings, D. Niyogi, W. Gilliam and C. Duke, (Eds), 2006a. Workshop on Agricultural Air Quality: State of the Science, North Carolina State University, Raleigh, NC 27695-7603, ISBN 0-9669770-4-1, pp. 1314.

Aneja V.P., W. Schlesinger, D. Niyogi, G. Jennings, W. Gilliam, R. Knighton, C. Duke, J. Blunden, and S. Krishnan, 2006b. Emerging national research needs for agricultural air quality. Eos - Transactions American Geophysical Union 87: 25-29.

Barthelmie, R.J. and S. Pryor, 1998. Implications of ammonia emissions for fine aerosol formation and visibility impairment- A case study from the Lower Fraser Valley, British Columbia. Atmospheric Environment 32: 345-352.

Battye, R., W. Battye, C. Overcash and S. Fudge, 1994. Development and Selection of Ammonia Emission Factors. EPA-600/R-94-190, U.S. Environmental Protection Agency Atmospheric Research and Exposure Assessment Laboratory, Research Triangle Park, North Carolina.

Battye, W., V. Aneja, P. Roelle, 2003, Evaluation and improvement of ammonia emissions inventories. Atmospheric Environment 37: 3873-3883.

Berg, W.E., 2006. Effects of acidifying liquid cattle manure with nitric or lactic acid on gaseous emissions. The Workshop on Agricultural Air Quality: State of Science, 4 June 2006, The Ecological Society of America. 2006. pp. 492 – 498.

Braam, C.R., J. Ketelaars and M. Smits, 1997. Effects of floor design on floor cleaning on ammonia emission from cubicle houses for dairy cows. NJAS Wageningen Journal of Life Sciences 45(1): 49-64.

Donaldson, K. and W. MacNee, 1998, The mechanisms of lung injury caused by PM_{10}. In: Hester, R.E. and Harrison, R.M. (Eds), Issues in Environmental Science and Technology 10, pp. 21-32.

Frank, B. and C. Swensson, 2002. Relationship between content of crude protein in rations for dairy cows and milk yield, concentration of urea in milk and ammonia emissions. Journal of Dairy Science 85: 1829-1838.

Galloway, J.N., J. Aber, J. Erisman, S. Seitzinger, R. Howarth, E. Cowling and B. Cosby, 2003. The nitrogen cascade. BioScience 53: 341-356.

Hansen, M.J. and T. Jensen, 2006. A biotrickling filter for removing ammonia and odour in ventilation air from a unit with growing-finishing pigs. The Workshop on Agricultural Air Quality: State of Science, 4 June 2006. The Ecological Society of America. 2006. pp. 844-846.

Hartung, E., 2006. NH_3 emission from pig husbandry in relation to ventilation control and indoor air cooling. The Workshop on Agricultural Air Quality: State of Science, 4 June 2006. The Ecological Society of America. 2006. pp. 657-664.

Heber, A.J., J. Ni, T. Lim, C. Diehi, A. Sutton, R. Duggirala, B. Haymore, D. Kelley and V. Adamchuk, 2000. Effect of a manure additive on ammonia emission from swine finishing buildings. Transactions of the ASAE 43(6): 1895-1902.

Hillman, P., K. Gebremedhin and R. Warner, 1992. Ventilation system to minimize airborne bacteria, dust, humidity, and ammonia in calf nurseries. Journal of Dairy Sciences 75: 1305-1312.

Irwin, J.G. and M. Williams, 1988. Acid rain: chemistry and transport. Environmental Pollution 50: 29-59.

Kronberg, S.L., 2006. Condensed tannin in drinking water of cattle and sheep to reduce their urine excretion and subsequent ammonia pollution. The Workshop on Agricultural Air Quality: State of Science, 4 June 2006. The Ecological Society of America. 2006. pp. 835-837.

Krupa, S., 2003. Effects of atmospheric ammonia (NH_3) on terrestrial vegetation: a review. Environmental Pollution 124: 179-221.

Lefcourt, A.M. and J. Meisinger, 2001. Effect of adding alum or zeolite to dairy slurry on ammonia volatilization and chemical composition. Journal of Dairy Science 84: 1814-1821.

Lippmann, M., 1998. The 1997 US EPA standards for particulate matter and ozone. In: Hester, R.E. and Harrison, R.M. (Eds), Issues in Environmental Science and Technology 10, pp. 75-79.

Loyon, L., F. Beline, F. Guiziou, P. Peu, S. Picard and P. Saint Cast, 2006. Assessment and comparison of annual gaseous emissions of three biological treatments of pig slurry with a storage – spreading system. The Workshop on Agricultural Air Quality: State of Science, 4 June 2006. The Ecological Society of America. 2006. pp. 863-867.

Model, A., 2006. Ammonia and trace gas emissions from organic fertilizers amended with gypsum. The Workshop on Agricultural Air Quality: State of Science, 4 June 2006. The Ecological Society of America. 2006. pp. 923-929.

Paerl, H.W., 1995. Coastal eutrophication in relation to atmospheric nitrogen deposition: current perspectives. Ophelia 41: 237-259.

Pederson, S., 2006. Agricultural best management practices in denmark. The Workshop on Agricultural Air Quality: State of Science, 4 June 2006. The Ecological Society of America. 2006. pp 56-67.

Schlesinger, W.H. and A. Hartley, 1992. A global budget for atmospheric NH_3. Biogeochemistry 15: 191-211.

Sutton M.A., D. Fowler and J. Moncrieff, 1993. The exchange of atmospheric ammonia with vegetated surfaces. I. Unfertilized vegetation. Quarterly Journal of the Royal Meteorological Society 119: 1023–1045.

Szögi, A., 2006. Reduction of ammonia emissions from swine lagoons using alternative wastewater technologies. The Workshop on Agricultural Air Quality: State of Science, 4 June 2006. The Ecological Society of America. 2006. pp. 1155-1160.

US EPA, 2005. National emissions inventory, http://www.epa.gov/ttn/chief/net/2002inventory.html, United States Environmental Protection Agency, Washington D.C.

Van Breeman, N., P. Burrough, E. Velthorst, H. Van Dobben, T. de Wit, T. Ridder and H. Reijnders, 1982. Soil acidification from atmospheric ammonium sulphate in forest canopy throughfall. Nature 299: 548-550.

Vanotti, M., 2006. Greenhouse gas emission reduction and credits from implementation of aerobic manure treatment systems in swine farms. The Workshop on Agricultural Air Quality: State of Science, 4 June 2006. The Ecological Society of America. 2006. pp. 1178-1185.

Warneck, P., 2000. Chemistry of the Natural Atmosphere. Second edition, Academic Press, Inc., New York, NY, pp. 484-485, 511-528.

Zapletal, M. and P. Chroust, 2006. Spatial distribution of ammonia emissions on the territory of the Czech Republic. The Workshop on Agricultural Air Quality: State of Science, 4 June 2006. The Ecological Society of America. 2006. pp. 1258-1262.

Atmospheric ammonia: detecting emission changes and environmental impacts – summary report of the UNECE Edinburgh workshop

M.A. Sutton[1], E. Giles[1] and S. Baker[2]
[1]Centre for Ecology and Hydrology, Edinburgh Research Station, Bush Estate, Penicuik, Midlothian, EH26 0QB, Scotland, United Kingdom
[2]Department for Environment Food and Rural Affairs, Air Quality Division, Ashdown House, 123 Victoria Street, London, United Kingdom

Background

The Workshop on Atmospheric Ammonia: Detecting Emission Changes and Environmental Impacts was held on 4–6 December 2006 in Edinburgh (United Kingdom). It was organised and supported by the Centre for Ecology and Hydrology (CEH), the Department for Environment, Food and Rural Affairs (DEFRA), the Scottish Executive Rural Affairs Department (SEERAD), COST 729 and the NitroEurope Integrated Project (NEU). Background documents and presentations are available at www.ammonia-ws.ceh.ac.uk/documents.html. The full official report of the on which this summary is based can be found at http://www.unece.org/env/wgs/docs39th%20session.htm ,where it is also available in French and Russian. The full proceedings will be published in due course.

The workshop was attended by 80 experts from the following 19 Parties to the Convention: Austria, Canada, Croatia, the Czech Republic, Denmark, France, Germany, Hungary, Ireland, Italy, The Netherlands, Norway, Poland, Portugal, Slovakia, Sweden, Switzerland, the United Kingdom and the United States. The European Commission, the EMEP (Cooperative Programme for Monitoring and Evaluation of the Long-range Transmission of Air Pollutants) Meteorological Synthesising Centre – West (MSC-W), the International Institute for Applied Systems Analysis and the secretariat were represented.

Objectives of the workshop

The objectives of the workshop were to:

1. Assess the extent to which the existing critical thresholds for ammonia reflect current scientific understanding by:
    a. Examining the case for setting new ammonia critical threshold(s) based on current evidence of direct impacts of ammonia on different receptors.
    b. Discussing the extent to which vegetation and sensitive ecosystems appeared to be differentially sensitive to ammonia versus other forms of reactive nitrogen (N).
    c. Debating the case for establishing indicative air concentration limits for indirect effects of ammonia which would be consistent with current critical loads for N.
2. Assess the extent to which independent atmospheric measurements can verify where regional changes in ammonia (NH_3) emissions have and have not occurred by:
    a. Quantifying the extent to which estimated regional changes in NH_3 emissions have been reflected in measurements of atmospheric NH_3 and ammonium.
    b. Distinguishing cases where the estimated changes in NH_3 emission are due to altered sectoral activity or the implementation of abatement policies, and thereby assess the

extent to which atmospheric measurements verify the effectiveness of NH_3 abatement policies.
   c. Making recommendations for future air monitoring and systems for assessing the national implementation of NH_3 abatement policies and considering the implications of any non-linearities for integrated assessment models.
3. Review approaches for downscaling transboundary assessments to deal with ammonia hot spots in relation to operational modelling and monitoring by:
   a. Reviewing current emission and atmospheric dispersion modelling methods for downscaling NH_3 dispersion and deposition in hot spots.
   b. Examining the status of methods for effect assessment and monitoring in hot spots.
   c. Recommending broad principles for assessment approaches in ammonia hot spots, including spatial approaches and interactions between transboundary NH_3 emission reduction targets and other policy measures.
4. Review mesoscale atmospheric transport and chemistry models in relation to their formulation and results for NH_3 by:
   a. Reviewing emission parameterisations used in the models, establishing comparability, spatial and temporal resolution and uncertainties.
   b. Reviewing dispersion, air chemistry and deposition formulations identifying key differences and uncertainties.
   c. Assessing the overall performance of the models against measurements and against a common reference, and thereupon making recommendations for improving mesoscale models of NH_3 transport and deposition, including the implications of any non-linearities for source-receptor matrices and integrated assessment models.

Conclusions

Critical levels for gaseous ammonia

The current NH_3 critical levels (CLEs) for vegetation under the Convention, agreed in Egham (United Kingdom) in 1992, were based on measurements and observations from the 1980s, mostly from The Netherlands, and were set at 3,300 $\mu g\ m^{-3}$ (hourly), 270 $\mu g\ m^{-3}$ (daily), 23 $\mu g\ m^{-3}$ (monthly) and 8 $\mu g\ m^{-3}$ (annual). The workshop concluded that these levels required revision in light of new evidence from field-based experiments and surveys.

The existing annual CLE (8 $\mu g\ NH_3\ m^{-3}$), when expressed as an equivalent deposition of N to an ecosystem, was less protective than the current critical load for most, if not all, European ecosystems and habitats. Field-based evidence relating effects on vegetation to NH_3 concentrations measured over one year or longer showed that the current annual CLE was too high.

A new long-term CLE for the most sensitive vegetation types (lichens and bryophytes) and the associated habitats was proposed, based on observed changes to species composition in the field. Most of the evidence came from studies in the United Kingdom, but there was corroborative evidence from Italy, Portugal and Switzerland. The proposed long-term CLE for NH_3 for (a) sensitive lichen communities and bryophytes, and (b) ecosystems where sensitive lichens and bryophytes were an important part of the ecosystem integrity was set at 1 $\mu g\ NH_3\ m^{-3}$.

There was less evidence available to quantify the concentrations at which long-term effects of NH_3 caused species changes in communities of higher plants. The workshop proposed a long-term CLE for higher plants of 3 μg NH_3 m^{-3}. This value was set for higher plants in general, but was particularly based on data from heathlands and forest ground flora. Given the larger uncertainties in this estimate, an uncertainty range was proposed of 2 to 4 μg m^{-3}, depending on the degree of precaution appropriate to different contexts.

On the basis of current knowledge, it could not be assumed that each of these new long-term CLE values would be protective for periods longer than 20–30 years. No assumptions had been made on the mechanism by which NH_3 exposure led to changes in species composition. Further details could be found on the website of the workshop. By emphasising long-term rather than daily NH_3 concentrations, the NH_3 critical level was concluded to have the advantage of providing a practical tool complementing the critical loads approach which was simple to apply for cost-effective regulation and monitoring of NH_3 specific measures.

Detecting changes in atmospheric ammonia

The workshop discussed progress in the state of knowledge in deriving trends from measurements and their use to verify abatement measures or other causes for decrease in emissions of NH_3 to the atmosphere. The workshop identified clear progress in closing the gap between the observed and expected values for reduced N, as well as a better understanding of the reasons behind this.

The long-term measurements available followed the emission trend. Current measurements made it possible to evaluate policy progress on NH_3 emission abatement. In those countries where there were big (>25%) changes in emissions, such as in The Netherlands and Denmark, the trend followed closely, especially when meteorology was taken into account. In other countries, such as the United Kingdom, the trend was much smaller, but there was no significant gap between measurements and model estimates. In the Netherlands, there was still an NH_3 gap – a significant (30%) difference between emissions-based NH_3 concentrations and measurements – but the temporal trend was the same. The difference might be due to either an underestimation of the emission or an overestimation of the dry deposition.

On the European scale it was difficult to match the emission change, both because of lack of measurements, especially in the eastern part of Europe, and because of the confounding factor of the SO_2 emission reductions, which affect the ammonium concentrations in aerosol and rain water.

Assessment methods for ammonia hot spots

The workshop agreed that accounting for hot spots for either upscaling of fluxes or risk assessment for nearby ecosystems required a precise description of all processes involved. Hot-spot assessment should also account for background concentrations and deposition history.

The key uncertainties in the models were emissions and dry deposition. Sufficient local input data were required for making effects assessments and landscape analyses. For dry deposition, this required better knowledge of NH_3 compensation points and surface resistances for different ecosystems, their dependence on climatic variables and the deposition history for NH_3 and other pollutants.

The workshop concluded that using different models allowed the analysis of landscape interactions between sources and receptors with sufficient accuracy for a range of conditions to consider real cases and scenarios. It also allowed the assessment of local, tailored abatement measures.

The workshop agreed that scenarios from local-scale modelling could be used in a statistical way to provide estimates of within-grid cell recapture for national- and regional-scale models, linked with global descriptors of the spatial variability in land cover.

Regional modelling of atmospheric NH_3 transport and deposition

A range of chemical transport models was used across the Convention to model the emission, transport and deposition of atmospheric NH_3 on the national and regional scales. These models had been developed from a range of historical backgrounds and with different purposes. Six models were considered, ranging from the national scale up to the full European scale. The models differed in concept, particularly in their chemical scheme and in scale, ranging from Lagrangian models on the national scale via Eulerian models on the European scale to nested models coupling the European scale with the local scale.

Key uncertainties in the modelling of atmospheric NH_3 were linked to emissions (absolute level and spatial and temporal allocation) dry deposition parameterisation, spatial resolution of the model and the description of vertical diffusion. All European-scale models (including the EMEP model) currently underestimated the measured NH_3 concentration. National models generally found better agreement with NH_3 measurements. The main reasons for the observed differences between the measured and modelled NH_3 concentrations were the spatial resolution of the models and the parameterisation of the dry deposition process.

The concentration of ammonium aerosol was fairly well described by all models. However, both under- and overestimates of measured concentrations were found. The magnitude of the wet deposition of ammonium was in general reproduced well by all models.

None of the models routinely used the compensation point (bidirectional exchange scheme) as a parameterisation in the dry deposition process of NH_3. This was thought to be one of the reasons why some models tended to underestimate concentrations, particularly in summer. The main reason for not taking this process into account was the lack of a generalised database for the compensation point with respect to the main land cover types used in the models.

The siting of measurements played an important role in the comparison with modelled concentrations. Some stations in agricultural areas should not be used for verification of the Eulerian models with large grid size (50 km), because of the significant contribution of sources close to the measurement stations that cannot be simulated by the models on this spatial scale.

Reliability of NH_3 emission estimates and abatement efficiencies

Few countries had considered uncertainty in NH_3 emissions in detail. Results indicated that national estimates may be accurate to within ±20%. For countries that had created inventories using emission factors (EFs) measured elsewhere, the uncertainty may be around 100%. The greatest uncertainty was likely to be for emission estimates for regions within countries. Sensitivity analysis of the United Kingdom inventory showed that activity data and other information on a range of relevant farming practices were the inputs for which the system was most sensitive. Cattle diets, especially grass-based ones, were considered particularly uncertain.

The United Kingdom and Denmark reported a high level of agreement between modelled and measured NH_3 concentrations, while models still underestimated measurements in the Netherlands. A detailed discussion of the Dutch 'ammonia gap' suggested that the EFs used in the Dutch inventory were accurate. The discrepancy was considered to result either from overestimation of abatement efficiencies or from overestimation of dry deposition velocities. Adjustment of either could eliminate the gap, but it was not yet known which was responsible.

The abatement efficiencies in the guidance document on ammonia to the 1999 Gothenburg Protocol were considered robust. While averages did not reflect the variability in data, quoting ranges may create uncertainty regarding which point in the range is most appropriate to use. Since data were obtained almost exclusively from Northern and North-Western Europe, abatement efficiencies could not be assumed to be applicable across the whole UNECE region. Only a brief statement is given in the guidance document on the impacts of reducing emissions of NH_3 following spreading on losses of other N pollutants, because nitrate leaching and nitrous oxide emissions tend to be site- and season-specific.

'Soft' approaches to NH_3 abatement were those implemented using basic facilities and simple management approaches (e.g. applying manure during weather conditions associated with little emission). While these offered an economically attractive method of reducing NH_3 emissions, it was often difficult to know their uptake by farmers and their efficiency, and therefore to convince environmental authorities of their effectiveness or to measure the achievement in national reporting.

Experience from the adoption of abatement technologies in other areas, suggested that *ex ante* cost assessments tend to overestimate the cost of implementation. However, taking emerging technologies into the industry could lead to a reduction in abatement efficiency. A number of emerging abatement options would be discussed in the full report of the workshop, together with a summary of other developments that may affect NH_3 emissions.

Ammonia policy context and future challenges

Ammonia emissions are major contributors to eutrophication and acidification of ecosystems and secondary $PM_{2.5}$ concentrations in Europe. Reduction of NH_3 emissions in Europe has been on the agenda for more than a decade, first on a national scale (e.g. in The Netherlands) and more recently through international efforts. The latter include the Convention's Protocols and EU directives and strategies.

The workshop considered the policy context of the NH_3 problem, including socio-economic, environmental, institutional and technological aspects, and the potential role of policy options in mitigating the ecosystem and health impacts of NH_3 emissions. The need to adapt tools used in policy analysis, such as integrated assessment models, and to consequently evaluate policies in view of new findings was also considered.

Ammonia policies were becoming interlinked with a number of other environmental and agricultural policies. In order to avoid the problem of 'pollution swapping', future policies needed to consider these interactions.

The workshop noted that, in responding to some of the policies like the EU Nitrates Directive (91/676/EC) or biodiversity-related directives, farmers in certain areas adjusted agricultural practices (e.g. by shifting application of manures from autumn to spring). This led to different seasonal patterns of NH_3 concentrations, although there was little knowledge of the environmental consequences.

Recommendations

A series of recommendations were made based on these conclusions. These are specified in full in the official UNECE report (url given above). The recommendations included:

a. specifying the new values for critical levels, including the future research needs;
b. exploring further the gap between ammonia measurements and model estimates, including the need for a model intercomparison;
c. fully implementing the EMEP monitoring strategy to provide the necessary comprehensive measurements of gaseous and aerosol NH_x;
d. further developing dynamic ammonia emission models to estimate diurnal and seasonal changes in emission strengths;
e. syntheising information from available databases to identify reference cases for assessment of ammonia hot-spots, including investigation of landscape level spatial scenarios;
f. developing a more generalised parametrisation to use in ammonia compensation point modelling in relation to the main land cover types;
g. conducting a coordinated intercomparison of mesoscale models for ammonia using a common model domain;
h. devoting effort to estimating the uncertainty of regional and national NH_3 emission inventories, including international collaboration to obtain better activity data regarding agricultural management practices across Europe;
i. to further examining the quantitative synergies and trade-offs that occur in abating different forms of nitrogen emission (NH_3, nitrous oxide, nitrate leaching);
j. putting further putting further research effort into methods to quantify the achievement of 'soft measures' to reduce ammonia, related to good agricultural practice and other novel strategies, so that the benefits can be considered within the Convention.

Finally, considering the N trade-offs, the workshop recommended the extension of currently used tools, verification of specific elements of the models, adaptation of monitoring networks, targeted measurement programmes, and possible revision of legislation in order to close existing loopholes and increase synergies in addressing nitrogen pollution at large. Priority should be given to measures aiming at reducing all kinds of nitrogen losses at farm level. Ammonia emission reduction policies must be analysed in a multi-effect (human health, greenhouse balance, acidification and eutrophication and related biodiversity loss), multi-media (air, water, soil), multi-scale (hot spots, regional, European, global) framework.
Considering the recommendation to lower critical levels for NH_3, there was a need for careful evaluation of the representativeness of EMEP modelling results for NH_3 concentration. It was also recommended to give further consideration to whether, and if so, how, the new critical levels would be used in addition to critical loads in formulating air pollution targets, especially on local or regional levels in areas with spatially variable NH_3 emissions and concentrations.
Considering the increase in springtime NH_3 emissions that had occurred in implementing some policies such as the EU Nitrates Directive, further research was recommended to quantify the seasonal dependence of environmental impacts of NH_3. More attention was also needed on how to monitor and incorporate impacts of other N-related policies in modelling tools.
It was recommended to explore possibilities of considering local Biodiversity Action Plans in larger scale modelling. Strategies existed to integrate them into the European scale, e.g. via the Flora Fauna and Habitats Directive and Natura 2000 network. However, the role of air pollution effects was often not explicitly taken into account even though N inputs had a large effect on biodiversity; there was room for improvement on local, national and European levels.

Reduced nitrogen in ecology and the environment

J.W. Erisman[1] and M. Sutton[2]
[1]Energy research Centre of the Netherlands (ECN), The Netherlands
[2]Centre for Ecology & Hydrology (CEH), United Kingdom

Introduction

Nitrogen, contained in amino acids, proteins, and DNA, is necessary for life. While there is an abundance of nitrogen in nature, almost all is in an unreactive form (gaseous nitrogen, N_2) that is not usable by most organisms. In the absence of human intervention, the supply of reactive nitrogen in the environment is not sufficient to sustain the current abundance of human life. Thus humans learned in the early 20th century how to convert gaseous N_2 into forms that could sustain food production. Over 40% of the world's population is here today because of that capability.

There are two major problems with nitrogen: some regions of the world do not have enough reactive nitrogen to sustain human life, resulting in hunger and malnutrition, while other regions have too much nitrogen (due mainly to the burning of fossil fuel and to the inefficient incorporation of nitrogen into food products) resulting in a large number of major human health and ecological effects. The rate of change of the problem is tremendous, probably greater than that for any other major ecological problem. For example, half of the synthetic nitrogen fertiliser ever used on Earth has been used in just the last 15 to 20 years. A prerequisite to reducing these problems is the development of a sound scientific base on which to begin to discuss policy options. This paper is a summary of the most important conclusions of the ESF-FWF Conference in Partnership with LFUI on Reduced Nitrogen in Ecology and the Environment, organised in the Universitätszentrum Obergurgl, Austria in 14-18 October 2006 (www.esf.org/conferences/lc06203). The scientific basis for reduced nitrogen in ecology and the environment needs to be strengthened. This conference was the first major step to develop this basis.

The least known part of the nitrogen cycle is the reduced nitrogen form. Reduced nitrogen, such as ammonia, ammonium and amines are essential in food production, in ecology and also in the environment. So far, reduced nitrogen has not been regarded as a serious issue for the environment and only relevant on the local scale because of the short atmospheric lifetime. The production and use of fertiliser, industrial activities and traffic (three-way catalyst) lead to emissions of reduced nitrogen to the environment. The biggest source, however, is related to animal excretion in extensive livestock breeding. Emissions of ammonia to the atmosphere can contribute to particulate matter formation affecting human health, decreasing biodiversity, contributing to eutrophication and acidification of aquatic and terrestrial ecosystems after deposition and to nitrous oxide formation, contributing to the greenhouse effect. The increased application rates of reduced nitrogen can cascade through the environment and contribute to different effects until in the cascade the reactive nitrogen is stored or converted back into N_2. The scale of these environmental effects has been extended from a few hotspots regions such as Denmark, The Netherlands, parts of France, the Po Valley, to the whole of Europe. There are clear signals that the effects of reduced forms of nitrogen in the environment can be different from the oxidised forms especially in relation to ecosystem effects and biodiversity. The most important findings of the Conference are summarised for the most important issues:

- atmosphere – biosphere interactions;
- agriculture and other emissions of ammonia;
- plants, soils and reduced nitrogen;
- environmental impacts;
- ammonia and air pollution;
- ammonia and coastal areas.

Atmosphere-biosphere interactions

Ammonia is essential for life and we industrially create 142 MTonnes annually to be used in a very inefficient system to sustain the relevant amino acids and other reduced nitrogen compounds in food.

The origin of ammonia was traced back to c. 100 AD in China and before that, in the first centuries BC, in Central Asia. Important factors determining the routes to Europe are the Silk routes, and the interest in ammonium chloride by Islamic alchemists, which probably developed from expertise in dyeing textiles and metal smithing as well as mystery rituals in the Hellenistic orient. The subsequent history of ammonia was traced through to the development industry in the 18th century and the developments of understanding reduced nitrogen in the atmosphere through the 19th century, which provided the foundations for the present study of reduced nitrogen in both atmospheric chemistry and agriculture. However, it was shown that the 19th chemists analysing for NH_3 appear to have worked in atmospheres strongly contaminated by local NH_3 sources, perhaps of human origin, so care is needed in interpreting these early datasets.

Through an enquiry the conference identified the 5 most important developments in ammonia research these past years through a ranking.

It has been demonstrated in the laboratory that plants can grow solely on atmospheric NH_3. Root uptake of NO_3 is lowered during these exposures. Very high exposure of atmospheric NH_3 can be dealt with by the plant until toxicity is reached. There are fewer indications of direct toxicity at long-term exposure with low to moderate concentrations of NH_3, although significant species changes do occur in response to chronic ammonia exposure.

Monitoring data are becoming available to show changes in concentration and deposition as the result of changes in emissions, but a more rigorous analysis is necessary. Furthermore, a monitoring and modeling strategy is necessary, especially because of the large spatial and temporal variations in NH_3 emissions and the relatively short atmospheric lifetime.

Organic N compounds are not well quantified in biogeochemical nutrient cycling. There are indications of methylamine emissions along with ammonia emissions. Methylamines are much more reactive and provide more basic neutralisation the atmosphere. The deposition of organic N and its contribution to N cycling in ecosystems needs further quantification.

There are differences in aerosol neutralisation in Europe and North America leading to different appreciations of the role of NH_3 in atmospheric chemistry. In Europe, sulphate is mostly neutralised by ammonia and nitrate equilibrium changes, influencing the transport distance of ammonium. In North America acid aerosol is generally not fully neutralised by ammonia. The neutralisation factor of aerosol and precipitation might be a good indicator for the relevance and emissions of NH_3.

Within the quantification of the atmosphere-biosphere interactions of ammonia, models have been developed these past years. The uncertainty in these models is dominated by the cuticle (and water layer) uptake and release of ammonia and the apoplast concentrations regulating the stomatal uptake.

Agriculture and other emissions of ammonia

There is still a big gap in estimated nitrogen balances at different scales. Much farm-level research has shown that the gap between input and output of N can be up to 40%. It is suspected that de-nitrification is the most important process that can explain the gap, even though organic nitrogen, which is often not fully included in N balances, might play a role.
Nitrogen efficiency in crop and/or meat production never reaches 100%. We always have to take into account losses to the environment. Depending on the system, the maximum nitrogen efficiency feasible is 40-50%. There is a need of a comprehensive assessment of reduced nitrogen Best Management Practices and how technology can be brought to lead to further reductions of reduced gases.
Currently volcanic emissions are not included in the emission inventories and might contribute to the global balance. Other natural emissions come from large colonies of seals or penguins. These emissions emissions from colony animals appear not to be a significant contribution to the overall global budget, but they contribute locally to massive nitrogen enrichment, especially since they occur in remote areas with few other fixed nitrogen sources. Because they represent isolated hotspots, they are ideal to study ammonia emissions, transport and effects.
Bat guoano deposits of several meters in caves were also shown to represent a research opportunity to examine concentration and deposition at locations with remote background. The caves also provide valuable information on the toxicity to a range of organisms, possibly including humans.
Inventories of agricultural emissions are becoming available for different areas of Europe and North America. The spatial scale needs to be refined in order to meet the requirements of dispersion models to assess the transport and deposition. Furthermore, the temporal variation in emission factors can be very large both from day to day as well as during the seasons. When using emission inventories the temporal variation need to be addressed separately.

Similarly, there remain significant industrial and energy and transport sources of reduced nitrogen and N_2O, as well as the well known sources of NO_x. For example, energy production and vehicle technologies to reduce NO_x emissions are leading to increases in both N_2O and NH_3 emissions.

Plants soils and reduced nitrogen

The increased use of biofuels is likely to be a counterproductive approach to mitigate global warming because the fuel energy gained from different biofuel crops might be offset against the nitrogen inputs and associated nitrous oxide emissions from these crops. N_2O is a 300 times more effective greenhouse gas than CO_2 and therefore, a small increase in N_2O emissions resulting from additional fertiliser use can easily offset large CO_2 reductions through the replacement of fossil fuels by biofuels.
Per unit of N deposition it was shown that gaseous NH_3 deposition was having substantial effects on both the shrub *Calluna vulgaris*, lichens and mosses, while effects of wet deposition were difficult to detect or so far small. A key finding was that chronic effects of NH_3 occurred over periods longer than one year, with the zone of dead heather and lichens gradually spreading,

so that the threshold for effects (critical level) decreased with time, on with a response of the function log [NH_3 CL] = - a time + b. For the wet treatments, although little damage to vegetation had so far been recorded, foliar chemistry and 'vitality' measurements indicate a larger response to NH_4 compared with similar doses of NO_3.
Experiments where reduced N had been added to a forest in Sweden were studied for many years after addition stopped and it was shown that while some ground flora species recovery occurred, this was non complete even in one experiment after 50 years. This demonstrates the long term effects of nitrogen.
The importance of organic nitrogen in atmospheric inputs to forests and other ecosystems is not well quantified and therefore a priority for future research. The role of organic nitrogen in nutrient cycling should be focused on qualifying uptake by different species, the sources of organic N and the output of the systems. Furthermore, it is a challenge to quantify N inputs and the differences between that to different receptors (e.g. ground flora vs. overall forest canopy).
While there is some information on N_2O emissions from Mediterranean shrublands, there is a complete lack of data on NH_3 emissions from semi-natural leguminous systems. Given the global significance of the amounts of nitrogen fixed in this way, this must be a priority for future research. There is at present a wide gap between detailed ecosystem process models to interpret field data of reduced nitrogen exchange with the atmosphere and approaches currently used for global upscaling. Bridging this gap should be a priority for the future.

Other environmental impacts

It was noted that there are major differences in sensitivity of different ecosystems to nitrogen inputs and differences between nitrogen form applied. For example, sub-artic vegetation is very sensitive to even moderate nitrogen loads.
Reduced N is seen to be much more harmful for (semi)natural vegetation compared to oxidised N, not only in heath lands, grasslands and soft-water lakes but also in ecosystems which are not sensitive to acidification like mires, fens and coastal waters. There are also observations of severe effects of reduced N on vascular plants and lichens in the Mediterranean region. Underlying biogeochemical cycles of nitrogen in Mediterranean ecosystems appears closely coupled to the plant functional types with the largest floral biodiversity in very limited NH_4 availability in sites.
Bog species are during relatively short term exposure much more sensitive to NH_3 compared to NH_4. It is known that harmful effect of wet deposited NH_4 also occur after long-term exposure, after saturation of the bog with N.

Ammonia and air pollution

Responses in ambient concentration and deposition to change in NH_3 emission are seasonally well understood. Where substantial reductions have been made, the benefits are clear from decreases in aerosol ammonium concentrations and ammonium in precipitation However, reductions in gaseous ammonia were in several instances shown to be much less than expected. These differences may in part be explained by changes in atmospheric chemistry over recent decades.
For example, interactions between SO_2, NO_x and NH_x have created a change in the main form of NH_x from $(NH_4)_2SO_4$ to NH_4NO_3, associated with reduced nitrogen occurring with a higher

ratio of gaseous NH_3 to aerosol NH_4. Increases in ambient NH_3 and NH_4 in precipitation appear to have occurred in remote areas of Europe, and these require quantitative explanation and more detailed measurements to demonstrate the cause. In addition, short term changes in emissions are not well quantified, nor in the emissions, nor in the responses in concentrations and deposition. Models and input data are still inadequate to show these changes.

The role of cuticular resistances in NH_3 fluxes to semi-natural vegetation has been shown to be the most important in regulating the deposition flux. This resistance determines which NH_3 concentration and which responses have key implications and more work is necessary.

Local assessments in Europe and North America suggested that dry deposition of NH_3 close to major sources contributes typically 4 to 10% of the total emissions within 500 m of the source, although these fractions are shown to depend on downwind canopy types and climate, and are currently very uncertain.

Ammonia and coastal areas

New concepts of nitrogen removal from the oceans were presented, in particular the occurrence of the anamox bacterial process for denitrification of ammonium and nitrite to dinitrogen in marine sediments. It remains doubtful, however, whether this newly recognised process is of major significance for marine nitrogen budgets.

Nitrous oxide emissions from the oceans, particularly in coastal and near-shore areas, are suspected of being underestimated due to insufficient spatial (and temporal) sampling/ measurement. There is a need for studies and establishment of concentration fields in both key coastal and oceanic areas.

Previously, it conventional wisdom was that oceans everywhere could represent a source of ammonia to the atmosphere. It appears that this is an oversimplification. Measurements have shown that low latitude (warm) waters can be regarded as sources, whereas high latitude (cold) waters are generally sinks. This is explained by strong temperature dependence of the Henry's Law constant for ammonia and leads to a significant reduction in the global marine flux of ammonia to the atmosphere.

Uncertainty in the size and sign of the spatial distribution of the air-sea flux of ammonia is mainly due to the poorly defined concentration driving force (difference between normalised air and water concentrations). There a few reliable measurements. This situation is in urgent need of attention if we are to understand the role that ammonia from the oceans plays in particle formation and in controlling the pH of rain and particles in the remote marine atmosphere.

Methylamines are stronger bases than ammonia and so, although their fluxes across the air-sea interface are only a few per cent of those for ammonia, they can therefore constitute 10-20% of the atmospheric supply of volatile bases in offshore marine environments. There is however little research on methylamines and it is advised to foster more research in this area.

Conclusions

The major conclusions and directions for scientific research can be summarised as follows:

1. *Global and regional cycles and reservoirs:* The ESF conference showed clearly that there is a need for a new thorough updating of the N cycle of at least the anthropogenic part, improving data on fluxes, reservoirs and also on uncertainties. This should be done for regions and globally. Already at the conference some new insights were presented and scientific unknowns were discussed. Among the most uncertain factors identified were:

- Missing sources for N_2O.
- Volcanic emissions of ammonia.
- The importance of amanox reaction as a sink for fixed nitrogen.
- The influence of climate change on the storage capacity of nitrogen.
- The role of the oceans in the N cycle.
- The long term stability of N in various terrestrial systems (soils).

2. *Process understanding:* New processes, which need to be further explored:
   - Amanox.
   - Flows of organic nitrogen.
   - Plant physiology: de-toxification.
   - Ocean emissions and uptake of NH_3.

   New data on known processes, which might help progress in these areas:
   - Temporal variation in agricultural emissions.
   - Atmospheric chemistry: aerosol formation.
   - Transport distance of ammonia: deposition processes.
   - Farmgate nutrient balances.
   - Cycling of reduced N in Mediterranean ecosystems.
3. Effects in biological systems:
   - Biodiversity: N induced vegetation deceases. Effects of low doses of N. New critical loads for nitrogen (separate for reduced and oxidised N). Ecosystem studies. There is a distinct difference between the effect of reduced and oxidised forms of nitrogen that need to be further explored.
   - Carbon sequestration: Nitrogen plays an important role in carbon sequestration in aquatic and terrestrial ecosystems, stimulating growth and thereby CO_2 uptake. However this relationship is not linear and varies in space, time and type of system, with overall effects on soil carbon rather uncertain. This needs to be further explored. Next to this, the additional nitrous oxide emissions resulting from increasing fertiliser use for biofuel production remains a key trade-off that must be considered. There is large uncertainty on the direct and indirect emissions of N_2O and thus doubts remain concerning the carbon neutrality of biofuels.
4. *Anthropogenic systems and N control:* New research items were identified:
   - Nitrogen and bioenergy.
   - The use of organic N as a fertiliser.
   - Amanox as a way of denitrification.
   - Control options.

References

This overview is based on the presentations by the different contributors to the conference. Their presentations can be found on: www.nine-esf.org.

Comprehensive reporting of agricultural emissions

U. Dämmgen[1], N.J. Hutchings[2] and J. Webb[3]
[1]*Federal Agricultural Research Centre, Institute of Agroecology, Braunschweig, Germany*
[2]*Dept of Agroecology, University of Aarhus, Research Centre Foulum, Tjele, Denmark*
[3]*AEA Energy & Environment, Harwell, United Kingdom*

Introduction

Accurate inventories of gaseous emissions are required to calculate total national emissions, identify the major sources and hence to develop effective abatement policies. In addition emission inventories may be used to describe past and present emissions (emission reporting), describe processes generating emissions and develop scenarios of future emissions (policy planning).

The first inventories of ammonia (NH_3) emissions from livestock production were calculated by multiplying livestock numbers by emission factors (EFs) per animal (e.g. Buijsman *et al.*, 1987). This approach did not allow for significant differences in the potential for NH_3 emissions due to differences in diet and hence N excretion, or differences in livestock and manure management between countries and regions. Subsequent inventories have replaced EFs per animal with specific EFs for the different phases of manure management, i.e. during animal housing, during manure storage, after manure spreading, and for different housing, storage or spreading systems, etc. (e.g. Misselbrook *et al.*, 2000). Such completeness of an inventory is state of the art. However, complete inventories using standard procedures are normally not 'mechanistic', i.e. they do not 'trace' a chemical element. In addition, increasing the number of EFs to account for emissions at each stage of manure management and discriminating between systems and abatement measures, makes the calculation of the interactions between abatement measures complicated. In particular, such an approach may fail to recognise that introducing abatement at an early stage of manure management, e.g. housing, will, by conserving NH_3-N, increase the potential size of NH_3 emissions later, i.e. during storage or after spreading. Thus a mass-flow approach is needed, and is particularly important when attempting to rank the costs of introducing measures to reduce NH_3 emissions.

Materials and methods

Complete is not synonymous with comprehensive. Comprehensive treatment allows for mass conservation and in many cases reduces calculated emissions. Comprehensive treatment, by accounting for N at each stage of the manure management process, detects side effects and estimates interactions.

In order to account for all losses and transformations of N, so that full account of changes in TAN during manure management, estimates need to be made of other gaseous emissions, mineralisation and, for litter-based manures, immobilisation. Examples include significant losses of N_2O emissions from poultry housing and from stored FYM and poultry manure (Chadwick *et al.*, 1999). When conditions favourable to both nitrification and denitrification are created, as is the case in aerated slurry stores and in FYM heaps, N_2 emissions may take place (e.g. Petersen *et al.*, 1998). Some of the N put into FYM stores is lost by effluent leaching (Amon *et al.*, 1997). And some of the TAN in urine may be immobilised in litter (Kirchmann and Witter, 1989).

Results and discussion

Thus, if in order to accurately model emissions of NH_3, it is necessary to quantify other gaseous emissions and transformations of TAN, then the logical step is to create mass flow models that act as inventories for all N species. Indeed, it is logical to go further and embrace the calculation of emissions of C species as well, since livestock excreta and manures are also sources of methane (CH_4)(Chadwick *et al.*, 1999) and non-methane VOCs (Hobbs *et al.*, 2004). So, while complete inventories treat each single chemical species on its own, comprehensive inventories identify homogeneous source categories (with respect to excretion, etc.), treat the categories with regard to their energy and mass flows (nitrogen and carbon), include options to incorporate future technical developments and avoid black boxes as far as possible. Inputs and outputs can be coupled with other models (environmental restrictions, political targets, herd management, cost calculations).

Conclusions

Comprehensive treatment may result in modelling agricural systems using a Mass Flow Approach for N and C species, dealing with the flows of: total N, TAN, total C, volatile solids: combined with the flow of energy. Comprehensive reporting – a dream or a nightmare?

References

Amon, B., J. Boxberger, T.H. Amon, A. Zaussinger and A. Pöllinger, 1997. Emission data of NH_3, CH_4 and N_2O from fattening bulls, milking cows, and during different ways of storing solid manure. In: Voermans, J.A.M. and Monteny, G.J. (Eds.), Ammonia and Odour Control from Animal Production Facilities. Elsevier, pp. 397-404.

Buijsman, E., J.F.M. Maas and W.A.H. Asman, 1987. Anthropogenic Ammonia Emissions in Europe. Atmospheric Environment 21: 1009-1022.

Chadwick, D.R., R.W. Sneath, V.R. Phillips and B.F. Pain, 1999. A UK inventory of nitrous oxide emissions from farmed livestock. Atmospheric Environment 33: 3345-3354.

Hobbs, P.J., J. Webb, T.T. Mottram, B. Grant and T.M. Misselbrook, 2004. Emissions of volatile organic compounds originating from UK livestock agriculture. Journal of the Science of Food and Agriculture 84: 1414-1420.

Kirchmann, H. And E. Witter, 1989. Ammonia volatilization during aerobic and anaerobic manure decomposition. Plant and Soil 115: 35-41.

Misselbrook, T.H., T.J. Van der Weerden, B.F. Pain, S.C. Jarvis, B.J. Chambers, K.A. Smith, V.R. Phillips and T.G.M. Demmers, 2000. Ammonia emission factors for UK agriculture. Atmospheric Environment 34: 871-880.

Petersen, S.O., A.-M. Lind and S.G. Sommer, 1998. Nitrogen and organic matter losses during storage of cattle and pig manure. Journal of Agricultural Science 130: 69-79.

Feeding and management: pigs

Nutritional options to reduce ammonia emission from excreta of pigs

A.W. Jongbloed, A.J.A. Aarnink and C.M.C. van der Peet-Schwering
Wageningen UR, Animal Sciences Group, P.O. Box 65, 8200 AB Lelystad, The Netherlands

Introduction

The main part of ammonia emission from pig houses originates from manure storage and fouled floors. The urea concentration of urine and the pH of faeces and urine are important characteristics of excreta determining the rate of ammonia emission. The urea concentration is highly dependent on protein nutrition and can be altered by changing dietary protein content. Other relevant dietary factors are (fermentable) non-starch polysaccharides (NSP) and acidifying salts. Based on research of the last decade at our institute, the results of monofactorial effects of these factors and their combined effects are described. Effects were quantified using an *in vitro* laboratory set up (Canh, 1998; Le, 2006).

Effect of dietary protein level

Several studies show that lowering dietary protein levels has a profound effect on ammonia emission (e.g. Canh, 1998; Le, 2006; Figure 1). Figure 1 shows a much lower NH_3 emission at lowered protein contents. As a rule of thumb, each percentage-unit lower protein content reduces ammonia emission by 10%. Lowered protein content, without compromising animal performance, can be achieved by supplementing free amino acids although there are limits on supplementary levels.

Effect of non-starch polysaccharides

Non-starch polysaccharides are not digested in the small intestine but can be fermented by microbes in the large intestine of pigs. Microbes use NSP (as a source of energy) and undigested protein (as a protein source) to synthesise microbial mass. Surplus of undigested protein can be used as energy source as well. In that case protein is broken down and the released ammonia is absorbed into the blood circulation and excreted as urea via urine. The result is a decreased ratio of urine-N and faecal-N (Figure 2 left; 8 experiments). Thus, increased dietary fermentable NSP

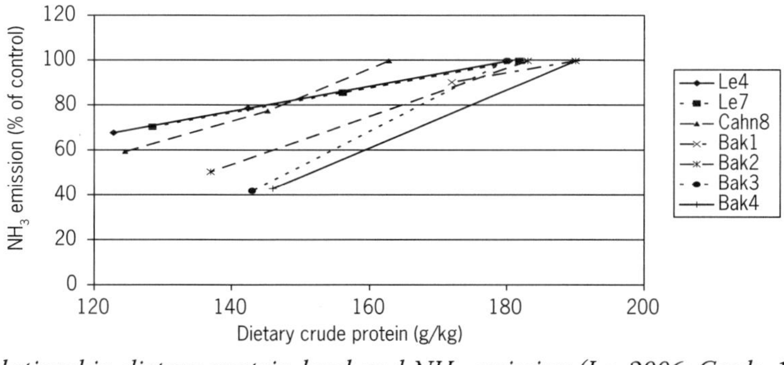

Figure 1. Relationship dietary protein level and NH_3 emission (Le, 2006; Canh, 1998; Bakker, 2004 pers. comm.; number after name in legend indicates chapter or treatment).

concentrations lead to a reduced NH_3 emission (Figure 2 right). In practical feeding of growing pigs, the problem is the lowered net energy content of most NSP-rich raw materials, resulting in lowered energy intake.

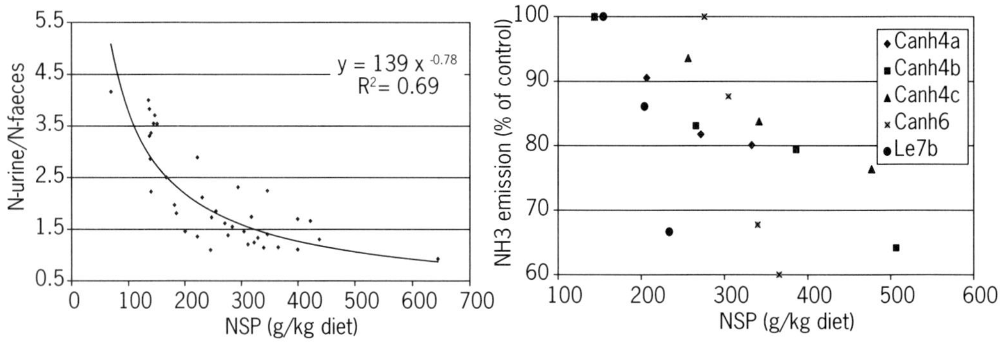

Figure 2. Effect of NSP content on N-urine/N-faeces ratio (left) and on NH_3 emission (right).

Effect of acidifying salts

Acidifying salts affect urinary pH, and therefore, may reduce NH_3 emission from fouled floors and from the manure pit. This was clearly shown by Canh (1998; Figure 3) using $CaCO_3$, $CaSO_4$, Ca-Benzoate or $CaCl_2$ at two supplementation levels to obtain 7 and 10 g Ca/kg diet, except for $CaCO_3$, which was also used at 4 g Ca/kg. Diets with a high dietary electrolyte balance (dEB, meq/kg; $Na^+ + K^+ - Cl^-$) and with a low dEB were used.

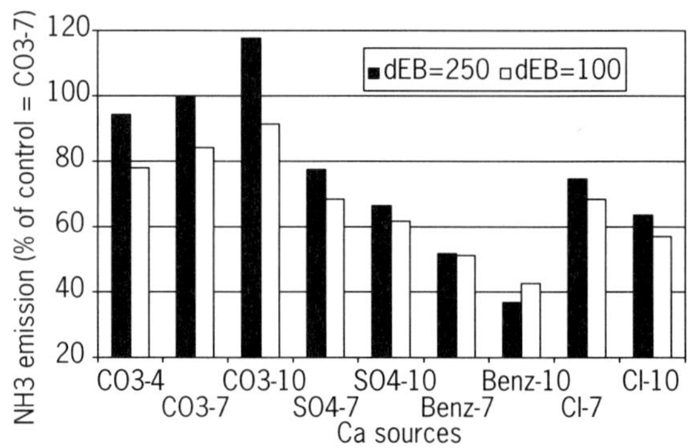

Figure 3. Effect of acidifying salts on NH_3 emission (Canh, 1998).

Combined effect of dietary factors and conclusion

In a large scale balance experiment using a central composite design, 26 different diets were tested differing in levels of protein (142, 161 and 180 g/kg), $CaSO_4$ (added: 0, 9 and 18 g/kg), NSP (140, 210 and 280 g/kg) or fermentable NSP (60, 90 and 120 g/kg) to evaluate possible interactive effects of the dietary factors imposed on NH_3 emission. Results showed that there was no interaction among the treatments. This means that the effects on NH_3 emission were additive. Cumulative NH_3 emission could be estimated by the following equation:

NH_3 (g/7d) = -5.347 + 0.056 x protein – 0.050 x acidifying salt – 0.010 x NSP, R^2 = 93.2; protein, acidifying salt and NSP are in g/kg.

It is concluded that dietary factors imposed to reduce NH_3 emission from pig houses are additive within the ranges studied. The *in vitro* model corroborated well with *in vivo* studies.

References

Canh, T.T., 1998. Ammonia emission from excreta of growing pigs as affected by dietary composition. PhD thesis, WIAS, Wageningen, The Netherlands, 163 pp.

Le, D.P., 2006. Odour from pig production: Its relation to diet. PhD thesis, WIAS, Wageningen, The Netherlands, 206 pp.

Effect of different feeding strategies on the ammonia emission from a fattening pig house

H. Cnockaert[1], M.J. Van Oeckel[2], D.L. De Brabander[2] and B. Sonck[1]
[1]Institute for Agricultural and Fisheries Research, Technology & Food Agricultural Engineering (TV-AT), B. Van Gansberghelaan 115, 9820 Merelbeke, Belgium
[2]Institute for Agricultural and Fisheries Research, Animal Science Unit, Scheldeweg 68, 9090 Melle, Belgium

Introduction

In Flanders it is since October 2003 enforced by law that new pig and poultry houses must be built according to a 'List of Building Techniques for the Reduction of Ammonia Emission'. The application of an adapted feeding strategy to reduce the NH_3 emission is not included, especially because of the lack of sufficient reliable scientific research reports. So, the objective of this investigation is to examine the effect of various feeding strategies on the NH_3 emission from a fattening pig house in practice. Four feed additives are involved for which it is proven in laboratory or experimental circumstances that they can reduce the NH_3 emission by different mechanisms: benzoic acid leads to a lowered urinary pH; enzymes result in a better digestion; clay minerals bind NH_3; and saponines inhibit the urease activity. In this paper the results of the first fattening period are given. The same additives will also be tested during a second and third fattening period but the results are not yet available. Therefore, the names of the additives are not mentioned and declared as treatment A, B, C and D.

Materials and methods

Fattening pig house and dietary treatments: The fattening pig house consists of seven identical compartments (channel ventilation) with eight pens each. The floor of each pen (3.0 m by 3.0 m) is a fully conventional concrete slatted floor. The dry feed troughs are placed in a corner of the pen. The nipple installation is attached to the troughs. Feed and water are provided ad libitum. Five dietary treatments were applied: a standard diet with a nutritional composition representive for Belgian practice in 2005 and four test diets based on the standard diet with one of the four feed additives A, B, C and D at commercial inclusion levels, ranging from 0.02 to 2%. The protein content was formulated to be 17% for the first phase diets (20 to 40 kg) and 16% for the second phase diets (40 to 110 kg). The 1.40 m deep slurry pit under the slatted floor of the pens is separated for each compartment.
Measurements: The measurements are performed with a photo-acoustic multigas monitor (Innova 1312 and 1314) combined with an 8-channel multisampler. The NH_3 sampling points in each compartment are situated 10 cm underneath the ventilation rate sensor. The concentration is measured every 45 minutes per sampling point. Ventilation rate and indoor temperature are recorded every three minutes. Every measured NH_3 concentration per day is multiplied by the average ventilation rate during the quarter of an hour before the moment on which the NH_3 concentration measurement is performed. The mean of all these measurements during a day is used to determine the daily NH_3 emission. Statistical calculations of the data were carried out with the program package SPSS 12.0 using the 'Duncan's multiple range test.

Results and discussion

The results showed that the course of the ventilation rate differed for two compartments compared with the other compartments because of the presence of a 'cold' side wall. The result is a lower ventilation rate with low outside temperatures and a higher ventilation rate with high outside temperatures. Because of this, only the results of the five remaining compartments are reported. The pigs (74 per compartment) were put into the compartments with an average starting weight of 21.9 kg (±0.5 kg). The average slaughtering weight was 117.5 kg (±2.5 kg). For all treatments optimal performance results were obtained with an average daily feed intake of 2.10 kg (±0.11 kg), a daily gain of 767 g (±19 g) and a feed conversion ratio of 2.74 (±0.17). The amount of fouling of each pen with faeces was assessed visually each week, and during the whole fattening period no difference in fouling was found between the compartments.

Table 1 presents an overview of the daily mean temperature, ventilation rate, NH_3 concentration and NH_3 emission based on the measurements performed during the period 25 May-17 July 2006. The course of the ventilation rate was the same for each treatment and no significant differences were found between the treatments. However, the NH_3 concentration was significantly higher for the standard treatment and treatment A in comparison with the other three treatments. The highest emission rates were detected for the standard feed and a significant difference was found with the other treatments. During the whole measuring period the treatments reduce the NH_3 emission with 13 to 24% compared to the standard feed.

Table 1. Daily average temperature (°C), ventilation rate (m^3/h), NH_3 concentration (ppm) and NH_3 emission (g/day) during the period 25 May – 17 July 2006.

	Treatment									
	Standard		A		B		C		D	
	Mean	Std	Mean	Std	Mean	Std	Mean	Std	Mean	Std
Temperature (°C)	24.2^a	1.3	24.2^a	1.3	24.4^a	1.5	24.1^a	1.2	24.4^a	1.3
Ventilation rate (m^3/h)	3,462^a	1,756	3,347^a	1,819	3,654^a	1,876	3,298^a	1,671	3,712^a	1,719
NH_3-concentration (ppm)	48.1^b	14.6	45.4^b	14.5	36.6^a	13.7	37.7^a	11.1	39.4^a	10.0
NH_3-emission (g/day)	2107^d	294	1,828^c	242	1,681^{ab}	366	1,612^a	201	1,780^{bc}	263
Reduction compared to standard (%)	-		13.2		20.2		23.5		15.5	

$^{a-d}$Numbers in the same row with different superscripts are significantly different (P<0.05).

Conclusions

The results of one pig fattening period during the summer period show that it is possible to reduce the ammonia emission with different feed additives in the fattening diet with 13 to 24%, while the performance results and pen fouling level remained optimal. Replicates with a second and third fattening period are necessary to confirm these conclusions.

Effect of diet on air emissions from pigs

W. Powers[1], S. Zamzow[2] and B. Kerr[3]
[1]Michigan State University, 2209 Anthony Hall, East Lansing, MI 48824, USA
[2]Iowa State University, USA
[3]USDA ARS Ames, IA 50011, USA

Introduction

Worldwide, animal production is under increasing scrutiny to reduce the emission of potentially harmful compounds such as NH_3 and H_2S. At this time, data relating to the typical levels emitted from livestock operations are limited. Differences that exist from one site to another in diet, manure handling strategies, animal numbers, ventilation, climate and weather make estimating gaseous emissions from animal facilities, as an industry, difficult. However, determining the 'typical' concentrations and emission rates from similar facilities are necessary before the impact of regulations and the need for regulation can be established. Emission factors from various operational components are often used to develop site emissions estimates. Diet modification to reduce nitrogen inputs into the animal by reducing dietary crude protein (CP) without negatively impacting performance is a proven method of reducing nitrogen excretion. In general, there has been limited research conducted that directly measures emissions following diet modification; most work reports emissions estimates based on nitrogen excretion measures and an assumption of NH_3 volatilisation. The impact of diet modification on other gases, including H_2S emissions, is poorly documented. The objective of this study was to evaluate the impact of feeding reduced crude protein diets, supplemented with exogenous amino acids, to swine throughout the grow-finish phase and to quantify gaseous emissions, nutrient excretions, and animal performance.

Materials and methods

Crossbred barrows (six per chamber at the start of the project; initial BW = 20.1 kg) were housed in eight indirect calorimeters at Iowa State University. Pigs were allocated to each chamber by weight in order to minimise body weight differences within each chamber. The pigs were penned in a 3.05×1.52 m raised deck with Tenderfoot® flooring. Swinging nipple waters were located above the middle of the pen and a two-hole feeder was located at one end of each pen. Chamber temperatures (18.3 °C to 25.6 °C) were adjusted weekly based on the average body weight within the chamber so as to remain in the thermoneutral zone of the animal. Fluorescent lighting was programmed to come on at 07:00 h and go off at 18:00 h. Barrows were fed a common mashed starter diet during a 13-d chamber acclimation period. Four feeding phases followed: Grower Phase 1 (G1; beginning at 24.5 kg BW), Grower Phase 2 (G2; 55.3 kg BW), Finisher Phase 1 (F1; 87.2 kg BW), and Finisher Phase 2 (F2; 111.4 to 122.7 kg BW). Pigs were offered one of three pelleted diets during each phase: a control diet (C), a low crude protein diet (LCP) and an ultra low crude protein diet (ULCP). Dietary crude protein (CP) was reduced between the three dietary treatments by adding supplemental amino acids to the diet such that the amino acid needs of the barrows were achieved with concomitant reduction in dietary N. Diets were formulated to contain similar lysine and energy content. Formulated CP of G1 was 22.5%, 20.0%, and 18.4% for the C diet, LCP diet and the ULCP diet, respectively. Progressing through each feeding phase, there was a decrease in the amount of formulated CP such that F2 was formulated to contain 16.6%, 15.4%, and 13.8% CP in the C, LCP and ULCP diet, respectively. Inclusion amounts of exogenous sources of amino acids in each of the dietary treatments changed as the pigs aged and the feeding phases progressed. Diets C and LCP were offered in three of the eight chambers during the four feeding periods and diet ULCP was offered in two of the eight chambers. Barrows were provided *ad libitum* access to feed and water. New feed was offered daily between 06:00 and 09:00 h. Feed data were recorded daily and remaining feed was removed and weighed from the feeders at the end of each feeding phase from which average daily intake was calculated. Diets were assigned, randomly, to groups in each of the eight chambers at the start of each feeding phase. Diets were sampled twice weekly and pooled together at the end of the feeding phase for proximate and amino acid analyses. At the end of G2, one pig was removed from each chamber such that five pigs remained in each pen for the finisher phases (F1 and F2). Exhaust air concentrations of NH_3 and H_2S were measured continuously from each chamber. All gases were measured simultaneously within a sample stream. Temperature and humidity were measured and recorded every 2 sec. Airflow rates into and out of each chamber were measured continuously, allowing for calculation of gas emission rates.

Results and discussion

Dietary treatment had no significant effect on weight gain, feed intake, or feed conversion measures. A diet effect was observed for average daily ammonia concentrations ($P<0.0001$). Exhaust (NH_3) concentration in rooms where pigs were fed the LCP diet were 16% less than the C diet (3.86 vs. 4.57 ppm). Ammonia concentrations were reduced 25% (2.93 ppm) in the ULCP diets compared to the LCP diet and 36% compared to the C. Average daily NH_3 emission rates were 26.8, 21.0, and 14.5 mg min^{-1} for the C, LCP, and ULCP diets, respectively, corresponding to a daily mass of NH_3 emitted per kg of animal live weight, of 88.0, 68.9, and 46.0 mg kg^{-1}. This represents a 22 and 48% reduction in NH_3 emissions for the LCP and ULCP diets, respectively, when compared to the C diet. Values reported here are similar to other values reported for pit-recharge and pull plug systems in the U.S. Hydrogen sulfide concentration and emissions were not different between rooms offered the different dietary treatments ($P<0.05$). Emission factors observed in this study (0.35 to 0.47 mg g^{-1} BW) were on the low end or less than that reported by others, approximating values reorted for pull plug systems.

Conclusions

In considering the implementation of reduced crude protein diets, diet cost must be considered as well as effects on animal performance. Feeding the LCP diet is practiced by a portion of the U.S. swine industry today. The ULCP diet would add considerable cost even when soybean meal prices are high. However, the cost of diet changes must be weighed against the cost of alternative emission control practices, most of which are likely engineered practices. Furthermore, in the event that swine operations are at some point accountable for N production, in general, point source control measures such as diet modification may be the feasible solution, regardless of the ULCP diet cost as compared to the C diet cost.

Effect of piglet diet on gas emissions

C. Pineiro[1], J. Morales[1], S. Godbout[2], S.P. Lemay[2], M. Belzile[2], J. Feddes[3], P. Illescas[4], G. Montalvo[4] and M. Bigeriego[5]
[1]PigCHAMP Pro Europa, S.A., Ctra. San Rafael, 42, 40006, Segovia, Spain
[2]IRDA, 3300 Rue Sicootte, C.P. 480 Saint-Hyacinthe, Québec, Canada
[3]University of Alberta, Edmonton, Alberta, Canada
[4]Tragsega, S.A. Conde de Peñalver, 84, 8 planta, 28006, Madrid, Spain
[5]Ministry of Agriculture, Fisheries and Food, Alfonso XII, 62, 28071, Madrid, Spain

Introduction

Livestock production in confinement facilities result in gas emissions such as ammonia (NH_3), methane (CH_4), nitrous oxide (N_2O) and carbon dioxide (CO_2). An effective way of reducing gas emissions is through the diet. During last years, different dietary strategies have been successfully assessed in this sense. The most important ways are reducing dietary N concentration, inclusion of acidifiers, and fibre supplementation. Reducing dietary N is based on higher protein requirements adjustments of animals combined with synthetic amino acids supplementation. Productive performance is not affected while ammonia emissions are significantly decreased. Dietary acidifiers have been effective in studies including benzoic acid in growing-finishing pigs through decreasing urinary pH. Urinary pH has a major impact on the volatilisation potential of ammonia. As pH is reduced, the ammonium concentration is increased and ammonia concentration is decreased within a solution. Finally, diets containing beet pulp have been demonstrating to decrease the manure pH and NH_3-N, probably due to the gut microbiota modification. The objective of this study was to assess the effects of different dietary strategies in piglets on gas emissions: crude protein reduction, benzoic acid inclusion, beet pulp supplementation and their combination.

Material and methods

A total of 80 piglets were allotted in ten environmentally-controlled chambers, each housing eight piglets of 13.1 kg of initial body weight (BW). They were used to monitor performance, water consumption, temperature, relative humidity, airflow rate, and NH_3, CH_4, and N_2O concentrations. The flooring was fully slatted with the manure being stored in a shallow pit located under the floor. The manure was removed by a vacuum pump and weighed. Methane, CO_2 and N_2O were analysed by a gas chromatograph. The NH_3 was analysed by a photo-acoustic infrared spectroscopy gas analyser (Innova model 1314, Analytical Instrument, California, USA). Piglets were fed on five different isoenergetic diets as follows: control diet, low protein (LP), sugar beet pulp (SBP), benzoic acid (BA) or the combination of all (LP+SBP+BA). The diet formulations are listed in Table 1. Each diet was replicated twice, and the pen of 8 piglets was the experimental unit. Average daily gain (ADG), feed intake (FI) and feed gain ratio (FGR) were controlled every week. Data were analysed as repeated measures using the GLM procedure of SAS, including dietary treatment and time (week) as main effects. Diet x time interaction effect was not significant and then removed from the statistical analysis.

Table 1. Diet formulation

Ingredients (g/kg)	Control	Beet pulp (SBP)	Low protein (LP)	Benzoic acid (BA)	LP+SBP+BA
Cereals[1]	692	590	768	681	659
Soybean meal 48%	120	120	71	120	60
Extruded soybean, full fat	96	94	69	97	85
Fish meal	38	38	20	38	20
Soy oil	22	21	34	22	28
Sugar beet pulp	-	100	-	-	100
Benzoic acid	-	-	-	5	5
L-Lysine HCl, 78%	6.1	6.1	8.0	6.2	8.1
Methionine-OH, 88%	0.85	1.17	1.70	0.88	2.06
L-tryptophan	0.92	0.97	1.90	0.90	2.04
L-threonine	0.63	0.76	1.20	0.62	1.30
Others[2]	23.5	28.0	25.2	28.4	28.6
Calculated nutrient concentration (%)					
Crude protein	19.7	19.7	16.6	19.7	16.6
Crude fibre	3.8	5.5	3.8	3.8	5.5
Digestible Energy (MJ/kg)	14.64	14.56	14.64	14.56	14.56
Digestible lysine	1.14	1.14	1.14	1.14	1.14

[1]Barley, corn and wheat.

[2]Others include dicalcium phosphate, calcium carbonate, salt, choline, enzymes and vit-min premix.

Results and discussion

Ammonia emissions were 0.9 mg/kg/h from the control diet, and similar to BA diet, but the LP, SBP and LP+SBP+BA diets had emission rates of about 50%. Similar results were observed for CH_4, where the control diet showed an emission of 0.59 mg/h/kg and the LP diet decreased emission rates about 40%. However, N_2O emission rate was higher (0.028 vs. 0.017 mg/h/kg, $P<0.05$) for the LP+SBP+BA when compared to the control diet. BA treatment also reduced ammonia emissions (9%) but not significantly. Other authors found higher effect associated to BA employing growing-finishing pigs instead of piglets. Differences in diet characteristics between both ages, like in protein content, might be affecting the mechanism of action of BA in the hindgut, which needs further research.

Regarding piglet performance, LP did not affect productive performance, according to many studies in the literature. Interestingly, piglets fed SBP or BA showed a higher final BW, 33.13 and 33.38 vs. 31.78 kg, $P<0.05$). The effect observed of both treatments on productive performance could be associated with a positive effect on the gastrointestinal microbiota. And piglets fed with the combination of all main effects (LP+SBP+BA) showed lower final BW (30.0 vs. 31.8 kg) because of a lower growth rate (0.53 vs. 0.58 kg/pig/day, $P<0.05$) in comparison with control diet. These results show the efficiency of nutrition on the control of emissions to the atmosphere, but also its implication on animal performance which always have to be balanced.

Dietary composition influences odour and ammonia emissions from pig manure

P.D. Le, A.J.A. Aarnink and M.W.A. Verstegen
Wageningen-UR, Animal Sciences Group, P.O. Box 65, 8200 AB Lelystad, The Netherlands

Introduction

Odour and ammonia (NH_3) emitted from pig manure cause nuisance for people living near pig farms, and environmental pollution, respectively. Odour and NH_3 are mainly produced by the microbial conversion of feed components in the large intestine of pigs and by microbial conversion of urinary and faecal excreta in manure. Protein (CP) and fermentable carbohydrates (FC) are the main precursors for odour production. Sulphurous compounds, indoles and phenols, are considered most important for odour nuisance from pig production facilities. Tryptophan (Trp), phenylalanine (Phe) and tyrosine (Tyr) are main substrates for the synthesis of indolic and phenolic compounds. The S-containing amino acids, methionine (Met) and cystine (Cys) are main substrates for the synthesis of S-compounds. Manipulating dietary factors of CP and FC was shown to be very effective in decreasing NH_3 emission from pig manure (Canh *et al.*, 1998, 1999). There are, however, few studies on effects of dietary manipulation on odour emission and the correlation between odour and NH_3 emissions. The main goal of the present studies was to assess the potential impact of CP and FC levels and amino acid (AA) types on odour emission. Furthermore, the correlation between odour and NH_3 emissions was assessed.

Materials and methods

In three different experiments, dietary factors were investigated for their effects on odour and NH_3 emissions from manure of growing and finishing pigs. In the 1st experiment, three dietary CP levels of 12, 15 and 18% were investigated. In the 2nd experiment, effects of specific crystalline AA supplementation to the diet were studied. Treatment groups were (1) 15% crude protein basal diet with three times the requirement of sulphur-containing amino acids (14.2 g Met + Cys/kg diet); (2) basal diet with two times the requirement of Trp and Phe + Tyr (2.9 and 20.4 g/kg diet, respectively); and (3) basal diet with AA supplementation to levels sufficient for maximum protein gain. The two former mentioned experiments were conducted on growing pigs and used a randomised complete block arrangement having three treatments in six blocks (n = 18/experiment). In the 3rd experiment, interactive effects of dietary CP and FC levels on odour and NH_3 emissions were studied. The experiment was conducted with finishing pigs (n = 36) in a 2x3 factorial randomised complete block arrangement with 6 treatment combinations in 6 blocks. There were two CP levels (low 12% and high 18%) and three digestible FC levels: (low 95.5, medium 145.5, and high 195.5 g kg^{-1} feed). Nutrients of experimental diets, except tested nutrients, were made equal.

Faeces and urine of each pig were accumulated together in a separate manure pit under the slatted floor. In the 5th week of collection, air samples for odour and NH_3 analyses were collected directly from each manure pit. Odour emission was measured by olfactometry according to the European standard (CEN standard 13725, 2003). Ammonia emission was determined by its concentration in two NH_3 traps each containing about 20 mL 0.5 M HNO_3.

Results and discussion

Lowering dietary CP level from 18 to 12% reduced odour emission (P=0.04) by 80% (from 5.76 to 1.18 $ou_E\ s^{-1}\ m^{-2}$) and NH_3 emission (P=0.01) by 53% (from 0.017 to 0.008 mg $s^{-1}m^{-2}$). Supplementing crystalline S-containing amino acids (Met and Cys) three times the animal requirement increased odour emission ($P<0.001$) by about 7-fold (from 2.33 to 16.50 $ou_E\ s^{-1}\ m^{-2}$). Supplementing crystalline Trp, Tyr, and Phe two times the requirement did not affect odour emission. Diets with different types of AA did not affect NH_3 emission ($P>0.05$). Dietary CP and FC levels showed to have an interactive effect on odour emission at P=0.06 but not on NH_3 emission ($P>0.05$). At a high dietary CP level, an increased FC level decreased odour emission from pig manure, while at a low CP level increased FC level increased odour emission. Ammonia emission from pig manure can be reduced substantially by decreasing dietary CP and by increasing FC. Lowering dietary CP from 18 to 12% decreased NH_3 emission ($P<0.001$) by 62% (from 0.03 to 0.011 mg $s^{-1}m^{-2}$). High FC diets resulted in low NH_3 emission (P=0.01). On average, for each 100 g increase in FC/ kg diet, as fed, NH_3 emission reduced by 29%. The correlation between odour emission and NH_3 emission was low (0.1, -0.3 and -0.1, respectively, for the 1st, 2nd and 3rd experiment) and deemed non-significant.

Dietary CP is the most important precursors for odour and NH_3 production thus reducing dietary CP levels decreases odour and NH_3 emissions from pig manure. Excess crystalline S-containing AA in the diet substantially increased odour emission. This means that the excess crystalline AA or their metabolic products provided precursors for odour production in manure. Interactive effects between CP and FC levels on odour emission were expected because in the large intestine of pigs and in manure stores, microbiota uses protein as a nitrogen source and obtain energy from FC for their biomass synthesis. When the FC to CP ratio is too low, microbiota uses CP as an energy source, causing the formation of odorous compounds. A low and non-significant correlation between NH_3 and odour emissions means that dietary strategies, which can reduce NH_3 emission effectively, may not work in the same manner for odour emission. In conclusion, dietary manipulation can be very effective in reducing odour and NH_3 emissions from pig manure. However, they may not work the same for both NH_3 and odour emissions.

References

Canh, T.T., A.J.A. Aarnink, J.B. Schutte, A. Sutton, D.J. Langhout and M.W.A. Verstegen, 1998. Dietary protein affects nitrogen excretion and ammonia emission from slurry of growing–finishing pigs. Livest. Prod. Sci. 56: 181-191.

Canh, T.T., M.W.A. Verstegen, N.B. Mui, A.J.A. Aarnink, J.W. Schrama, C.E. Van 't Klooster and N.K. Duong, 1999. Effect of non-starch polysaccharide-rich by-product diets on nitrogen excretion and nitrogen losses from slurry of growing-finishing pigs. Asian-Aust. J. Anim. Sci. 12: 573-578.

CEN standard 13725, 2003. Air Quality - Determination of Odour concentration by Dynamic Olfactometry. European Committee for Standardization, Brussels, Belgium.

Feeding and management: poultry

Effect of diet on air emissions from laying hens

W. Wu[1], W. Powers[2], R. Angel[3], C.E. Hale III[4] and T. Applegate[5]
[1]*Iowa State University, USA*
[2]*Michigan State University, 2209 Anthony Hall, East Lansing, MI 48854, USA*
[3]*University of Maryland, USA*
[4]*Rose Acre Farms, USA*
[5]*Purdue University, USA*

Introduction

It is well documented that gaseous emissions from laying hen feeding operations can have potential negative impacts on the environment and on human and bird health. Feeding diets formulated to reduce excess crude protein (CP) inputs help to reduce nitrogen (N) excreted resulting in lower ammonia (NH_3) emission. Acidogenic materials reduce manure pH resulting in the protonation of NH_3 to ammonium. Gypsum (calcium sulphate) is an acidogenic compound that can serve as a partial replacement for limestone as a calcium (Ca) source. A study conducted by Hale (2005) showed that using a reduced CP diet in combination with acidogenic materials such as gypsum and nitrogenous binding compounds like zeolite decreased NH_3 emission from laying hen excreta. However, the effectiveness of feeding such a diet on all gaseous emissions has not been reported. The objective of the current study was to evaluate the effectiveness of feeding a reduced emissions diet (R) containing 6.9% of a gypsum-zeolite mixture which replaced 35% of the limestone and slightly reduced CP to laying hens of different ages on egg production and emission of NH_3, H_2S, NO, NO_2, CO_2, CH_4 and non-methane total hydrocarbon as compared to feeding a commercial diet (C).

Materials and methods

The study consisted of 3 trials utilising birds which were initially 21, 38, 59 wks old. During each trial, 640 Hy-line W36 hens (BW = 1.36, 1.47 and 1.52 kg in trials 1, 2 and 3, respectively) were randomly assigned to one of eight air emissions chambers for a 3-wk period. Between each trial, chambers were completely cleaned. In each chamber, eight cages of 10 birds per cage were used (3.55 m^2 per bird). Feed, in mash form, and water were available for *ad libitum* consumption. On an analysed basis, the C diet contained 18.0, 17.0 and 16.2% CP and 0.25, 0.20 and 0.20% S in trials 1, 2 and 3, respectively, while the R diet contained 17.0, 15.5 and 15.6% CP and 0.99, 1.20 and 1.10% S in trials 1, 2 and 3, respectively. Diets were formulated to contain similar Ca and P concentrations. All diets were formulated to meet National Research Council (1994) nutrient recommendations.

Gaseous concentration and air flow were monitored from each chamber in a sequential manner resulting in 10-11 daily observations per chamber. During each 15 minutes observation period, the concentration of NH_3, H_2S, NO, NO_2, CO_2, CH_4 and non-methane total hydrocarbon were recorded through computer.

Data were analysed using a mixed model with the day as random variable. (SAS v 8.0). Emission data were adjusted for number of birds. Significance was accepted at or below a $P<0.05$.

Results and discussion

Average daily egg weight (ADEW; 57.4 g), average daily egg production (ADEP; 82.5%), average daily feed intake (ADFI; 92.6 g) and BW change (BWC; 24.3 g), across ages, were unaffected by diet. Age affected ADEW (52.1, 58.9 and 61.2 g), ADEP (86.7, 87.1 and 73.7%), ADFI (86.8, 96.2 and 94.6g) and BWC (65.2, 17.3 and -9.7 g) in trials 1, 2 and 3, respectively.

In trials 1, 2 and 3, daily NH_3 emissions from hens fed the R diets (185.5, 312.2 and 333.5 mg $bird^{-1}$) were lower than those of hens fed the commercial diet (255.0, 560.5 and 616.3 mg $bird^{-1}$). It was showed that age affected the NH_3 emission, daily emission was greater from older hens than from younger hens. Daily H_2S emissions from hens fed the R diets (1.6, 7.1 and 3.7 mg $bird^{-1}$) were lower than those of hens fed the commercial diet (0.5, 1.9 and 0.8 mg $bird^{-1}$). Daily CO_2 emissions from hens fed R diet (64,626, 79,799 and 80,335 mg $bird^{-1}$) were less than those of hens fed the commercial diet (65,627, 86,243 and 82,694 mg $bird^{-1}$) and an age affect existed for CO_2 emissions. Daily methane emissions from hens fed R diet (138.5, 28.2 and 9.4 mg bird-1) were lower than those of hens fed the commercial diet (153.1, 43.8 and 11 mg $bird^{-1}$). Age affected methane emission with younger birds producing more methane than older birds. Daily emissions of NO from hens fed R diets (0.71, 0.11 and 0.21 mg $bird^{-1}$) were less than those from control diet (0.94, 0.2 and 0.28 mg $bird^{-1}$). No diet or age effects were observed for NO_2 and non-methane total hydrocarbons.

Diet acidification, CP reduction and zeolite supplementation reduced the emissions of NH_3 (by 39%), CO_2 (by 5%), CH_4 (by 17%) and NO (by 48%). Diet acidification increased the gaseous emissions of H_2S mainly because of the high concentration of sulphate in the acidifying agent used. More research is needed to address the increased H_2S emissions, long term effects on animal performance and health and to explore the impact of graded concentrations of the additive on emissions.

Conclusions

The U.S. Environmental Protection Agency Emergency Planning and Community Right-to-Know Act (EPCRA) and Comprehensive Environmental Response, Compensation, and Liability Act (CERCLA) reporting requirements allow daily emissions 45 kg of both NH_3 and H_2S from poultry farms. The current study illustrated that feeding a reduced emission diet could increase hen population by 40% without exceeding the requirements. Even though gypsum addition increases H_2S emissions considerably, these concentrations did not surpass reporting limits even for the larger laying hen complexes. It will be NH_3 that drives bird numbers with respect to reporting requirements for CERCLA/EPCRA. Long-term feeding of the diet was evaluated in a separate study with results unavailable at this time.

References

Hale E.C., 2005. Reduction of Ammonia Emission and Phosphorus Excretion in Laying Hen Manure Through Feed Manipulation. Symposium State of the Science Animal Manure and Waste Management, San Antonio, Texas.

National Research Council, 1994. Nutrient Requirements of Poultry, 9th rev. ed. National Academy Press, Washington, DC.

Effect of diet on air emissions from broiler chickens

W. Powers[1], R. Angel[2], and T. Applegate[3]
[1]Michigan State University, 2209 Anthony Hall, East Lansing, MI 48854, USA
[2]University of Maryland, 4131 Animal Sciences Center, College Park, MD 20742, USA
[3]Purdue University, Lilly Hall of Life Sciences, West Lafayette, IN 47907, USA

Introduction

Public and regulatory concerns related to air emissions from livestock and poultry operations has increased in recent years. Of concern is the paucity of data on emissions from different livestock and poultry operations. Because there is very limited information in the published literature on the actual air emissions from broiler operations, a study was conducted at a new air emissions chamber facility at Iowa State University to determine baseline air emissions of ammonia (NH_3), hydrogen sulfide (H_2S), nitric oxide (NO), nitrite (NO_2), carbon dioxide (CO_2), methane (CH_4) and non-CH_4 total hydrocarbon from broiler flocks grown on a 'typical' industry phase feeding (control, C) and management program and on a reduced protein (LP) feeding program.

Materials and methods

Ross 308 hatchling male broiler chickens were allocated to one of eight air emission chambers during each of three sequential 42-d flocks (50 chicks per chamber per flock). Clean wood shavings were sampled, placed in pans and weighed prior to the start of the first flock, only, and cake was mixed into the litter between flocks. Chambers were randomly allocated at the start of each flock to one of two dietary treatment: a control treatment (C) that consisted of four feed phases: a 17-d starter (St, d 0 - 16), a 13-d grower (Gr, d 17 – 29), a 6-d finisher (Fn, d 30 - 35) and a 6-d withdrawal (Wd, d 36 – 42) and a reduced crude protein treatment (LP) that consisted of six feed phases: a 7-d pre-starter (PreSt, d 0 - 6), a 10-d starter (d 7 - 16), a 7-d grower 1 (Gr1, d 17 – 23), a grower 2 (Gr2, d 24 – 29), a 6-d finisher (d 30 - 35) and a 6-d withdrawal (d 36 – 41). Chamber body weight was measured and recorded at the start of each flock and on d 7, 17, 24, 30, 36, and 42. Feed disappearance was determined for each phase allowing for feed efficiency (feed to gain, FCR) calculations. Diets, formulated to be isocaloric, by phase, were primarily corn and soybean based, but differed in protein content as follows: 22.1, 20.0, 17.2, and 16.6% for the C St, Gr, Fn, and Wd diets, respectively while those for the LP PreSt, St, Gr1, Gr2, Fn, and Wd diets were 22.0, 18.6, 18.1, 17.3, 15.8, and 15%, respectively. To maintain minimum amino acid (lysine, methionine, total sulfur amino acids, threonine, arginine, isoleucine, tryptophan, and valine) required concentrations in the diets synthetic amino acids were used. In C diets only synthetic lysine and methionine had to be used while in the LP diets lysine, methionine, threonine, arginine, tryptophan, valine and isoleucine were included in the diet.

At the end of flock 3, 20 broilers per chamber were randomly selected and sampled for breast yield determinations. Feed was removed 12 hours prior to sampling. Broilers were weighed individually, the intestinal tract (from crop to cloaca) and abdominal fat pad removed for the determination of carcass weight. The breast was then removed and weighed.

Throughout each flock, emissions of CO_2, NH_3, NO, NO_2, SO_2, H_2S, CH_4 and non-methane total hydrocarbons were measured and average concentration of each gas during the last 5 min of each 15-min sampling period was recorded.

Results and discussion

Dietary treatment had an effect on final body weight (42 d) with broilers fed the LP treatment diets being 87 g lighter than those fed the C treatment diets (2.780 and 2.693 kg, respectively over the three flocks). Broilers fed the LP treatment diets were lighter at 42 d than those fed the C treatment diets in all three flocks but the weight differences were greatest in flock 1 (2.705 and 2.861, 2.841 and 2.907, and 2.534 and 2.573 kg for broilers fed the LP and C treatment diets in flocks 1, 2, and 3, respectively). Feed consumption was lower in broilers fed the LP treatment diets resulting in no difference in feed efficiency between treatments C and LP (1.89 and 1.91, respectively over the three flocks). Mortality was similar between treatments (6%) with no flock or treatment effect observed. Others have observed decreases in body weight when diet protein is reduced but it has been speculated that the decreases in performance observed by these researchers have been caused by not supplementing back with sufficient amounts of limiting amino acids other than methionine and lysine. On a practical basis, however, bird performance can be hindered by excessive lowering of CP in diets due to a number of factors: reduced dietary potassium levels, altered dietary ionic balance, deficiency of nonessential amino acids, imbalances among certain amino acids (e.g. branched chain amino acids), and/or potential toxic concentrations of certain amino acids. These issues did not appear to be the cause for the lower body weight in the current study. Diet amino acid analysis showed that formulated and analysed concentrations were similar. When looking at when differences started to appear and the proportional differences at the different ages, it appears that the decrease in protein in Gr1 diet in the LP treatment may have been too severe.

In flock 3, when birds were sampled for yield determinations, treatment did not affect body weight, carcass weight, breast weight, or dress percent and breast yields. Numerical differences in body weight were similar to those seen in the three flock performance data. Body weight of sampled birds was 2.721 and 2.667 kg for broilers fed the C and LP treatment diets but because replication was low (one flock, four replicates of 20 birds per treatment) differences that were significant in the performance data were not significant in this portion of the study. Breast weights, however, were not different between broilers fed the C and LP treatment diets (542.9 and 540.7 g, respectively) and breast yields were 22.75 and 23.13% for broilers fed the C and LP treatment diets. Thus dietary treatment had no effect on breast weight or yield.

Dietary treatment affected NH_3 emissions with the LP treatment resulting in lower daily NH_3 emissions (26.5 mg kg^{-1} d^{-1} in flocks 1) compared to emissions from broilers fed the C treatments diets (33.8 mg kg^{-1} d^{-1} in flocks 1). NO and NO_2 were also lower in the LP vs the C treatment. There was no effect of treatment on NO, NO_2, SO_2, H_2S, CO_2, CH_4, or non-CH_4 hydrocarbons.

Conclusions

Lowering dietary protein while maintaining minimum concentrations of amino acids resulted in substantial (22%) reductions in daily emissions of NH_3 and no impact on breast weight or yield.

Combined effect of dietary acidogens and cation exchangers on laying hen manure ammonia emissions and egg production costs

E.C. Hale III
EcoCal Products, P.O. Box 179, Brownstown, Indiana, 47220, USA

Introduction

Dietary acidogens are materials which, when fed, act to reduce excreta pH. Examples of dietary acidogens are calcium sulfate (gypsum), monobasic calcium phosphate, sodium bisulfate, and phosphoric acid. Various materials, including zeolites and humates, have a high cation exchange capacity, and will pass undigested through the gut of laying hens. When laying hens are fed combinations of acidogens to reduce excreta pH below 7, and indigestible cation exchangers, ammonia (NH_3) formed from catalytic degradation of uric acid is immediately protonated to form ammonium (NH_4^+), which is subsequently bound by the cation exchanger, and retained in the manure.

Materials and methods

Experiment 1: The effect of an ammonia emission reducing diet containing 1.25% zeolite and 5.25% gypsum on manure ammonia emission rates in high-rise laying houses was determined. Two high rise layer houses, one containing 164,000 hens, the other containing 173,000 hens, with different ages and nutritional requirements, were monitored during this study.
Between 11-1-2004 and 11-30-2005, the daily emission rate for both houses was measured to quantify differences resulting from differing populations, ages, and diets. Pairs of daily emission rates were converted to a ratio (H2/H1), and the average ratio determined. The diet in House 2 was changed to the test diet on 12-1-04. House 1 acted as a control. Monitoring for the effect of the diet was performed between 12-16-2004 and 1-31-2004. The observed test period emission rates in House 1 were multiplied by the average emission rate ratio to reduce the effect of inter-house differences. The results are illustrated in Table 1 (Heber, *et al.*, 2006, unpublished report to Ohio Fresh Egg).
Experiment 2: The effect of an ammonia emission reducing diet containing 1.25% zeolite and 5.75% gypsum on manure nitrogen (N), phosphorus (P) and potassium (K) content was determined over an 11-month period. Two high-rise houses with initial populations of 125,000 hens were used in this study. One house was fed the test diet, the other acted as a control. Manure was allowed to accumulate in the basement of the house for the entire 11-month period. At the end of the study, 20 manure samples were collected from each house at varying locations and depths in the manure pile. The samples were individually analysed for total N, P, and K content. Average manure N, P, and K content for the test and control house is reported in Table 2.
Experiment 3: The effect of an ammonia emission reducing diet containing 1.0% zeolite and 2.5% gypsum on egg production, egg grade, hen mortality, and production costs was determined over a 14-week period. A total of 375,000 hens were fed the test diet, 375,000 additional hens acted as a control. Egg production and grade figures, and production costs are presented in Table 3.

Results

Table 1. Ammonia emission rates, reported in average g/AU/day.

	House 1	House 2	Avg. Ratio, H2/H1
Baseline Emission Rate, 11-1-04 to 11-30-04	309	366	1.44
Test Emission Rate, 12-16-04 to 1-31-05	411	202	
Corrected Test Period Emission Rate	592	202	
Difference, H1/H2		-66%	

Table 2. Manure N, P, and K content, reported in kg/metric ton.

	Total N	Total P	Total K
Standard Diet	38.3	7.7	10.3
Amended Diet	27.9	16.7	26.2
Difference, %	+37%	-54%	-61%

Table 3. Average weekly production, test diet vs. control diet.

	Test	Control	% Diff
Total Production, eggs/week	1516560	1444574	4.98
Mortality, hens/week	494	693	-28.68
Grade A Large +, eggs/week	1377108	1311483	5.00
Undergrade Eggs, eggs/week	108923	141907	-23.24
Feed Cost, $/ton	104.12	101.95	2.12
Consumption, lbs feed/week	451966	444655	1.64
Production Cost, cents/egg	1.57	1.62	-3.22

Conclusions

Manure excreted by laying hens fed diets containing gypsum and zeolite exhibits significantly reduced ammonia emissions, increased N content, and reduced P and K content compared to control. Improvements in the production environment translate to improved production, reduced mortality, and reduced production costs.

Feeding and management: cattle

Prediction of fecal and urinary nitrogen excretion by dairy cattle

F.K. van Evert, H.G. van der Meer and J.J.M.H. Ketelaars
Plant Research International, P.O. Box 16, 6700 AA Wageningen, The Netherlands

Introduction

When quantifying nitrogen flows on a dairy farm, both fecal and urinary excretion of N must be determined. Just knowing total excretion is not sufficient, because fecal N is relatively stable, whereas urinary N is sensitive to leaching and volatilisation immediately after excretion and may quickly be lost to the environment. Nitrogen ingested in excess of the requirement for milk production and growth of the animal, is excreted. In practice, N content of the diet can be varied widely (between approx. 25 and 35 g (N) / kg (dm)) without appreciable effect on milk production, milk N content, or growth of the animal. The N balance of a lactating dairy cow can be expressed as (all quantities in kg):

$$N_{intake} = N_{milk} + N_{growth} + N_{feces} + N_{urine} \qquad (1)$$

where N_{milk} and N_{growth} are given by the productivity level of the cow and the N content of the product. Because fecal nitrogen consists in part of N in undigested feed particles, the fraction N_{feces}/N_{intake} is sometimes called the 'apparently undigestible' fraction of diet N. Boekholt (1976) proposed a simple relationship between a diet's N content and the 'apparent digestibility' of the N. When this relationship is used to determine fecal N excretion, urinary N excretion becomes the only unknown term in Equation (1) and can thus be calculated. The objective of our study was to evaluate Boekholt's relation using independent data for dairy cattle, beef cattle and young stock.

Materials and methods

We used data from 12 feeding experiments from a number of countries, in which dairy cows at productivity levels between 5,500 and 10,000 kg fat- and protein-corrected milk per year were represented, and which also included young stock and beef cattle.

Results and conclusions

Boekholt (1976) found linear relationships between the N content of the diet (g (N) / kg (dm)) and the apparent digestibility of the N, albeit with different parameter values for different types of diets. For diets consisting of hay and concentrates, based on data collected from 362 lactating dairy cows, an intercept of -5.3 and a slope of 0.833 was found; for diets consisting of fresh grass, he found an intercept of -5.6 and a slope of 0.929. We used the first equation for diets consisting of less than 70% fresh grass, and we used the second equation for grass-based diets. From the plot of predicted against measured fecal N excretion (Figure 1), it is clear that Boekholt's relation holds for dairy cows at a wide range of productivities, for young stock, and for beef cattle.

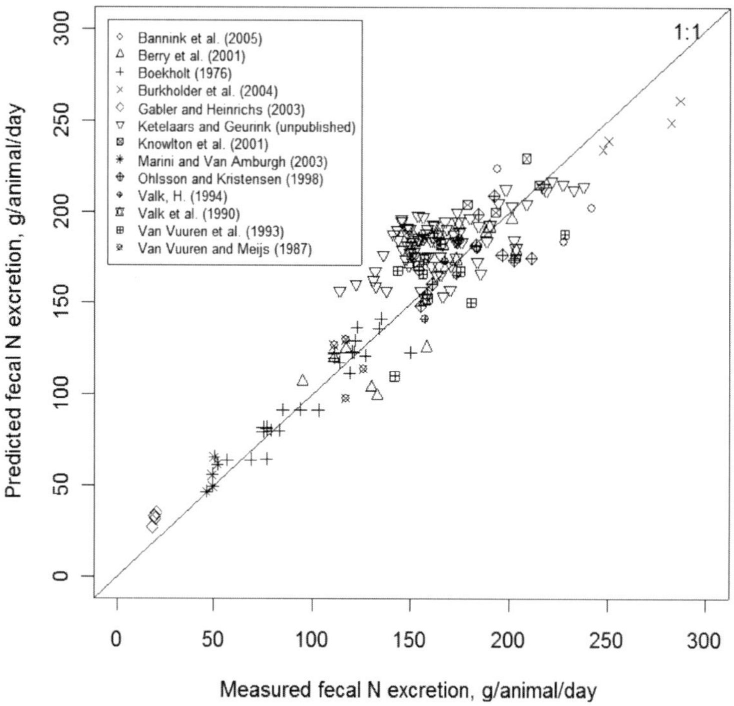

Figure 1. Measured and predicted faecal nitrogen excretion. For explanation and references, see text.

References

Bannink, A., H. Valk and J. Dijkstra, 2005. Een nadere analyse van de invloed van voeding op de productieresultaten en de mestsamenstelling op biologische melkveehouderijbedrijven binnen BIOVEEM I. ASG Rapport 04/I00565. Animal Sciences Group, Lelystad.

Berry, N.R., P.L. Jewell, F. Sutter, P.J. Edwards and M. Kreuzer, 2001. Effect of concentrate on nitrogen turnover and excretion of P, K, Na, Ca and Mg in lactating cows rotationally grazed at high altitude. Livestock Production Science 71: 261-275.

Boekholt, H.A., 1976. De stikstofhuishouding van het melkrund en de betekenis van de gluconeogenese uit aminozuren. Ph.D. Thesis, Landbouwhogeschool Wageningen.

Burkholder, K.M., A.D. Guyton, J.M. McKinney and K.F. Knowlton, 2004. The effect of steam flaked or dry ground corn and supplemental phytic acid on nitrogen partitioning in lactating cows and ammonia emission from manure. Journal of Dairy Science 87: 2546-2553.

Gabler, M.T. and A.J. Heinrichs, 2003. Effects of increasing dietary protein on nutrient utilization in heifers. Journal of Dairy Science 86: 2170-2177.

Knowlton, K.F., J.H. Herbein, M.A. Meister-Weisbarth and W.A. Wark, 2001. Nitrogen and phosphorus partitioning in lactating Holstein cows fed different sources of dietary protein and phosphorus. Journal of Dairy Science 84: 1210-1217.

Marini, J.C. and M.E. Van Amburgh, 2003. Nitrogen metabolism and recycling in Holstein heifers. Journal of Animal Science 81: 545-552.

Ohlsson, C. and V.F. Kristensen, 1998. Reducing inputs and losses of nitrogen and energy on dairy farms. Report in EU project AIR3-CT-332

Valk, H., 1994. Effects of Partial Replacement of Herbage by Maize Silage on N-Utilization and Milk-Production of Dairy-Cows. Livestock Production Science 40: 241-250.

Valk, H., H.W.K. Poelhuis and H.J. Wentink, 1990. Effect of Fibrous and Starchy Carbohydrates in Concentrates as Supplements in a Herbage-Based Diet for High-Yielding Dairy-Cows. Netherlands Journal of Agricultural Science 38: 475-486.

Van Vuuren, A.M. and J.A.C. Meijs, 1987. Effects of herbage composition and supplement feeding on the excretion of nitrogen in dung and urine by grazing dairy cows. In: H.G. Van der Meer, Unwin, R.J., Van Dijk, T.A. and Ennik, G.C (Eds.), Animal manure on grassland and fodder crops. Fertilizer or waste? Martinus Nijhoff. pp. 17ff.

Van Vuuren, A.M., C.J. Van der Koelen, H. Valk and H. Devisser, 1993. Effects of Partial Replacement of Ryegrass by Low-Protein Feeds on Rumen Fermentation and Nitrogen Loss by Dairy-Cows. Journal of Dairy Science 76: 2982-2993.

Nutritional options to manipulate ammonia emission from excreta of dairy cattle

A. Bannink[1], J. Reijs[2] and J. Dijkstra[2]
[1]Wageningen UR, Animal Sciences Group, P.O. Box 65, 8200 AB Lelystad, The Netherlands
[2]Wageningen University, Animal Nutrition Group, P.O. Box 338, 6700 AH Wageningen, The Netherlands

Introduction

A main part of ammonia (NH_3) emission from dairy cattle operations originates from fouled floors and manure storage. This emission depends, first of all, on the quantity of urea that is excreted with urine, in turn dependent on protein nutrition and the number of cows present. This contribution gives an outline of recent work on modelling the effect of nutritional measures on N excretion by cows, and the expected consequences for NH_3 emission.

Quantifying NH_3 emission from excreta

In quantifying NH_3 emission, it is important to make a distinction between environmental and animal factors. This contribution focuses on nutritional factors which affect N excretion with urine and faeces by the animal, and excreta characteristics relevant for NH_3 emission.
Dynamic mechanistic models were developed (Dijkstra *et al.*, 1992; Bannink *et al.*, 2007) which represent the physiological mechanisms involved with feed digestion and nutrient utilisation by cows. Recently, the model was extended to quantify excreta composition and consequences for slurry N composition and NH_3 emission (J.W. Reijs, unpublished results). Also effects on excreta volumes and acidity were documented (Bannink *et al.*, 2007). From a nutritional point of view, the detailed representation of these models allows the evaluation of multiple physiological aspects on NH_3 emission in a more integrated manner.

Effects of nutritional measures

Feeding less (digestible) protein: Feeding less (digestible) protein without affecting milk yield leads to less N excretion with urine. This may be achieved by less protein in dietary dry matter or less digestible protein. Reijs and others (unpublished results) demonstrated that both measures are highly effective to reduce urine N. Lowering N fertilisation rate and harvesting grass in a more mature stage, and a 50% exchange by maize silage resulted in a maximum reduction of 15% faecal N and 75% of urinary N (reduction in urine N more than proportional to that in total N excretion). Assuming that a constant fraction of urea is emitted, NH_3 emission reduced from 13 to 5% of total N excreted. Several nutritional measures are available which affect N digestibility and urine N losses, leading to a more than proportional decrease in NH_3 loss. The percentage of excreted N lost as NH_3 may decease with a similar order of magnitude as total N excretion.
Lowering rumen protein balance: Model calculations indicate that more optimal use can be made of the capacity of recycling of urea from blood to rumen. In the Netherlands, a surplus of rumen degradable protein of 50 g N per day (OEB of 300 g per day) is recommended. A reduction to zero rumen N balance (N available from microbial N synthesis - microbial N synthesised) did not negatively affect microbial growth efficiency (J.W. Reijs, unpublished

results; Bannink *et al.*, 2007) as often presumed. This omission results in 10% less urine N, and a similar reduction in NH_3 emission. Although this measure is feasible through whole lactation, in early lactation (negative energy balance and peak milk yield) diets are usually supplemented with protein to sustain milk protein yield. These supplements may be partly degraded in the rumen and increases rumen N balance.
Stimulating hindgut fermentation: Increase of dry matter intake and milk yield increases the efficiency of N utilisation, reduces rumen digestibility (lower pH, faster passage) and stimulates fermentation in the hindgut. Also feeding specific by-products or more resistant starch-sources may shift fermentation of organic matter from rumen to hindgut. With the most extreme nutritional measures tested by Reijs and others (unpublished results) an increased hindgut fermentation led to a doubling of N captured in the form of microbial N. At a maximum, this leads to slightly more than 10% reduction in urine N and hence in NH_3 emission (excluding effects of altered excreta acidity).
Excreta volume and acidity: Although an increased urine volume decreases urea concentration, it will be accompanied by a higher frequency of urination as well and involves a more frequent refreshment of urine on floors and the manure top layer. This effect partly compensates the potential diminishing effect of reduced urea concentrations on NH_3 emission. Acidification of excreta reduces NH_3 emission as well. Urine acidity is more sensitive for salts additions to diets already low in cation content and pH values may approach 7 then. Faeces can be acidified to some extent by nutritional measures (hindgut fermentation) and reported pH values range from more than 7 to less than 6, probably depending on source and technological treatment of starch-supplements (Bannink *et al.*, 2007). Considering the sensitivity of ammonia fractions in ammoniacal N, nutritional measures that acidify urine or faeces may still have a considerable effect on NH_3 emission. The precise mechanisms involved will be investigated further.

Conclusions

Nutritional measures have multiple effects on N digestion, N excretion and excreta characteristics. Although still in its infancy, the strategic use of the dynamic mechanistic models as discussed here may prove worthwhile to evaluate the effectiveness of nutritional measures to reduce NH_3 emission from dairy cattle operations.

References

Bannink, A., J.W. Reijs and J. Dijkstra, J., 2007. Integrated approaches to evaluate nutritional strategies for dairy cows. In: J France and E. Kebreab (editors), Mathematical Modelling in Animal Nutrition. Centre of Nutrition Modelling, University of Guelph, CA (in press).

Dijkstra, J., H.D.St.C. Neal, D.E. Beever and J. France, 1992. Simulation of nutrient digestion, absorption and outflow in the rumen: model description. Journal of Nutrition 122: 2239-2256.

Milk urea concentration as an indicator for ammonia emission reduction from dairy barns by feeding measures

G. van Duinkerken and L.B.J. Šebek
Wageningen UR, Animal Sciences Group, P.O. Box 65, 8200 AB Lelystad, the Netherlands

Introduction

Emission of ammonia from the dairy barn depends on the cow's diet, barn design, out- and indoor climate and farm management, e.g. grazing regime (Monteny, 2000). Ammonia emission from dairy barns mainly originates from urea excreted with urine. The hydrolysis of urea is catalysed by urease, which is produced by microorganisms in faeces, on fouled floors and in slurry pits. Urinary urea concentration and temperature are main factors influencing the rate of ammonia volatilisation. As urinary urea levels are highly correlated with the diet, ammonia emission from dairy cow houses can significantly be reduced by implementing nutritional measures (Van Duinkerken *et al.*, 2005). In an experimental study it was demonstrated that urea concentration in bulk milk is a useful indicator for ammonia emission reduction from the dairy barn in situations with summer feeding (Van Duinkerken *et al.*, 2005). In addition, the correlation between bulk milk urea and ammonia emission was studied for a situations with rationed grazing and indoor supplemented feed. Furthermore, European studies on current milk urea levels were inventoried to discuss the potential for emission reduction by feeding measures.

Material and methods

The correlation between bulk milk urea and ammonia emission from a dairy barn was studied in an extensive experiment in a research facility (Van Duinkerken *et al.*, 2005). The experiment was in a 1x3 design with three adjusted levels for the factor bulk milk urea (15, 35 and 55 mg urea/100 g). All milk urea levels were in 3 repetitions in a randomised sequence; implicating that there were 9 consecutive periods. Each period had a 3-week duration[1] and cows grazed daily during an 8.5 hour period. Several tools were used to adjust the milk urea levels: (1) the level of nitrogen fertilisation of the pasture, (2) level and type of indoor supplemented feed and (3) herbage mass and regrowth age in a rotational grazing system (Van Duinkerken *et al.*, 2004). Measurement methods on ammonia emission and milk urea concentration were described by Van Duinkerken *et al.* (2005). The relationship between bulk milk urea concentration and ammonia emission from the barn was estimated using a dynamic regression model (Pankratz, 1991).

Results and discussion

Results showed that in the studied farm system with rationed grazing and indoor supplemented feed, the ammonia emission could be described with an emission model using temperature and milk urea concentration as input variables. Model calculations show that emission from the barn rises with 2.6% for 1 °C temperature increase. Ammonia emission rises exponentially with increasing milk urea concentrations. At a level of 20 mg urea/100 g milk, emission increases

[1] Period 9 (the last period) had a length of 2 instead of 3 weeks, because of bad weather and grazing conditions.

with about 2.5% when urea concentration is increased with 1 mg/100 g. At a level of 30 mg urea/100 g milk emission increases with about 3.5% when urea concentration is increased with 1 mg/100 g. These results were in accordance a similar study on summer feeding (Van Duinkerken *et al.*, 2005). The similarity of results between these studies indicates that feeding measures have high potential for emission reduction and that bulk milk urea is a useful indicator for ammonia reduction. In the Netherlands, the national averaged milk urea concentration decreased from 30 to 25 mg/100 g milk over the period 1998 to 2003, indicating that farmers were able to reduce ammonia emission from dairy houses with about 12% on average. Also as a consequence of the results of Monteny (2000) and Van Duinkerken *et al.* (2005), Dutch government and dairy farming sector have agreed upon a voluntary effort to reduce emission. If the sector succeeds to decrease the national average bulk milk urea concentration to 20 mg/100 g milk in 2010, no housing measures will be obligatory for dairy farms with a grazing system. In other European countries, there is a comparable potential for emission reduction by feeding measures. Current milk urea levels (Table 1) are clearly above 18 mg/100 g milk. This level is the benchmark for an optimum bulk milk urea concentration, when results of Schepers and Meijer (1998) are evaluated for a situation with feeding according energy and protein requirements and a rumen degradable protein balance of 0 g/day.

Table 1. Milk urea levels in several European studies with varying circumstances.

Country	Reference	Milk urea (mg/100 g)	Remark
Belgium	Frand *et al.*, 2003	25	Average of 14 herds
Netherlands	Van Duinkerken *et al.*, 2005	25.1	National average in 2001
Finland, Sweden	Nousiainen *et al.*, 2004	27.7	Average of 50 trials
Spain	González Rodríguez and Yánez, 2002	30.5	Supplemented grazing dairy cows in Galicia
United Kingdom	Cottrill *et al.*, 2002	23	Average of a group of UK-herds with Holstein cows

References

Cottrill, B., H.J. Biggadike, C. Collins and R.A. Laven, 2002. Relationship between milk urea concentration and the fertility of dairy cows. Vet Rec. 151: 413-416.

Frand, X., E. Froidmont, N. Bartiaux-Thill, V. Decruyenaere, A. Van Reusel and J. Fabry, 2003. Utilization of milk urea concentration as a tool to evaluate dairy herd management. Anim. Res. 52: 543-551.

González Rodríguez, A. and O.P. Vázquez Yáñez, 2002. Milk urea content on supplemented grazing dairy cows in Galicia. In: Fisher, G. and B. Frankow-Lindberg (Eds.), Lowland and Grasslands of Europe: Utilization and Development. FAO/CIHEAM Interregional and Cooperative Research and Development Network on Pastures and Fodder Crop Production; Lowland Grasslands Subnetwork. FAO, Rome.

Monteny, G.J., 2000. Modeling of ammonia emissions from dairy cow houses. Ph.D. Thesis, Wageningen University, The Netherlands.

Nousiainen, J., K.J. Shingfield and P. Huhtanen, 2004. Evaluation of milk urea nitrogen as a diagnostic of protein feeding. J. Dairy Sci. 87: 386-398.

Pankratz, A. 1991. Forecasting with Dynamic Regression Models. Wiley, New York, USA.

Schepers, A.J. and R.G.M. Meijer, 1998. Evaluation of the utilization of dietary nitrogen by dairy cows based on urea concentration in milk. J. Dairy Sci. 81: 579-584.

Van Duinkerken, G., M.C.J. Smits, L.B.J. Šebek, P.F.G. Vereijken, G. André and G.J. Monteny, 2004. Ammonia emission form dairy barns in a farm system with limited grazing and related to milk urea concentration. Report 46, Animal Sciences Group of Wageningen UR, Lelystad.

Van Duinkerken, G., G. André, M.C.J. Smits, G.J. Monteny and L.B.J. Šebek, 2005. Effect of rumen-degradable protein balance and forage type on bulk milk urea concentration and emission of ammonia form dairy cow houses. J. Dairy Sci. 88, 1099-1112.

Examples of dairy cows diets to high yielding dairy cows, fed for decreased ammonia emission

C. Swensson
Department of Agricultural Biosystems and Technology, Swedish University of Agricultural Sciences, P.O. Box 59, 23053 Alnarp, Sweden

Introduction

Dairy production produces not only high quality products like milk and meats but also wastes. One of the negative impacts from dairy production is ammonia emission from manure. Most important source of ammonia emission is the content of urea in the urine from cattle. One reason for the problem with ammonia emission is that 'ruminants are routinely overfed with nitrogen' (Pfeffer and Hristov, 2005). The purpose of this paper is to compare the outcome from a dairy cow feed experiment using whole crop silage as a measure to decrease ammonia emission with calculations made by the new Nordic feed evaluation system Norfor. The hypothesis is that there is a relationship between the content of crude protein, the urea content in urine and ammonia emission.

Material and method

The feeding experiment was carried out at the experimental dairy farm Mellangård, SLU Alnarp. The experimental design was a Latin square model with 3 replicates. 12 Swedish Holstein dairy cows were included in the experiment. 4 different feed rations were composed; two different roughages were included whole crop wheat silage (WS) or super-pressed beet pulp silage (BP). They were given in high or low amount. The type of roughage were combined with diets with high content of crude protein (CP) content (18%) or low content of CP (16%). Ammonia emission was estimated from the manure. A more detailed description of the experiment is given in Terez Persson *et al.* (2005). Results from this experiment were compared with the outcome from calculations from the Nordic feed evaluation system Norfor (Gustafsson *et al.*, 2005). The calculations were made with the Norfor Training model version 1.4 (add in version 1.9.0.1). In this model the nutritional value of the ration is calculated and the production of energy-corrected milk, kg ECM and kg protein is predicted (Sjaunja *et al.*, 1990). A nitrogen balance is also calculated on dairy cow level.

Preliminary results

In Table 1 the average consumption of feed in the experiment is presented. These values were used in the Norfor model and the results from the model were compared to the outcome from the experiment. The predicted milk yield, kg ECM, calculated by Norfor was lower for all diets. Predicted protein production had only minor deviations from actual production (Table 2). The predictions of N in manure according to Norfor showed a similar ranking between the different treatments. According to Norfor, diet WSHP had a high content of N in urine and therefore a higher risk of ammonia emission compared to the other diets. The ammonia from manure was measured in the experiment and came to the same conclusion.

Table 1. Daily consumption of the feedstuff (Modified from Terez Persson et al.*, 2005).*

	Diet			
	BPHP, kg DM	BPLP, kg DM	WSHP, kg DM	WLHP, kg DM
Hay	2.3	2.1	2.2	2.2
Mixed ration	8.3	7.5	8.2	8.3
BP	2.1	3.8		
WS			1.7	2.8
Concentrate A	3.0	3.0	2.2	4.1
Concentrate B	4.7	3.8	6.0	3.2

Table 2. Comparison between results from experiment and predictions made by Norfor (Modified from Terez Perssson et al.*, 2005).*

	Diet			
	BPHP	BPLP	WSHP	WLHP
Milk experiment, kg ECM /cow and day	34.4	32.2	33.7	32.1
Milk, prediction, kg ECM /cow and day	29.5	29.2	28.2	28.9
Protein experiment, g /cow and day	1148	1102	1127	1075
Protein prediction, g /cow and day	1067	1137	1132	1048
N in manure experiment, g/d	384	350	371	342
N in manure prediction, g/d	399	350	407	385
Ammonia from manure, ppm	3.8	3.3	4.0	3.2
N in urine, prediction	210	167	218	199

Discussion

There is an obvious need to develop tools in dairy production which demonstrates the potential of ammonia emission from dairy cow diets. The N content in urine is a good predictor of the potential of ammonia emission from the manure. Norfor, together with other feed evaluation system, are a way forward to avoid overfeeding nitrogen to dairy cattle. According to Van Amburgh *et al.* (2006) the overfeeding could be at least 10% due to lack of data when modelling.

References

Gustafsson, A-H., H. Volden, M. Mehlqvist, M. Larsen, G. Gudmundsson and O. Aaes, 2005. Norfor^{TM} – the new Nordic feed evaluation system for cattle. 56th Annual meeting of the EAAP, Uppsala, Sweden, June 2005: 8 pp.

Pfeffer, E. and A.N. Hristov, 2005. Interactions between Cattle and the Environment: a General Introduction. In: Pfeffer, E. and Hristov, A.N (Eds.), Nitrogen and phosphorus nutrition of cattle. CABI Publishing,Wallingford UK: pp. 1-12.

Sjaunja, L.-O., L. Baevre, L. Junkkarinen, J. Pedersen and J. Setälä, 1990. A Nordic proposal for an energycorrected milk (ECM) formula. 26th session of the ICRPMA.

Terez Persson, L., B. Frank and A.H. Tauson, 2005. Evaluation of whole crop wheat silage as feedstuff to high yielding dairy cows, fed for decreased ammonia emission. Report 136. Dept. of Agricultural Biosystems and Technology, SLU, Sweden, 68 pp.

Van Amburgh, M.E., T.R. Overton, L.E. Chase and E.B. Peterson, 2006. Protein requirements of lactating dairy cattle: A case study approach. In: Proceedings 2005 Cornell Nutrition Conference for Feed Manufacturers. Cornell University, Ithaca, NY., pp. 213-221.

Additives

Application of urease inhibitors in dairy facilities to reduce ammonia volatilisation

M. Leinker[1], A. Reinhardt-Hanisch[1], E. von Borell[2] and E. Hartung[3]
[1]Universität Hohenheim, Institut of Agricultural Engineering, Garbenstraße 9, 70599 Stuttgart, Germany
[2]Martin-Luther-Universität Halle-Wittenberg, Institute of Animal Breeding and Husbandry with Veterinary Clinic, Adam-Kuckhoff-Straße 35, 06099 Halle/Saale, Germany
[3]Christian-Albrechts-Universität zu Kiel, Institute of Agricultural Engineering, Max-Eyth-Straße 6, 24118 Kiel, Germany

Introduction and objectives

Agricultural animal husbandry is a substantial source of ammonia (NH_3) emissions in Europe. The largest amount of NH_3 is formed and emitted on top of housing floors due to the catalytic breakdown of urea (Braam and Swierstra, 1999). Effective urease inhibitors decrease urease activity (NH_4^+-formation). Thus they can prevent the breakdown of urea and the release of NH_3 effectively (Parker *et al.*, 2005). In the current research project the potential of new urease inhibitors was examined and an application strategy for on farm use was developed.

Material and methods

The effect of new urease inhibitors on minimising NH_4^+-formation and NH_3-release was tested on laboratory scale. In practice only the effect on minimising NH_4^+-formation was measured. The laboratory experiments were conducted with two standardised measurement systems (principle dynamic chamber; glass bottles filled with 2 l of slurry and wind tunnel with an emitting surface coated with a shallow layer of feaces). The level of urea breakdown in combination with NH_4^+-formation and NH_3-release was measured (Leinker *et al.*, 2005). The most effective inhibitor from the laboratory experiments was tested in a natural ventilated dairy barn for 65 lactating cows. In the practical trials urease activity was measured on top of the housing floor with mobile static chambers (Braam and Swierstra, 1999) in three different areas of walking alleys in plots without and with inhibitor treatment. Two different application strategies were tested (trial period June to November 2005 and May to June 2006): (1) within 4 day trials, urease inhibitor was applied once at the 1st day (1 times either 3, 30 and 300 mg/m^2); (2) within 4 day trials, urease inhibitor was applied on the 1st, 2nd und 3rd day (3 times either 2.5, 5 and 10 mg/m^2). The inhibitor was dissolved in demineralised water. With an amount of 0.2 litre m^{-2} this solution was applied on the floor.

Selected preliminary results

The potential to reduce NH_3-release by applying urease inhibitors was depended on type and temperature of substratum as well as type and concentration of the inhibitor (glass bottle measurement). The best effect to reduce NH_3-release was achieved at an applied inhibitor concentration of 0.1% of TKN (Total Kjeldahl Nitrogen) (Table 1). The new urease inhibitor type-D showed overall trials the best reduction of NH_3-release in cattle slurry by about 82 to 88% and in pig slurry by about 60 to 100%. Pollution swapping to N_2O is not expected because of an equal balance of input and output NH_4^+ and NH_3 nitrogen. The results in the wind tunnel experiments proved that the NH_3-emission can be reduced by inhibitor treatment of 2.5 mg/m^2

on average by about 46% (17 to 87%). Additionally NH_4^+-formation was diminished on average by about 76% (16 to 94%). The applied amount of 2.5 mg m^{-2} on the emitting surface in the wind tunnel is approximately a tenth less than the applied concentration of 0.1% of TKN in the glass bottle measurements. In the practical trials inhibitor type-D showed similar temperature and fouling related effects on reducing urease activity (NH_4^+-formation). Urease activity was reduced by about 71% (61 to 88%) with a favourite treatment of 3 times 2.5 mg/m^2 within 4 days (Table 2). The potential of inhibitor type-D to reduce NH_3-emission in practical scale is estimated by about 40 to 50% according to the wind tunnel experiments.

Table 1. Reduction of NH_3-release by using different urease inhibitors in laboratory (glass bottles).

type of slurry	substratum temperature [°C]	reduction of ammonia release by use of different urease inhibitors[1] C		D		E		F		G	
		%	n	%	n	%	n	%	n	%	n
cattle slurry	5	57	12	87	12	2	4	82	4	71	4
	15	43	12	88	12	14	4	45	4	23	4
	25	63	12	82	12	19	4	70	4	18	4
pig slurry	5	8	4	100	4	n.m.		n.m.		n.m.	
	15	26	4	100	8	n.m.		n.m.		n.m.	
	25	-11	8	60	12	-19	4	46	8	-16	4

[1]Concentration of urease inhibitors: 0.1% of Total Kjeldahl Nitrogen; measuring about 96 hours; n = number of repetitions; n.m. = not measured.

Conclusion and perspective

Systematic researches under laboratory and practical conditions have shown that the application of inhibitor type-D on emitting surfaces effectively reduced the catalytic breakdown of urea. Both the formation of NH_4^+ and the emisison of NH_3 were diminished. The reduction potential was high by applying minimised inhibitor concentration and high application frequency. Considering a dairy barn with 65 cows, a walking space of 5 m^2 per cow and an inhibitor treatment of 2.5 mg/m^2 per day, 297 g of urease inhibitor are needed per year. Additionally about 23.7 m^3 water per year is needed for application. Further aspects have to investigate for a standard practical implementation: for instance measuring long term effectiveness of the inhibitors when covering the complete housing floor as well as NH_3 measurements on top of the housing floor. In addition application techniques to get a homogeneous overlay of inhibitor solution have to be developed. Stationary spraying systems with scrapers or mobile systems on scrapers or robots are possible for the implementation.

References

Braam, C.R. and D. Swierstra, 1999. Volatilization of Ammonia from Dairy Housing Floors with Different Surface Characteristics. J. Agri. Eng. Res.72: 59–69.

Leinker, M., A. Reinhardt-Hanisch, E. Hartung and E. von Borell, 2005. Reducing Ammonia Emissions by Using Urease Inhibitors. Landtechnik 60, Heft 6: 342-343.

Parker, D.B., S. Pandrangi, L.W. Greene, L.K. Alamas, N.A. Cole, M.B. Rhoades and J.A. Koziel, 2005. Rate and frequency of urease inhibitor application for minimizing ammonia emissions from beef cattle feedyards. Transaction of the ASAE 48: 787-793.

Table 2. Urease activity in a dairy barn on top of a housing floor before and after inhibitor treatment (trial period June to November 2005 and May to June 2006).

treatment practice	inhibitor concentration		number of measurements[1]		urease activity[2] [mg NH_4^+-N m^{-2} h^{-1}] (reduction %)					
			reference plot	inhibitor plot	reference plots	inhibitor plots				
					day[3] 0	day 0	day 1	day 2	day 3	mean
1 x application (day 0)	1 x 3 mg m^{-2}	[3 mg m^{-2}]	41	74	1875 (100)	407 (78)	671 (64)	1776 (5)	1313 (30)	1042 (44)
	1 x 30 mg m^{-2}	[30 mg m^{-2}]	27	54	2443 (100)	197 (92)	414 (83)	696 (72)	1405 (42)	678 (72)
	1 x 300 mg m^{-2}	[300 mg m^{-2}]	11	27	2141 (100)	161 (92)	154 (93)	407 (81)	1088 (49)	453 (79)
3 x application (day 0-1-2)	3 x 2,5 mg m^{-2}	[7,5 mg m^{-2}]	48	268	1012 (100)	120 (88)	299 (70)	393 (61)	374 (63)	297 (71)
	3 x 5 mg m^{-2}	[15 mg m^{-2}]	13	33	1702 (100)	271 (84)	242 (86)	133 (92)	118 (93)	191 (89)
	3 x 10 mg m^{-2}	[30 mg m^{-2}]	21	60	1921 (100)	12 (99)	241 (87)	348 (82)	422 (72)	256 (87)

[1]Number of measurements at reference plots without inhibitor (day 0); number of measurements at inhibitor plots (day 0 to 3).

[2]With urease activity is meant the NH_4^+-N-formation over an duration of 30 minutes [mg NH_4^+-N m^{-2} h^{-1}] (Braam and Swierstra, 1999).

[3]Day where the measurements were accomplished.

Research of feed and manure additives utilisation for ammonia and greenhouse gases emissions abatement in livestock breeding

A. Jelinek[1], M. Dedina[1] and J. Dolejs[2]
[1]Research Institute of Agricultural Engineering, Drnovska 507, 161 01 Prague 6, Czech Republic
[2]Research Institute of Animal Production, Pratelstvi 815, P.O.Box 1, 104 01 Prague 114, Czech Republic

Introduction

Ammonia emissions originating from animal breeding are one of the biggest sources of atmospheric ammonia pollution in Europe (Buijsman *et al.*, 1987). Ammonia, after deposition in soil, can significantly increase its acidification, eutrophication, nitrate leaching (Schulze *et al.*, 1989) and can negatively influence surrounding biotopes (Roeloffs and Houdijk, 1991). In order to reduce ammonia and greenhouse gases emissions, the Czech Republic has adopted principles of Gothenburg and Kyoto Protocols. Acts and directives, fully compatible with the EU legislation have been incorporated in the Czech legislation. The integrated pollution prevention and control (IPPC) document and the Nitrate Directive are among those directives that have significant effect on air and water pollution. The EU has also set up a policy towards emission ceilings for member states. Limit for the Czech Republic is 80 kton NH_3 per year (to be reached in 2010). In case of greenhouse gases emissions, the Czech Republic is obliged to reduce these emissions by 8% every year until 2012 (in relation to emissions reached in 1980). To meet the above mentioned demands, it was necessary to implement ammonia emissions abatement techniques and relevant measures. They are presented either in the Best Available Techniques Reference Document (BREF) or in the UNECE guidance (UNECE, 1999). We have also found measures in use very expensive (Webb *et al.*, 2005) and economically unavailable for the Czech agriculture sector. In practice, manure and feed additives are able to meet the European directives demands for ammonia emissions abatement. An advantage of these additives is their immediate application without high investment costs. Additives are widely used in both large and extra large enterprises for pigs production in the Czech Republic.

Materials and methods

A photo-acoustic infrared gas monitor Innova 1312 is being used for ammonia and greenhouse gases emissions measurement in stables and slurry storages. This method of measurement is readily available but its application is relatively costly (Mosquera, 2005). The indoor air temperature, relative humidity, values of air pressure and ventilation parameters are continually measured and recorded by appropriate devices. Probes, drawing air samples, are placed in front of exhaust ventilators in the stream of exhaust air in accordance with the methodology (Jelinek *et al.*, 2004). A continual ammonia emissions measurement (running at least 24 hours without interruption) is carried out in the second third of the pig fattening period (usually after 120th day of the period). Comparative measurements are taken in hall with additives applied and in model hall without additive application. Samples of additives are applied into feeding mixture or spread on slated floor, according to suppliers´ instruction. More than 42 measurements were carried out, using 18 different manure and feed additives, in commercial pigs breeding farms during the past eight years.

Results and discussion

The objective of this study has been focused on assessment of efficiency of different additives for reducing ammonia emissions. Simultaneously with measurement of ammonia emissions were also carried out measurements of methane, carbon dioxide and hydrogen sulphide. A number of tested additives have proved a positive effect on abatement of greenhouse gases. An average reduction of methane by 28.1%, carbon dioxide by 28.5%, hydrogen sulphide by 22.4% and ammonia by 24.0% were observed. Though an effectivity of many additives, intended for ammonia emissions reduction are currently known, discussion is still being carried on (Amon *et al.*, 1995, Kemme *et al.*, 1993). Presented results can contribute to the better knowledge of the ammonia emissions reduction processes.

Conclusions

The study of manure and feed additives for ammonia and greenhouse gases emissions abatement supports the new BAT determination. In addition, according to our results, an appropriate additive can not only decrease ammonia but also other greenhouse gases. The study was sponsored by the Ministry of Agriculture of the Czech Republic, project No. QF 3140 Reduction of greenhouse gases and ammonia emissions from agricultural activities.

References

Amon, M., M. Dobeic, T.M. Misselbrook, B.F. Pain, V.R. Phillips and R.W. Sneath. 1995. A farm scale study on the use of De-Odorase for reducing odour and ammonia emissions from intensive fattening piggeries. Bioresour. Technol. 51: 163–169.

Buijsman, E., J.F. Mass and W.A.H. Asman, 1987. Antropogenic NH_3 emissions in Europe. Atmos. Environ. 21: 1009-1022.

Jelinek, A., M. Dedina, P. Pliva and J. Soucek, 2004. Research of biological agents effects on reduction of ammonia concentration in stables of intensive farm animals breeding. Res. Agr. Eng. 50: 43-53.

Kemme, P.A., A.W. Jongloed, B.M. Dellaert, and F. Krol-Kramer. 1993. The use of *Yucca schidigera* extract as a urease inhibitor in pig slurry. In: Proc. of the 1st Int. Symp. on Nitrogen Flow in Pig Production and Environ. Consequenses, Wageningen, the Netherlands, pp. 330–335.

Mosquera, J., G.J. Monteny and J.W. Erisman, 2005. Overview and assessment of techniques to measure ammonia emissions from animal houses: the case of the Netherlands. Environ. Pollut. 135, 381-388.

Roeloffs, J.G.L and A.L.M. Houdijk, 1991. Ecological effects of ammonia. In: Nielson, V.C., B.F. Pain and J. Harting (Eds), Ammonia and Odour Emissions from Livestock Production, Elsevier, Barking, UK, pp. 10-16.

Schulze, E.D., W. de Vries, M. Hauhs, K. Rosen, L. Rasmusen, S.O. Tamm and J. Nilsson, 1989. Critical loads for nitrogen deposition on forest ecosystems. Water, Air and Soil Pollution 48: 451-456.

United Nations Economic Commission for Europe (UNECE), 1999. Control techniques for Preventing and Abating Emissions of Ammonia. EB.AIR/WG.5/199/8/Rev.1. UNECE, Geneva, Switzerland, 37 pp.

Webb, J., H. Menzi, B.F. Pain, T.H. Misselbrook, U. Dämmgen, H. Hendrichs and H. Döhler, 2005. Managing ammonia emissions from livestock production in Europe. Environ. Pollut. 135: 399-406.

Reduction of ammonia emissions from pig housing, slurry storage and applied slurry through acidification of slurry

P. Kai[1], P. Pedersen[2], J.E. Jensen[3] and S.G. Sommer[1]
[1]Institute of Agricultural Engineering, Aarhus University, Schüttesvej 17, 8700 Horsens, Denmark
[2]The Department of Pig Housing and Production Systems, Danish Pig Production, Vinkelvej 11, 8620 Kjellerup, Denmark
[3]Extension Service Landbonord, Erhvervsparken 1, 9700 Broenderslev, Denmark

Introduction

Livestock production is the most important source of ammonia in the atmosphere in Europe Consequently, ceilings on the annual NH_3 emissions were included in the Gothenburg Protocol United Nations Convention on Long-range Transboundary Air Pollution, and in the EU National Emissions Ceilings Directive (NECD). Acidification of livestock slurry is an obvious treatment for the purpose of reducing NH_3 emissions from livestock production facilities (Clemens *et al.*, 2002; Vandre and Clemens, 1997). However, till now, development of the technology has failed due to risk of foaming and because of the potential hazards associated with the use of acids. The present study measured the whole farm effect of a newly developed commercial slurry acidification system on NH_3 emissions from pig housing, slurry storage, and applied slurry.

Materials and methods

The effect of acidification of slurry on the NH_3 emission was studied on a farm with 1200 growing/finishing pig places (25-100 kg) in four separate compartments. Acidification of the slurry was performed by pumping slurry from the slurry pits to a 20 m^3 treatment tank. Sulphuric acid (96% H_2SO_4) was added at a rate of 0.5% by weight to the bottom of the treatment tank to reduce pH to 5.5. Following acidification the slurry was aerated to reduce foaming (Infarm A/S, Aalborg, DK). Part of the slurry in the treatment tank was returned to the pig house resulting in about 15 cm of slurry in the pits. Surplus slurry from the treatment tank was transferred to a slurry store at the farm. The NH_3 emission per pig was calculated for each of 6 batches and was based on half hour measurements of NH_3 and estimations of the ventilation rate using the CO_2 balance method described by Pedersen and Sällvik (2002). Two slurry storage experiments were carried out with slurry stored in open face PVC containers (0.95 m high, 1.09 m^2, and 1.04 m^3) for the duration of 6 and 13 months, respectively. The NH_3 emission was determined by measuring the loss in total nitrogen (total-N). The effect of acidification on the NH_3 emission from applied pig slurry was investigated in a comparative crop response field study with band spread untreated and acidified slurry. Slurry was applied to circular plots with a diameter of 36 m. The NH_3 flux was measured during the course of 7 days using passive Leuning samplers (Leuning *et al.*, 1985) using the micrometeorological mass balance method (Wilson *et al.*, 1982).

Results and discussion

The average NH_3 emission from the control compartments amounted to 0.43±0.06 kg NH_3-N per pig, while the compartments with acidified slurry in average emitted 0.13±0.06 kg NH_3-N per pig equal to a reduction of 70%. This difference was statistical significant (P<0.001). No surface crust developed on the stored pig slurries during the two experimental periods. 5% of the total N was lost during 6 months storage of untreated slurry, while 45% was lost during the 13 storage period. The acidified slurry lost 0.3% of the N during 6 months storage and 5% during 13 month storage. After one week almost 50% of the applied TAN had volatilised from the untreated slurry. The NH_3 emission from acidified slurry was 67% lower than the control slurry equal to 16% of added TAN. The fertiliser efficiency, based on grain yields, of acidified slurry amounted to 81 kg mineral fertiliser per 100 kg manure N, while the untreated slurry corresponded to 46 kg mineral fertiliser. Using the Danish manure norm emission factors to estimate NH_3 emissions from untreated slurry, i.e. housing 16%, storage 9%, and application 10.5% of total N in the slurry (Poulsen *et al.*, 2001; Sommer and Hansen, 2004) it can estimated that of 100 kg excreted N 68 kg N is available for the crop (Figure 1). For acidified slurry the corresponding emission factors are 4.8%, 1%, and 3.5% leaving 91 kg N available for the crop. The commercial slurry acidification system is officially approved Best Available Technique (BAT) in Denmark.

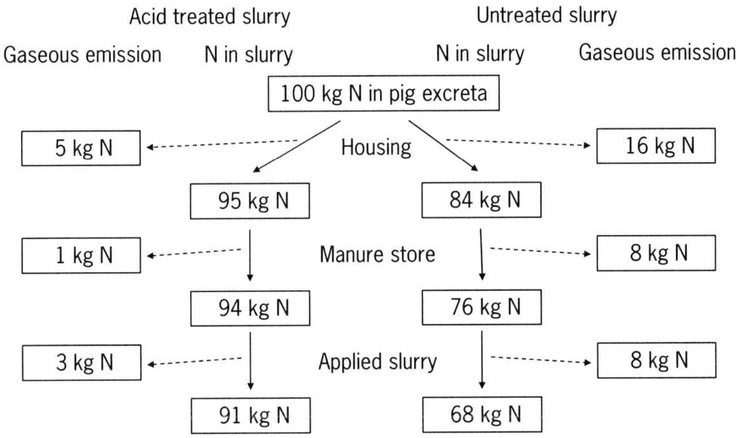

Figure 1. N flow during manure management of 100 kg nitrogen in pig slurry with and without acid treatment.

References

Clemens, J., S. Bergmann and R. Vandre, 2002. Reduced ammonia emissions from slurry after self-acidification with organic supplements. Environmental Technology 23: 429-435.

Leuning, R., J.R. Freney, O.T. Denmead and J.R. Simpson, 1985. A sampler for measuring atmospheric ammonia flux. Atmos. Environ. 19: 1117-1124.

Pedersen, S. and K. Sällvik, 2002. Climatization of Animal Houses. CIGR Working Group Report on: Heat and Moisture Production at Animal and House Level. Danish Institute of Agricultural Sciences, Denmark.

Poulsen, H.D., C.F. Børsting, H.B. Rom, and S.G. Sommer, 2001. Kvælstof, fosfor og kalium i husdyrgødning – normtal 2000 (Norm values of nitrogen, phosphorous and potassium in animal manure). DJF-Report – Husdyrgødning nr. 36. Miljøministeriet for fødevarer, landbrug og fiskeri. Danmarks Jordbrugsforskning, Tjele. pp. 152.

Sommer, S.G. and M.N. Hansen, 2004. Ammoniakfordampning fra udbragt gylle (Ammonia emission from applied animal slurry). Grøn Viden, Markbrug No. 303, pp. 6.

Stevens, R.J., R.J. Laughlin and J.P. Frost. 1992. Effects of separation, dilution, washing and acidification on ammonia volatilization from surface-applied cattle slurry. J. Agric. Sci. 119, 383-389.

Vandre, R. and J. Clemens. 1997. Studies on the relationship between slurry pH, volatilization processes and the influence of acidifying additives. Nutr. Cycl. Agroecosys. 47: 157-165.

Wilson, J.D., G.W. Thurtell, G.E. Kidd and E.G. Beauchamp. 1982. Estimation of the rate of gaseous mass transfer from a surface source plot to the atmosphere. Atmos. Environ. 16: 1861-1867.

Reduction of ammonia emission and improved production efficiency in pig houses by acidification of slurry

S. Levi and S. Daren
Daren Laboratories & Scientific Consultants Ltd., Hamazmerah, 13, 74047 Nes Ziona, Israel

Introduction

Ammonia emission from pig houses mainly originates from the sub-floor storage of liquid manure. One of the options to control the release of ammonia from the accumulated manure is to lower the pH below seven. After rejecting the use of inorganic acids such as nitric, phosphoric and sulphuric, themselves potential pollutants, to achieve this, an aqueous biodegradable acid/polymer based slurry additive formulation was developed and tested, both under laboratory conditions and in-situ. The *in-situ* experiments, including an economic evaluation, were financed by the EC as a CRAFT project (QLK5-CT-2001-70429, RAPID). Previous laboratory studies had already revealed that the formulation effectively reduced ammonia emissions from stored pig slurry, after adding and mixing, or after spraying the formulation on top of the slurry. During experiments in the framework of the CRAFT project, optimal spraying strategies were developed.

Materials and methods

The *in-situ* experiments were carried out in compartments of the research station for pig husbandry in Sterksel, the Netherlands. An automatic spraying system with under floor nozzles was developed for this purpose, to enable the testing of different spraying regimes (various volumes per spray, spraying intervals etc.). The spraying was automatically actuated by an ammonia sensor when the ammonia concentration in the slurry pit reached a preset threshold.

Results and discussion

Results showed that at an optimal spraying regime, the 20 - 25 ppm ammonia emission from the ventilated pig house was reduced by 50% compared to a reference treatment in an untreated compartment. Spraying could successfully and automatically be carried out using a threshold for the ammonia concentration in the air beneath the floor (in the pit). Under the optimal spraying regime, animal production efficiency – expressed in growth per day and feed conversion ratio – was increased significantly. This was most likely due to the improved indoor air quality. It means that the investment and operational costs, estimated to amount to 15 € per pig place per year, will be repaid by improved animal production. The parallel implication of this treatment is that since ammonia is not being emitted, the nitrogen remains in the slurry where it remains available as a fertiliser when the slurry is spread on soil intended for crops. In another study in the CRAFT programme, performed in PRI, The Netherlands, treated cow manure was spread on test plots in which grass was grown. The yield of grass on plots that received formulation treated manure was10-30% greater than a similar plot that received untreated cow manure. Another series of tests using the RAPID formulation was performed by CRPA, Italy, on chicken manure.

Odour abatement

Almost all livestock farming produces odours. Increasingly, people from non-farming backgrounds are moving into the countryside from towns and cities, and often complain about livestock odours. It is arguable that odour reduction could make a farm or manure handling business easier to run and more socially acceptable, especially when planning permission is needed or when the business is sold or expanded. Exceptionally, regulatory authorities can enforce odour abatement measures and impose fines when no action is taken. Odour abatement pressures are not applied uniformly to farms, but are context specific. *In extremis,* effective odour abatement could be essential for the continuation of a business. The results from insitu experiments showed that the formulation could be a useful tool in such situations. For example, the studies of its use in poultry farming, completed in Italy, suggested that odour abatement correlates with the abatement of ammonia emission.

The application technology

The simplest system would involve the delivery of the solution in reusable 250 or 1,000 litre containers with the necessary dilution then being performed on site. The results showed that once diluted to appropriate concentrations, conventional crop spraying systems could be used to apply the formulation to slurry or solid manure stores, or during land spreading of slurries or solid manures The small-scale spraying apparatus used for dairy slurry experiments in the UK, delivered 6 litres per minute of diluted formulation (1:4 dilution) over $5m^2$. These data were compared with an estimate of the full economic cost of commercial crop spraying in the UK, which is approximately €0.09 per litre of diluted material. At dilution rates in the range of 1:2 to 1:4, the RAPID formulation costs are of a similar order of magnitude. For application to solid manures from poultry, the results indicated application costs of €0.10-0.15 per laying hen place and €0.08-0.09 per broiler place for large-scale systems.

Cattle housing

Ammonia emissions from a low-profile cross-ventilated dairy barn

R.E. Sheffield[1], M. De Haro Marti[1], J.F. Smith[2] and J.P. Harner[3]
[1]Department of Biological and Agricultural Engineering, University of Idaho, USA
[2]Department of Animal Science and Industry, Kansas State University, USA
[3]Department of Biological and Agricultural Engineering, Kansas State University, USA

Methodology

This study was conducted on a 128 x 64 m 800-head LPCV barn, with a 0.5/12 roof pitch in Milnor, ND (Figure 1). Tests were conducted during three series of three randomly assigned, pre-selected ventilation [low = 20 fans (198.22 m^3/s); medium = 40 fans (364.34 m^3/s); high 78 = fans (537.45 m^3/s)] periods lasting 2 hours. The herd was comprised of crossbred and Holstein cows milked 3X and housed in freestalls with sand bedding. Manure was scraped from the barn, to a flush-flume collection pit on the north-end of the barn. Collected manure was then processed through a sand-manure separator before collected manure and parlor wastewater were transferred to an earthen manure collection basin. Gas emission rates were estimated using an open-path ultraviolet (UV) spectrometer system. An open-path UV Spectrometer, known as a UV Sentry (Cerex Environmental, Atlanta, GA), was placed inside the barn adjacent to the exhaust fans. Sample UV spectra were recorded every minute during the eighteen sampling periods and ammonia (NH_3) data was analysed. Emission rates were calculated from the product of the gas concentration, gas molecular weight, and air velocity. Differences between groups were tested for significance ($P < 0.05$) using Differences in Least Squared Means test of the PROCMIXED procedure of SAS (SAS 9.1).

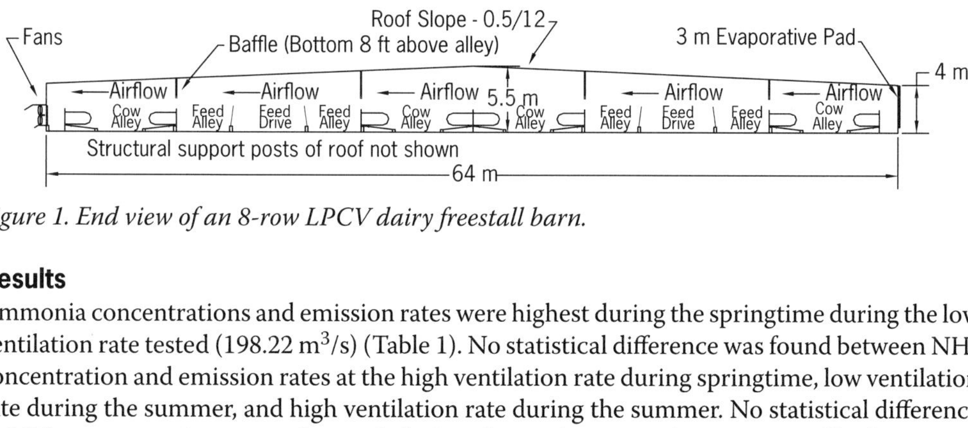

Figure 1. End view of an 8-row LPCV dairy freestall barn.

Results

Ammonia concentrations and emission rates were highest during the springtime during the low ventilation rate tested (198.22 m^3/s) (Table 1). No statistical difference was found between NH_3 concentration and emission rates at the high ventilation rate during springtime, low ventilation rate during the summer, and high ventilation rate during the summer. No statistical difference in NH_3 concentrations was observed during the medium ventilation rates of both seasons. Average concentrations of NH_3 observed here (spring = 1,219±5 ppb; summer = 1,117±4 ppb) are lower than those reported by Zhoa *et al.* (2005) and Mutula *et al.* (2004) of 300–3,000 ppb and 36,000–51,000 ppb, from naturally ventilated freestall barns in Ohio and Texas, respectively. Springtime NH_3 emissions from the LPCV barn were found to be lower than those calculated during studies of naturally ventilated freestall barns in Europe. During this study, NH_3 emissions at the low ventilation rate were found to be 856 mg/h/500 kg live weight during the spring and 678 mg/h/500 kg live weight during the summer, compared to those measured in England (1048 mg/h/500 kg), The Netherlands (1,769 mg/h/500 kg), Germany (1,168 mg/h/500 kg), and Denmark (843 mg/h/500-kg) by Groot Koerkamp *et al.* (1998). Additionally, Schmidt *et al.* (2002) measured NH_3 emission rates from naturally freestall dairies of 224 mg/h/500 kg live weight during the winter and 481 mg/h/500 kg live weight during the summer in Minnesota. These differences between reported values are likely due to differences in the measurement techniques and the methods used for quantifying the ventilation rate from each respective barn, and differences between emission rates due to barn configuration, manure management, and ventilation rate. Further research is needed to investigate if gaseous emissions observed in this study using sand bedding would differ if organic bedding such as dried manure solids or sawdust were used.

Table 1. Ammonia concentration and emission rate from an 800-cow LPCV dairy barn.

Season	Ventilation Rate	Concentration (ppb)		Emission Rate (μg/s)	
		Mean	Standard Error	Mean	Standard Error
Spring	Low	1,370	10.3	172,248	2,464
	Medium	1,181[b]	8.2	273,133[a]	1,962
	High	1,108[a]	8.2	377,874[b]	1,962
Summer	Low	1,084[a]	7.0	136,426	1,676
	Medium	1,157[b]	7.0	268,596[a]	1,679
	High	1,112[a]	7.1	379,190[b]	1,681

[abc]Within a column, means without a common superscript differ ($P<0.05$) using differences in Least Squares Means.

Conclusions

Gaseous emissions were found to be dominated by nitrogen-based compounds during emissions studies at an 800-cow low-profile cross-ventilated dairy freestall barn in Southeastern North Dakota. Indoor ammonia concentrations were found to be considerably less than those reported in naturally ventilated freestall barns during previous studies. Ammonia emission rates were similar to freestall barns in Denmark, and less than those measured in England, The Netherlands, and Germany and greater than those observed in Minnesota, USA.

References

Groot Koerkamp, P.W.G., J.H.M. Metz, G.H. Uenk, V.R. Phillips, M.R. Holden, R.W. Sneath, J.L. Short, R.P. White, J. Hartung and J. Seedorf, 1998. Concentrations and emissions of ammonia in livestock buildings in Northern Europe. J. Agric. Eng. Res. 70: 79-95.

Mutulu, A., S. Mukhtar, S.C. Capareda, C.N. Boriack, R.E. Lacey, B.W. Shaw and C.B. Parnell, 2004. A process-based approach for ammonia emission measurements at a freestall dairy. Presented at the ASAE/CSAE Annual International Meeting, Ottawa, Canada 1-4 August 2004.

Schmidt, D.R., L.D. Jacobson and K.A. Janni, 2002. Continuous monitoring of ammonia, hydrogen sulfide and dust emissions from swine, dairy and poultry barns. Presented at the ASAE Annual International Meeting/CIGR XVth World Congress, Chicago, USA July 28-31 2002.

Zhao, L.Y., M. Brugger, R. Manuzon, G. Arnold and E. Imerman, 2005. Study of air quality of air quality spatial and temporal distributions on large dairy farms in Ohio. Proceedings of the Seventh International Symposium, 18-20 May 2005, Beijing, China. ASABE. St. Joseph, Michigan USA.

Ammonia emission from cow houses within the Dutch 'Cows & Opportunities' project

M.C.J. Smits and J.W.H. Huis in t Veld
Wageningen UR, Animal Sciences group, PO Box 65 8200 AB Lelystad, The Netherlands

Introduction

In total 17 farmers spread over the Netherlands were participating in a Duch project Cows & Opportunities (C&O) with the aim to develop and show a durable way of farming on commercial farms. A few years after this project started, short measurements of ammonia emissions of cow houses were done within a selection of these farms while the farmers were already improving nitrogen management. The effect of ventilation rate on ammonia emission has not been studied before in naturally ventilated commercial Dutch dairy farms. Bulk milk urea concentration (BMU) has been suggested as an indicatory tool for monitoring cow house ammonia emissions (Duinkerken *et al.*, 2005). The main aims of this paper are to study the variations in cow house emissions between the C&O herds and its relationship with bulk milk urea and ventilation rate.

Materials and methods

Farms: Ammonia emissions of 12 selected cow houses were measured occasionally during a short period (approx. 1 week) in summer or mild winter. On 7 selected farms this was repeated in a second short period and in two of these farms also in a third period. Measurements were also performed on an applied research farm De Marke. This farm is developing and demonstrating a durable way of farming already for more than a decade. All cow houses were naturally ventilated loose housing systems with cubicles. On the selected C&O farms slatted floors with underfloor slurry pits were used. De Marke was using a special solid floor with parallel grooves and a scraper.

Tracergas ratio method: A tracergas SF6 was injected at a known fixed rate near the slatted floors in the naturally ventilated cow houses. Concentrations of SF6 and NH_3 were measured in the air inlet and air outlet openings. The source strength of ammonia, being the ammonia flux, was calculated from the ratio of measured concentrations and the known source strength of SF6.

Correcting for grazing hours and temperature: At low temperatures and during grazing hours, emission rates from the cow houses are lower. For comparison between farms, the measured daily emissions were corrected towards zero grazing hours and a temperature of 15 °C, by adding 2.4% per grazing hour and approx. 2.7% per degree Celsius that the Temperature differed from 15 °C respectively. These corrections were based on Monteny (2000) and Duinkerken *et al.* (2005) respectively.

Results

On average 10.1% of nitrogen excreted in the cow house was emitted as ammonia on C&O farms (range 5.0-23%). BMU concentrations were low and its variations within and between farms were small and could only explain a small percentage of the variation in emission. A large part (67%) of the variation between farms was explained by the ventilation rate (Figure 1).

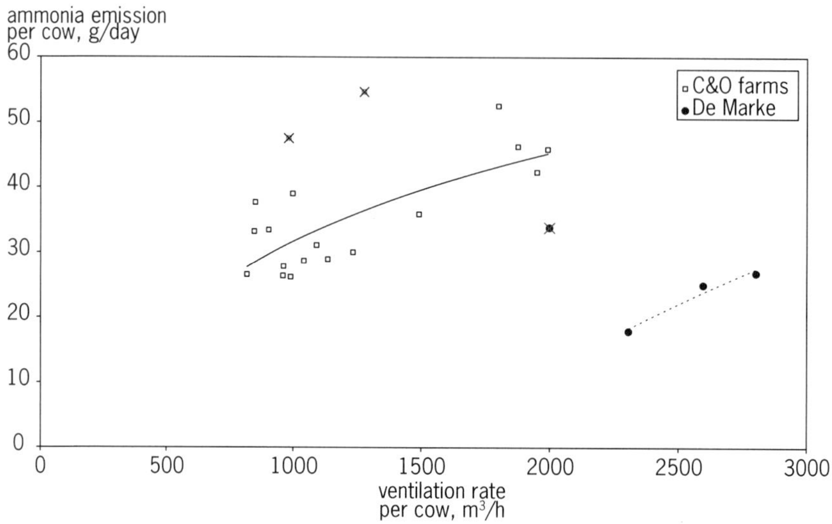

Figure 1. Relationship between average ventilation rate and average ammonia emission per short measuring period. Outliers (crosses) are ignored in the regression lines. De Marke is different because of its special floor and management.

Discussion

Although bulk milk urea on the farms was low, large variations in emissions were found due to varying ventilations rates. At high wind speeds and large ventilation openings even much higher emissions could have been found if the nitrogen flows on these farms, reflected in the low BMU concentrations, would not have been managed so well. At higher wind speeds, large ventilation openings should be avoided. Automatically controlled natural ventilation openings are recommended to prevent heat stress on hot windless days and to reduce high ventilation and emission rates, on non-hot summer days. A solid floor may also help to reduce air exchange from the pit and emission from the cow house especially at high ventilation rates.

References

Van Duinkerken, G., G. André, M.C.J. Smits, G.J. Monteny and L.B.J. Sebek, 2005. Effect of Rumen-Degradable Protein Balance and Forage Type on BMU Concentration and Emission of Ammonia from Dairy Cow Houses. J. Dairy Sci. 88: 1099-1112.

Monteny, G.J., 2000. Modelling of ammonia emissions from dairy cow houses. PhD-thesis, Wageningen University, the Netherlands, 155 pp.

Measurements of ammonia emission from naturally ventilated dairy cattle buildings

G. Zhang, J.S. Strøm, H.B. Rom and P. Kai
Institute of Agricultural Engineering, Research Centre Bygholm, Aarhus University, Schüttesvej 17, 8700-Horsens, Denmark

Introduction

A major part of the global NH_3 (ammonia) emissions is recognised from agriculture. In Denmark, the ammonia emissions from livestock buildings make up about 40% of the total ammonia emission from agriculture operations (Andersen *et al.*, 2001). It is crucial to find technical solutions to reduce the NH_3 emission. However, practical methods for evaluation of the emission from naturally ventilated buildings and data on the emissions are still needed. The average emission rates of ammonia and other contaminant gases from naturally ventilated dairy cattle buildings with different types of floor and different manure handling systems were presented and discussed by Zhang *et al.* (2005). The objectives of this paper are to address the diurnal variation of ammonia emission from naturally ventilated dairy cattle buildings and the differences between the actual and average values for the emission rates during longer measurement periods.

Materials and methods

Measurements were carried out in nine selected free-stall dairy cattle buildings with different type floors and different manure handling systems (Figure 1). All buildings were naturally ventilated with side-wall and ridge openings. The selected buildings were newer type dairy cattle buildings with a milking centre at one end and the animal occupied area in the rest of the building (Zhang *et al.* 2005). In order to achieve reliable averages of the emission data and

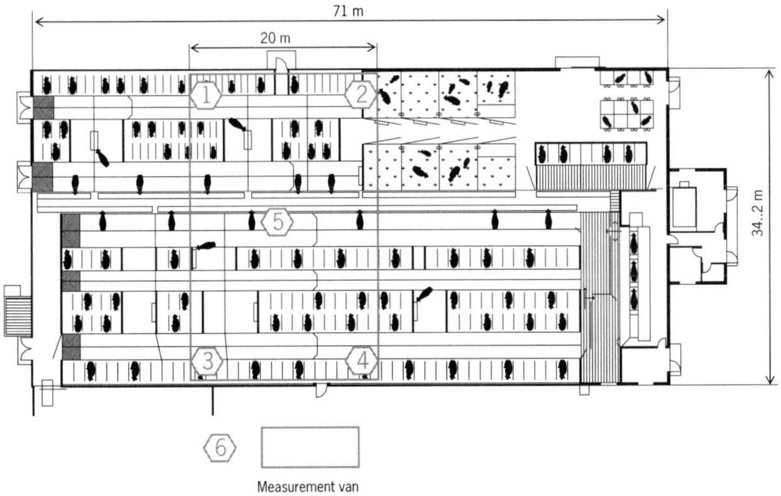

Figure 1. Measurement section and sensor positions.

the variations in concentration, air exchange rate and emission, air sampling was done for a minimum period of three days was selected with five days as a preference.

Results and discussion

Estimation of air exchange rates: Both tracer gas method and a model aided CO_2 balance method was used for estimation of the air exchange rate. Based on the CO_2 model, the air exchange rate can be determined by the mass balance of CO_2. However, due to the imperfect mixing and the non-uniform wind pressure distribution, there is still uncertainty in the method, but the results did show much difference than the tracer gas method.
Variation of ammonia emission: Examples of the variation in the NH_3 emission rates throughout a measurement period are shown in Figure 2 for two of the dairy buildings. The data are one-hour moving averages. The emission rate is given as g HPU-1 d-1 where HPU is defined as 1000 W total heat produced by the livestock at an environmental temperature of 20 °C. It is seen that the emission levels varied considerably during the measurement period. The lowest emission rates during a 24-hour period tended to occur after midnight, which are possibly caused by the low activity level of the cows, and the low indoor air and floor temperatures. There were a few large spikes or peak values during a 24-hour period, however, they were probably due to different management schedules such as feeding, floor scraping and milking, etc. In most cases, the highest emission rates occurred in afternoon. Generally, the diurnal patterns were found in all the measurements.
Air exchange rates, wind speeds and variation of ammonia emission rates: A general trend was that the air exchange rates varied according to the wind speeds, and the emission rates varied in proportion to the air exchange rates. It was noticed that the ammonia emission patterns were very similar to the air exchange rates as well as to the wind speeds. This implies that the air exchange rate is a crucial factor that may affect the emission. The air exchange rate can be recognised as an indicator that provides the momentum of the airflow passing over the contaminant surfaces in the buildings, i.e. the floor and the slurry channels. The higher the air exchange rates, the higher an air speed can be created in occupied zones (Strøm *et al.*, 2002).

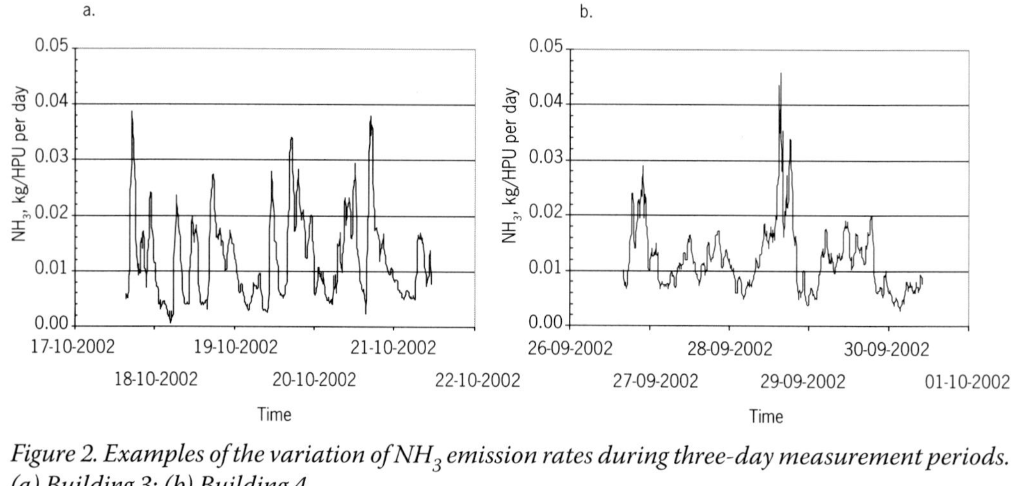

Figure 2. Examples of the variation of NH_3 emission rates during three-day measurement periods. (a) Building 3; (b) Building 4.

Conclusion

The results showed that the emission levels varied considerably during the measurement period. The variation of ammonia emission rates could be from 2 to 40 g/HPU/day in a building during a 24-hour period. The lowest emission rates tended to occur after midnight. The air exchange rates in the buildings affect emission rates significantly. The results on the diurnal patterns indicate that in order to make a steady-state estimation of the gas emission from naturally ventilated cattle buildings, the measurement period should be at least 24 hours.

References

Andersen, J.M., H.D. Poulsen, C.F. Børsting, H.B. Rom, S.G. Sommer, N.J. Hutchings, 2001. Ammonia emission from agriculture since the middle of the 80'es (Danish text with English summary, Danish title: Ammoniakemission fra landbruget siden midten af 80'erne). Report No. 353. National Environmental Research Institute, Denmark.

Strøm, J.S., G. Zhang and S. Morsing, 2002. Predicting near-floor air velocities for a slot-inlet ventilated building by jet velocity decay principles. The Transactions of the ASAE 45: 407-413.

Zhang, G. J.S. Strøm, B. Li, H.B. Rom, S. Morsing, P. Dahl and C. Wang, 2005. Emission of ammonia and other contaminant gases from naturally ventilated dairy cattle buildings. Biosys. Eng. 92: 355-364.

Monitoring ammonia emissions from cattle houses using Ferm tubes

C H Burton[1,2], T.H. Misselbrook[3], D.C. Welch[2] and C Hampson[2]
[1]Cemagref, 17 Avenue de Cucillé, 35044 Rennes, France
[2]Formerly of Silsoe Research Institute, Wrest Park, Silsoe, Bedford, United Kingdom
[3]Institute of Grassland and Environmental Research, North Wyke, Okehampton, Devon, United Kingdom

Introduction

Livestock farming, especially cattle, contributes to over half of the ammonia emissions produced across the UK and most of this comes from the animal buildings (Defra, 2002). There is a range of equipment that can measure ammonia concentration within a livestock building (Phillips *et al.*, 2000; Phillips *et al.*, 2001) but many of these methods require information of the ventilation rate which is a problem where there is natural rather than forced ventilation. Ferm tubes directly measure the ammonia emission flux. They were chosen as a suitable technology for monitoring a modern 100 animal cattle building over two successive winter confinements. The study sought both to evaluate the technology and to follow emission factors throughout the seasons.

Materials and methods

'Ferm tubes' are passive samplers for measuring ammonia flux (Scholtens *et al.*, 2003). The general construction comprises two lengths of tube each lined with acidified paper. The sections are joined to make a single length but separated by an orifice plate the hole being around 1mm in diameter. This greatly constricts the flow of air through the tubes and thus allows long periods of exposure without saturation. The whole assembly is mounted in the building orifice to be monitored such that the tubes are perpendicular to the opening and thus in line with the expected airflow. Depending on the direction of the air flow, ammonia will be entrained on to one or the other of the adsorbing tubes; the *net* flux will thus be the difference. The total emission from the building is given by Equation 1:

$$\text{SUM } \{n = 1, n\} \ [(A_n/(K_s.a).(F_n/t_n)] \qquad (1)$$

where A_n is the area of the opening covered by the sampler n and a is the orifice area. Thus $A_n/(K_s.a)$ is the scale number allowing for the correction factor K_s. F_n is the net amount of ammonia entrained on the sampler; dividing this by the exposure time, t_n gives the ammonia flowrate passing through the sampler. Multiplying this by the scale up number give s the total flux for the opening. One last calculation is to divide the emission by the liveweight of animals in the building and so give the emission as g-ammonia per day per LU (where one LU is equivalent to a cow of 500 kg liveweight). A typical cattle building of modern design was selected, i.e. open boarded walls, cubicles for the animals and large entrances for vehicle access which allowed generous ventilation. The floor was scrapped daily directing manure to a store located 10 metres south of the building. 160 samplers were used with a concentration in the large more exposed spaces. All large openings (over 0.1 m^2) were monitored but in the case of small gaps such as boarded walls, representative openings were selected. Exposure was for 24 hours. The farm was visited a total of 13 times between September 2002 and April 2004.

Results and discussion

A summary of the results is set out in Table 1. In the first campaign, there was a seasonal trend with emissions falling from 72 g/day•LU in October to 20-24 for most of the winter before rising to 53 in March. Misselbrook *et al.* (2000) give a mean figure of 34 g/day•LU falling to half this value where litter is used. The second campaign is followed a similar pattern producing values of a similar order but no marked increase in the Spring.

Table 1. Overall emission factors for ammonia from monitoring campaign of cattle building.

	Sept	Oct	Nov	Dec	Jan	Feb	Mar	Apr
Campaign 1								
Emission g/LU•day		73	43	24	25	21	53	
Campaign 2								
Emission g/LU•day	70	24	31	23	43		19	25

Conclusions

Ferm tube technology is an effective and versatile method for monitoring emissions of ammonia from naturally ventilated buildings. The study revealed clear patterns. The effect of wind was mostly confined to influencing where the ammonia left the building. Over a short period of 2-3 hours, strong winds could accelerate emission but the average value over longer periods was less affected. The more significant parameter was building temperature which was greatly influenced by the seasonal average; plotting the latter against emission produced a weak but significant correlation underlining this effect.

References

DEFRA, 2002. Ammonia in the UK. Department for the Environment, Food and Rural Affairs. DEFRA publications, London, SW1A 2XX United Kingdom. 89 pp.

Misselbrook, T.H., T.J. Van der Weerden, B.F. Pain, S.C. Jarvis, B.J. Chambers, K.A. Smith, V.R. Phillips and T.G.M. Demmers, 2000. Ammonia emission factors for UK agriculture. Atm. Env. 34: 871-880.

Phillips, V.R., D.S. Lee, R. Scholtens, J.A. Garland and R.W. Sneath, 2000. A review of methods for measuring emission rates of ammonia from livestock buildings and slurry or manure stores, Part 1: assessment of basic approaches. J. Agr. Eng. Res. 77: 355-364.

Phillips, V.R., D.S. Lee, R. Scholtens, J.A Garland and R.W. Sneath, 2001. A review of methods for measuring emission rates of ammonia from livestock buildings and slurry or manure stores, Part 2: monitoring flux rates, concentrations and airflows. J. Agr. Eng. Res. 78:1-14.

Scholtens, R., J.M.G Hol, M.J.M. Wagemans and V.R. Phillips, 2003. Improved passive flux samplers for measuring ammonia emissions from animal houses, Part 1: basic principles. Biosys. Eng. 85: 95-100.

Inventory of reduction options for ammonia and methane emissions in Dutch cattle farming

H.J.C. van Dooren, M.C.J. Smits, N.W.M. Ogink and G.J. Monteny
Animal Sciences Group, Wageningen UR, P.O. Box 65, 8200 AB Lelystad, The Netherlands

Introduction

The part of agriculture to the total ammonia (NH_3) and methane (CH_4) emissions in the Netherlands was in 2004 120 kton (90%) and 404 kton (49%) respectively (Milieubalans, 2006). Cattle farming is the major source of both NH_3 and CH_4 emissions from agriculture in the Netherlands. In 2004 veal calves, beef cattle and dairy cattle were responsible for 5%, 6% and 46% respectively of the national NH_3 emission from animal husbandry. Endogenous emissions from enteric fermentation, mainly coming from cattle, amounted 72% of the national CH_4 emission from agriculture (Milieubalans, 2006). Given the considerable contribution of cattle to national emissions of NH_3 and CH_4 and the expected future national emission targets an overview of further possibilities to reduce emissions is necessary. Objective of this inventory was to create a basis for further development in emission reduction by summarising national and international available promising reduction options.

Dairy farming in the Netherlands

Within cattle farming as a whole, the dairy sector is the main contributing sector. Husbandry systems for beef cattle are comparable with those for dairy cattle. Reduction options for veal calves have already been proposed (Smits *et al.*, 2005). For these reasons the focus in this paper will be on dairy cattle. The average annual milk production per cow increased significantly in the last decades. As the national milk quotum remained constant the number of cows decrease from 4.3 million in 1980 to 2.6 million in 2005 (CBS, 2007). Due to increasing milk production the decrease in number of cow only slightly affected the total amount of slurry produced. Also the number of dairy farms decreased: from 67,167 in 1980 to 23,527 in 2005. This resulted in an increasing scale of dairy farming in The Netherlands and an increasing specialisation in terms of housing systems. The majority of the dairy cows (86%) are housed in a cubicle system with concrete slatted floors. Only 10% of the dairy cows are housed in an alternative system like the deep litter system and tying stall system. The rest (4%) is housed in a cubicle system with a solid concrete floor or an emission reducing grooved floor (CBS, 2007).

Available reduction options

As cattle is mainly housed in naturally (i.e. passively) ventilated buildings, integrated reduction options for NH_3 and CH_4 are difficult to implement. Reduction options developed in the past focuses mainly on different floor constructions. The list of options currently recognised as emission reducing contains mainly sloping solid floors with scraper, alternatively equipped with a washing system. The principle idea behind these systems is to minimise the emissions by covering the underfloor slurry pit and daining off the urine from the floor combined with separate removal of faeces. The systems however have serious constraints in respect to animal welfare as they become too slippery and are therefore rarely used. The grooved floor was based on the same principle but without a sloping floor. Other options like acidification or aeration of slurry, reducing slurry temperature, slurry dilution, reducing emitting surface, reducing ventilation and air movement or adding substances with chemical or absorbing properties all seized upon a different link in the chain of ammonia formation and volatilisation but were never widely implemented because of high costs or negative trade offs on other aspect of dairy farming. Adapting the diet of dairy cows by replacing grass with maize will reduce the urea concentration in urine and finally the NH_3 emission. This option can be monitored by the urea concentration in milk. It has never been incorporated in legislation but was implemented in 2002 as a convenant between govermant and farmers union: farmers voluntarily reduce the average urea level in bulk milk to 20 mg per 100 ml in 2010 and remain free from investments in NH_3 emission reducing housing system. Available reduction options for CH_4 emissions are removing the slurry from the housing as soon as possible and covering the external storage facility, collecting the passively emitted CH_4 for flaring or biological scrubbing or anaerobic digestion of slurry and conversion the CH_4 in a combined heat power unit (CHP-unit) to CO_2 simultaneously producing heat and electricity.

Further promising reduction options

Apart from the above mentioned reduction option that have been developed and assessed earlier there are some new possible promising reduction option that need further development and evaluation:

- Feeding strategies might also influence the CH_4 emission caused by rumen fermentation.
- Storage of all slurry outside animal housings may reduce both NH_3 and CH_4 emissions.
- Alternative housing systems for dairy cattle might have a lower NH_3 emission as they are based on organic litter. Measurements of NH_3 emission of deep litter systems for dairy cattle however show large varioation within and between measured farms (Mosquera *et al.*, 2005).
- Active cooling of slurry with groundwater or surpluss of heat from anaerobic digestion.
- Creating a relatively small but NH_3 rich air flow that can be clean in an air scrubber.
- The use of urease inhibitors.

Conclusions

After prioritising the mentioned reduction options on factors like reduction potential, cost effectiveness, feasibility of implementation and possible trade offs to other environmental aspects the most promising option need further exploration or development.

References

CBS, 2007. CBS Statline, http://statline.cbs.nl.

Milieubalans, 2006. Milieu en Natuur Planbureau, http://www.mnp.nl.

Mosquera, J., J.M.G. Hol and J.W.H. Huis in 't Veld, 2005. Onderzoek naar de emissies van een natuurlijk geventileerde potstal voor melkvee I. De stal, Agrotechnology and Food Innovation, Wageningen UR, Rapport 324, 58 pp.

Smits, M.C.J., R.W. Melse, A.C. Smits and N.W.M. Ogink, 2005. Bouwsteen Stallen. Quick scan van opties voor vermindering van ammoniak- en geuremissie uit vleeskalverstallen in de Agrarisch Enclave Uddel Elspeet, Agrotechnology and Food Innovation, Wageningen UR, Rapport 59, 52 pp.

Pig housing

Ammonia emission in organic pregnant sows with and without access to paddock

S. Ivanova-Peneva[1], A.J.A. Aarnink[2] and M.W.A Verstegen[2]
[1]Agricultural institute, 3 Simeon Veliki blvd., 9700 Shumen, Bulgaria
[2]Animal Sciences Group of Wageningen UR, P.O. Box 65, 8200 AB, Lelystad, The Netherlands

Introduction

In organic pig production most of the sows have access to a paddock. This may influence the excreting behaviour of the pigs. A part of the urine and faeces will be excreted on the paddock. From research in cows it is known that ammonia emission from a urine puddle in the pasture is a lot lower than the emission from a urine puddle on a concrete or slatted floor. Grass intake by the pigs from the paddock may influence nitrogen excretion and ammonia emission, as well. The aim of this study was to determine any differences in ammonia emission between sows with access to a paddock and sows without access to a paddock.

Material and methods

Two groups of 15 sows (Great York x Dutch Landrace) were raised in a confinement building in group straw pens with individual feeding stalls and outside concrete yard. One of the groups of sows had access to a paddock (paddock group), and the other had not (control group). In the first period sows had unlimited access to the paddock, while this was restricted from 9:00 h until 15:00 h during the second period. Both measuring periods were in summer, with 45 days in between. Sows were restrictedly fed once per day in the morning, with the amount of feed depending on weight, parity number, and stage of pregnancy. Diets consisted 8.7 MJ Net Energy, 153 gkg^{-1} Crude Protein, 80 gkg^{-1} Crude Fiber, 44 gkg^{-1} P, 89 gkg^{-1} K and 6.3 gkg^{-1} Lysine. Water was available *ad libitum*. There was one drinker for each group, positioned at the outside yard. Natural ventilation was provided and there was no heating system. Feed, urine, faeces and manure were sampled two times in the pregnancy period. Samples were analysed in a laboratory for total–N, total-P, total-K, NH_4^+-N, pH, dry matter and ash. Ammonia emissions were measured by the ventilated chamber technique at different locations inside the building and on the paved outside yard. The procedure and the way ammonia emissions was calculated was done as described in Ivanova-Peneva *et al.* (2006).

Results and discussion

There was a clear difference in N-concentration of urine between sows with and without access to pasture. Sows with access to the paddock had about twice higher total nitrogen content in the urine (8.96 gkg^{-1} versus 4.67 gkg^{-1} during the first measuring period and 9.07 gkg^{-1} versus 5.76 gkg^{-1} during the second measuring period; $P<0.05$). Probably the consumption of clover grass, which is rich in nitrogen, has influenced the nitrogen content of urine. The ammonia emission, however, did not differ between treatments (Table 1).
The data form the table show that there were differences in ammonia emissions mainly between the two measuring periods. Ammonia emission per m^2 did not differ significantly between treatments, but differed between periods and locations (inside and outside the building). The effect of both is highly significant. The effect of measuring period was probably caused by the

Table 1. Ammonia emissions in pens of sows with and without access to a paddock.

Period	NH_3 emission ($gm^{-2}d^{-1}$)		Effects		
	Control group	Paddock group		Probability	S.E.D.
First	6.078	6.578	Period	$P<0.001$	0.128
Second	3.196	3.043	Location	$P<0.001$	0.323

time the measurements were done compared to the moment of cleaning. The compartment of pregnant sows was cleaned every three weeks. The first measurements were done at the end of this period, while the second measurements were done closely after cleaning. As was shown before (Ivanova-Peneva *et al.*, 2006) manure management seems to have a big influence on ammonia emission in organic pig production (von Wachenfelt and Jeppsson, 2006). When measurements were done one week after introducing new straw in the compartment (second measurement), ammonia emissions were quite lower.

References

Ivanova-Peneva, S.G., A.J.A. Aarnink and M.W.A. Verstegen, 2006. Ammonia and mineral losses on Dutch organic farms with pregnant sows. Bios. Eng. 93: 221-235.

von Wachenfelt, H. and K.H. Jeppsson, 2006. Nitrogen Losses from Organic Housing Systems for Fattening Pigs. In: Proceedings of Workshop on Agricultural Air Quality, Washington, DC, USA, 5-8 June, 2006, pp. 1192-1199.

Heath risks of ammonia exposure for farm animals

J. Hartung
University of Veterinary Medicine Hannover, Foundation, Buenteweg 17p, 30559 Hannover, Germany

Abstract

Ammonia is one of the typical gases regularly present in the air of livestock buildings. It gives cause for concern for several reasons. First, exposure to higher concentrations of ammonia may damage directly mucous membranes of the respiratory tract and the conjunctiva of the eye of animals. Secondly, chronic exposure may exacerbate multi-factorial diseases, such as atrophic rhinitis in pigs and air way diseases such as bronchitis or pneumonia in animal and man. Third, ammonia is emitted into the ambient air from livestock buildings, manure stores and during land spreading causing odour nuisance in residents living in the neighbourhood of farm enterprises. Fourth, deposition of ammonia can directly damage plants and trees and contributes to acidification of soil and water as well as to the phenomenon of forest decay. While a bulk of work has been carried out in the last 20 years on the environmental impact of ammonia, the increase in knowledge on the role of ammonia in the development of diseases in different farm animal species is still limited. This is mainly due to the fact that ammonia under practical conditions in animal house air is always associated with other gases and pollutants such as micro-organisms or it is attached to dust particles which complicates the assessment of the heath effects caused solely by ammonia. The consequences are that the presently used thresholds of ammonia for farm animals (e.g. 20 ppm for pigs) are more oriented to what is technical possible in an animal house than what protects the health of the animals. Research indicate that present thresholds do not meet the needs of the animals and should be reduced which coincides with the intentions to protect the environment by reducing ammonia emissions from farm animal production.

Distribution of ammonia in swine houses at emission reduction

K.-H. Krause[1], S. Linke[1], H.-J. Müller[2] and M. Mußlick[3]
[1]Federal Agricultural Research Centre (FAL), Bundesallee 50, 38116 Braunschweig, Germany
[2]Leibniz-Institute for Agricultural Engineering e.V. (ATB), Max-Eyth-Allee 100, 14459 Potsdam, Germany
[3]Thuringian Ministry of Agricultural, Environment Protection and Environment, Beethovenstraße 3, 99096 Erfurt, Germany

Introduction

Ammonia emission reductions of animal houses, with mechanically forced ventilation, can be more easily realised with the installation of special biofilters. Regulatory agencies prefer such end-of-pipe-strategies because it is easier to verify and to control. The degree of effectiveness of biofilters increases with the growing supply of ammonia, e.g. by below-floor ventilation. In general the ammonia concentration is smaller in the breathing zone of the animal than at above-floor ventilation. The disadvantage of both methods is to suck the whole volume stream through the biofilter. For the example of a swine house, we will demonstrate how to combine air flow patterns with cleaning of outgoing air on a low level of ammonia mass streams with regard to animal welfare and environment protection.

Material and methods

The examples (a) and (b) in Figure 1 are well-known systems with one exhaust fan location. Example (c) offers a two-way flow distribution system in dependence on the forced pressure distribution. The performance of the below floor pit fans must have strong enough pressure characteristics to avoid a flow reverse airflow, so no fresh air is sucked over the manure surface into the animal house, which would means an increase of ammonia transport into the region over the slatted floor. During summer ventilation with large volumetric exchange rates such

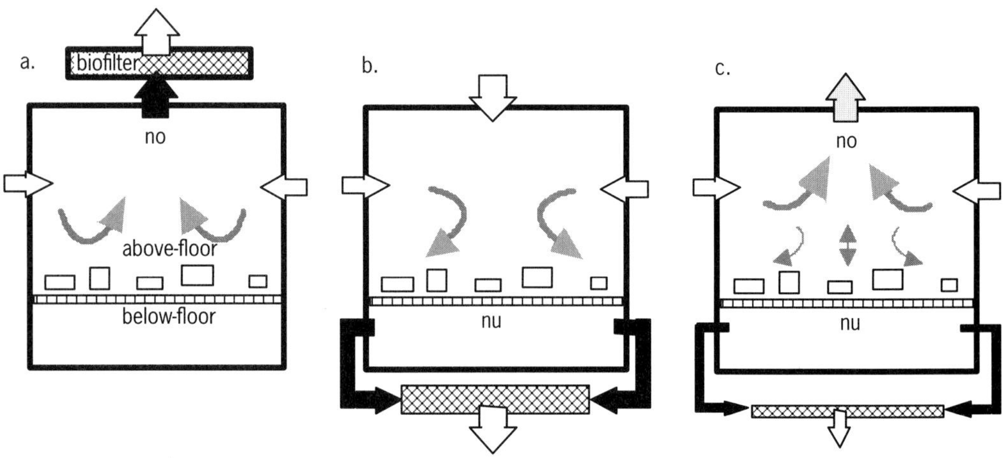

Figure 1. Different systems of emission reduction in a piggery: flow schemes.

a case could occur, if the fan is designed in a wrong way. The case (c) requires effective fans below-floor.

Results and discussion

Time series of concentrations Cu below-floor and Co above-floor were measured. At the same time the corresponding partial volumetric rates were recorded. With relation to the volume of the animal house the exchange rates nu (below floor airflow) and no (above floor airflow) describe the partial ventilation situation and the total exchange rate n is expressed by the sum of no and nu. Collecting the data that are not regularly spread over the area of interest (Cu/Co = f(n)) a contour map is produced with the isolines nu/no, see Figure 2, where the intervals between the contours are set to 0.02. We are interested in those cases where the ratio nu/no is low and the concentration ratio Cu/Co is high. The pre-determined configuration of the ventilation system in the swine house of concern shows an asymptotic behaviour at Cu/Co = 2 for the used pit fan. To reach a better ratio of volumetric exchange rate a low volume stream must overcome a great pressure resistance at a high exchange level.

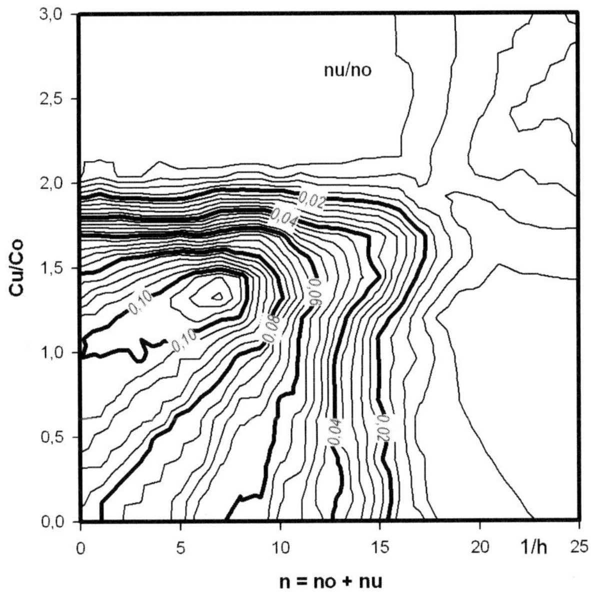

Figure 2 Isolines of constant exchange ratio nu/no in dependence on Cu/Co versus n.

Conclusion

The reduction of the ammonia output of a swine house by an intelligent air flow arrangement leads to small biofilters at below-floor ventilation.

References

Krause, K.-H., S. Linke, M. Mußlick, H-J. Müller and C. Weihs, 2006. A new ventilation system for reduction of odour and ammonia emissions from pig houses. In: World Congress: Agricultural Engineering for a Better World, Bonn, 03-07 September 2006, book of abstracts, Düsseldorf, VDI-Verlag, pp. 447-448.

Adjustment of the livestock ventilation system as a potential ammonia abatement technique

N. Kozlova, V. Vtory and N. Maximov
North-West Research Institute of Agricultural Engineering and Electrification (SZNIIMESH), P.O. Tiarlevo, Filtrovskoje Shosse 3, 196625 Saint-Petersburg, Russia

Introduction

Major environmental problems originates with the design and construction of the farm's ventilation system. So it seems important to supply design-oriented information, materials and methods to individual who design and build livestock buildings. Examination of a pig fattening house as a ventilation system, determined the ammonia emissions from various parts of the pen and the manure pit (Maximov, 2000; Kozlova *et al.*, 2005). These results were used as the basis for the developing decisions on ammonia abatement measures in animal housing when designing environmental control systems.

Materials and methods

In an animal house the ventilating air, through interaction with animals and manure affects ammonia emission to the atmosphere (Figure 1). Indoor air quality is regulated to comply with the animal welfare and productivity requirements. Outlet (exhaust) air concentrations and airflow rates are used to calculate all emissions to the atmosphere. In the animal housing system with extraction of dirty and contaminated air from the underfloor manure pits, which are an additional ammonia emission source, the concentration of ammonia and other unwanted gases in the outlet air increases. This is shown schematically by the dotted line. Since the quantity and quality of the inlet air are adjusted in air control systems, this might be considered an ammonia abatement measure.

Results and discussion

Data on the links between ammonia emissions, the ventilation rate and indoor temperature experimental data were found in the following literature: (Maximov, 2000, Gallman *et al.*, 2005,

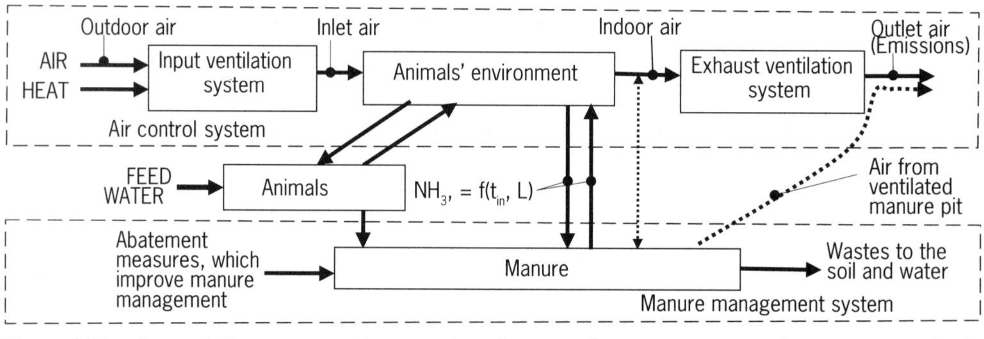

Figure 1.The place of climate control system in release and transportation of ammonia and other contaminants into the atmosphere as a result of interaction of two media – air and manure – in animal house.

Gustafsson *et al.*, 2003, Jeppsson, 2003). All these past research results differed in trial aims, experimental techniques, methods of data handling, and further interpretation of results. But they show that there exists a correlation between the ventilation rate and hazardous emission, and between temperature and emission, and this correlation is strong. Simulation of operating modes of the climate control systems in a dairy barn and a pig fattening house was used to identify the control factors for the ventilation rate and the indoor air temperature under different outdoor climate conditions, the type and weight of animals. In a pig fattening building for 600 pigs with the initial weight of 40 kg and 100 kg slaughter weight, the optimal ventilation rate must vary from 10,000 to 70,000 kg/h, and heat consumption varies from 0 to 150 kW with variation of animal weight and outdoor climate - to maintain the acceptable indoor air temperature and relative humidity levels under typical St. Petersburg climate conditions. In a livestock building, where the ventilation rate is not controlled, fattening periods with excess ventilation air and increased ammonia emissions occur. Natural ventilation systems with the controllable inlet and outlet units are more efficient, as in this case it is possible to decrease ventilation rates in winter, or when the animals are yet small. Simulation results have demonstrated that from the standpoint of ammonia emissions, evaporative cooling of animals should be preferred to avoid the heat stress of the dairy cow in summer instead of over-ventilating the animal house.

Conclusion

To reduce ammonia emission from the animal house it is necessary to control the operation factors of the ventilating and heating system in order to keep the ventilation rate, indoor air temperature and air flow over the manure surface as low as possible. At the same time, the ventilation system must maintain an acceptable indoor air quality to comply with the animal welfare requirements. Investigation outcomes have shown that optimisation of the livestock climate control system as ammonia emission abatement also can impact energy usage.

References

Gallmann E., E. Hartung and T. Jungbluth, 2003. Long-term study regarding the emission rates of ammonia and greenhouse gases from different housing systems for fattening pigs – final results. In: Proceedings of the international symposium on 'Gaseous and Odour Emissions from Animal Production Facilities, Horsen 1-4 June 2003, pp. 122-130.

Gustafsson, G, K-H. Jeppsson, J. Hultgren and J-O. Sannö, 2003. Techniques to Reduce the Ammonia Release from a Cowshed with Tied Dairy Cattle. In: Proceedings of the international symposium on 'Gaseous and Odour Emissions from Animal Production Facilities, Horsen 1-4 June 2003, pp. 239-248.

Jeppsson K.-H., 2003. Diurnal variation in ammonia, carbon dioxide and water vapour emission from a deep litter house for fattening pigs. In: Proceedings of the international symposium on 'Gaseous and Odour Emissions from Animal Production Facilities, Horsen 1-4 June 2003, pp. 131-139.

Kozlova, N., A. Afanasyev and N. Maximov, 2005. Some aspects of emissions abatement from agriculture in the Russian Federation. In: U. Dämmgen, T. Kuczyński, J. Webb and A. Myczko (Eds.), Emissions from European agriculture. Wageningen Academic Publishers, Wageningen, the Netherlands, pp. 293-299.

Maximov, N.V., 2000. Character of ammonia formation and emission in piggeries. 'Ecology and Agricultural Machinery'. V2. Ecological aspects of plant and livestock production technologies. [In Russian], SZNIIMESH publisher, St-Petersburg, pp. 243-247.

Determination of ammonia emissions in a pig house from ventilation flow estimated by two different methods

V. Blanes-Vidal[1], S. Calvet[1], M. Pablos[1], F. Estellés[1], M. Laínez[2] and A.G. Torres[1]
[1]Universidad Politécnica de Valencia. Departamento de Ciencia Animal. Camino de Vera s/n 46022, Valencia, Spain
[2]IVIA, Centro de Investigación y Tecnología Animal, Polígono de la Esperanza 100, 12400. Segorbe, Castellón, Spain

Introduction

Ammonia emission from mechanically ventilated animal houses can be estimated from the difference between the ventilation rate and NH_3 concentration in the indoor and outdoor air. The ventilation flow can be directly determined by measuring fans or indirectly estimated using different methods. One of these indirect methods consists in calculating the ventilation flow from indoor and outdoor CO_2 concentration measurements and CO_2 production in the animal house and carrying out a CO_2 balance (CIGR, 2002; Blanes and Pedersen, 2005; Müller *et al.*, 2006). The objectives of this work were: (1) To measure continuously NH_3 concentrations in a pig building for fatteners; (2) To determine NH_3 emission from an animal house, by estimating the ventilation flow by two indirect methods: obtained from the ventilation controller and estimated from the CO_2 balance in the building.

Materials and methods

The investigation was carried out with 80 fattening pigs, during 14 days, in an experimental building located in Segorbe (Valencian Community region, Spain). The pig building contained 16 pens, with a fully slatted concrete floor. No bedding was used. The house was provided with a negative pressure mechanical ventilation system. During the experiment, the slurry was stored in a pit under the slat, and it was removed at the end of the trial. The initial average body weight was 88.3 kg, and the final weight, 104.2 kg. Measurements included: temperature, and NH_3, CH_4, N_2O and CO_2 concentration in the indoor and the outdoor air. A photoacoustic gas monitor (Innova) was used for gas concentration measurements. The ventilation flow (V) was determined by two methods: (1) From the measured indoor temperature (T_i), which was related to the ventilation flow (V_T), and (2) from the CO_2 balance in the house (V_{CO2}). Regarding the first method, the ventilation flow in the room was automatically controlled according to T_i. The equation relating T_i and V_T was obtained from the calibration performed by the ventilation system manufacturer. In this equation, the actual V_T was expressed as a percentage of the maximum ventilation flow. For determining the maximum ventilation flow in the room, a preliminary test was carried out, in which the rotational speed of the fan was set at maximum, and the ventilation flow under this condition, was calculated from air velocity measurements at the outlet section. In the final experiment, measured T_i and the obtained equation were used for continuously calculating V_T. The ventilation flow was also indirectly estimated from the carbon dioxide balance in the animal house (V_{CO2}). In this work, a value of 0.185 $m^3\ h^{-1}\ hpu^{-1}$ for the total CO_2 production in the farm has been used for the CO_2 balance (CIGR, 2002), where 1 hpu (heat production unit) is equivalent to 1000 W of total heat produced by the animals at 20 °C. This value was adjusted considering the diurnal variation on CO_2 production (CIGR, 2002). On the basis of recorded data, a data set with averages from two hours periods was generated.

Results

NH_3 concentration in the animal house during the experiment period was on average 4.77 mg m^{-3}. Evaluation of NH_3 emissions from the pig building showed that the mean and standard deviation of NH_3 emissions were 3.86±1.38 g d^{-1} pig^{-1}, when the ventilation flow was determined from the indoor temperature and the ventilation controller (V_T), and 5.03±1.16 g d^{-1} pig^{-1}, when it was determined from the carbon dioxide balance (V_{CO2}). Figure 1 shows measured indoor temperature in the animal house during the experiment period and ammonia emission calculated from ventilation flow estimated by both methods. The correlation between ammonia emissions calculated from both methods was R^2 = 0.61. NH_3 emission varied during the day, and were stable and low between 22:00-10:00 and peaked at 18:00.

Conclusions

Ventilation flow in animal houses is a fundamental variable in the calculation of emission rates. In this paper, two different indirect methods for estimating the ventilation flow were used to calculate ammonia emissions from a pig house. The results showed a discrepancy in ammonia emission estimation of about 30%.

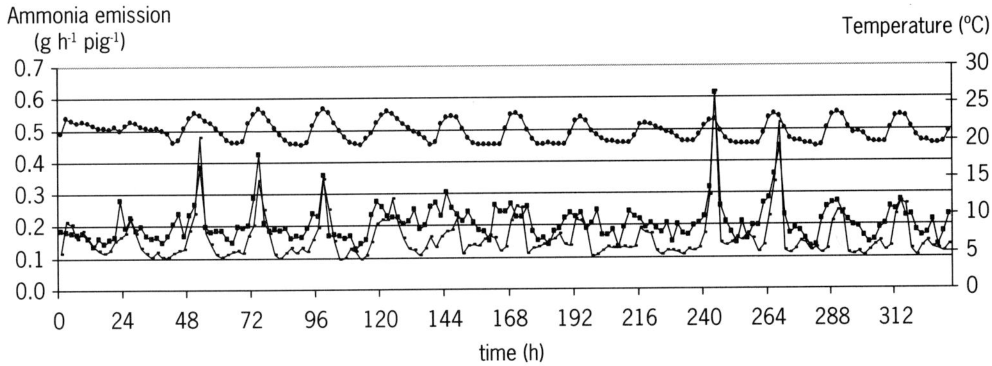

Figure 1. Measurements during the experiment: temperature (°C, bullets), NH_3 emissions calculated from V_T (g h^{-1} pig^{-1}, small blocks) and NH_3 emissions calculated from V_{CO2} (g h^{-1} pig^{-1}, large blocks).

Acknowledgements

Funded by the Spanish Ministry of Education and Science (project AGL 2005-07297) and by the 'Conselleria de Empresa, Universidad y Ciencia', co-funded by FEDER European funds.

References

Blanes V. and S. Pedersen, 2005. Ventilation Flow in Pig Houses measured and calculated by Carbon Dioxide, Moisture and Heat Balance Equations. Biosyst Eng. 92: 483-493.

CIGR, 2002. Climatization of animal houses. Heat and moisture production at animal and house levels. Pedersen S. & Sallvik K. 4th Report of Working Group.

Müller H.J., H.B. Rom and S. Pedersen, 2006. Comparison of methods of determination of ventilation flow in animal houses with different kinds of animals. Paper 937A. XVI CIGR World Congress.

Ammonia and odour from different surfaces in a house for fattening pigs

S.A. Nimmermark
Swedish Univ. Agric. Sciences, Dept. of Agr. Bios. and Technology, 230 53 Alnarp, Sweden

Introduction

Ammonia as well as odour can be a problem in houses for fattening pigs and also for the neighbourhood and the environment. Increased knowledge about the emission from different surfaces in a pig house can be a useful tool for identification of possible reduction techniques. Emissions from surfaces inside the building as well as emissions from ventilated manure culverts contribute to the total emission to the surrounding. The objective of the study was to compare ammonia and odour release from different surfaces in a house for fattening pigs and to compare the emissions from systems with over or under-floor air extraction.

Material and methods

Two different research facilities with fattening pigs were used in the study. A pig house with natural ventilation and room for 128 fattening pigs in 8 pens with partly slatted floor above a 1.0 m deep and 1.3 m wide manure culvert and a deep bedding area was used for estimating emissions from different surfaces. The ammonia and odour release were measured in the exhaust air from fan ventilated flux hoods covering six different surfaces in a pen and one surface in the walking alley. Measurement of release from each surface was made at three different days. Another building with 8 pens with partly slatted floor was used for studying ammonia and odour release from a manure culvert, i.e. over or under-floor air extraction. The manure culvert in this pig house is 1.2 m deep and 1.1 m wide. Ammonia and odour were measured in the exhaust air close to fans using a system with (1) wall mounted fans and (2) fans connected to the manure culvert. During the fattening period, measurements were made about 10 separate days using wall fans for ventilating the room and about 10 separate days using low evacuation through the manure culvert. Pig weight was estimated by the help of standard weight gain curves. Ammonia was measured by the help of gas detection tubes. Odour was sampled in nalophan bags and analysed for odour concentration by the help of an olfactometer (Ecoma TO7) and a qualified panel. Ventilation rate was measured by the help of impellors.

Results

Surfaces inside the building: Ammonia as well as odour release was high from wet dirty litter. Concentration differences between exhaust from the hoods and the room air is shown in Figure 1. Dirty surfaces on the slatted floor emitted a limited amount of ammonia and rather much odour.

Under-floor air extraction versus wall-mounted fans: Average odour emission was 9.5 $OU_E \cdot m^{-2} \cdot s^{-1}$ for wall-mounted fans and 20.2 $OU_E \cdot m^{-2} \cdot s^{-1}$ for evacuation through the manure culvert. The odour emission was significantly higher for the system ventilating the pig house through the manure culvert than for the system using only wall mounted fans. No significant difference was found for pig weight. Average ammonia emission was 35 $\mu g \cdot m^{-2} \cdot s^{-1}$ for wall-mounted fans and 46 $\mu g \cdot m^{-2} \cdot s^{-1}$ for evacuation through the manure culvert.

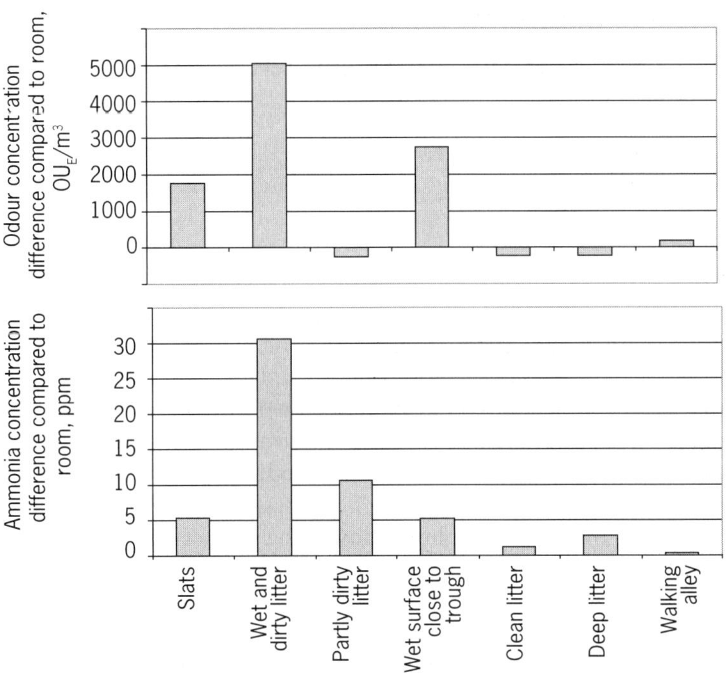

Figure 1. Ammonia and odour concentration difference between the exhaust air from the flux hoods and the air inside the room at air flow rate 150 $m^3 \cdot m^{-2} \cdot h^{-1}$ inside the hood.

Discussion and conclusions

Measured ammonia release from the slats is comparable to values found by Andersson *et al.* (1994). A very rough estimation of the emission contribution from the various types of surfaces suggests that 17% of the ammonia and 39% of the odour in the house is released from the slats. Corresponding values for the wet and dirty litter and for the deep litter are 37% / 44% and 16% / -10% respectively. Odorants seem to be adsorbed in clean litter. Ammonia and odour emissions found for the system with wall mounted fans are somewhat higher but of the same magnitude as values for ammonia found by Aarink *et al.* (1995) and values for odour found by Ogink and Koerkamp (2001). Release of odorants from the manure culvert seems to increase the emissions to the neighbourhood considerably when exhaust fans are evacuating the room through the manure culvert.

References

Aarnink, A.J.A., A. Keen, J.H.M. Metz, L. Speelman and M.W.A. Verstegen, 1995. Ammonia emission patterns during the growing periods of pigs housed on partially slatted floors. J. Agr. Eng. Res. 62: 105-116.

Andersson, M., K.-H. Jeppsson and E. von Wachenfelt, 1994. Ammonia emission from different surfaces in livestock buildings. In: AGENG '94, Milano, Report N. 94-C-001.

Ogink, N.W.M. and P. Koerkamp, 2001. Comparison of odour emissions from animal housing systems with low ammonia emission. Water Science and Technology 44: 245-252.

Pit and wall emission streams for H_2S, NH_3, CO_2, PM and odour from deep-pit pig finishing facilities

L.D. Jacobson, B.P. Hetchler and D.R. Schmidt
University of Minnesota, Biosystems & Agricultural Engineering Dept., 1390 Eckles Ave., St. Paul, MN, USA

Introduction

Pig production buildings generate and emit a variety of airborne pollutants and some of them are hazardous compounds, such as ammonia (NH_3) and hydrogen sulphide (H_2S), which may create chronic indoor air health concerns for workers and for the pigs housed. Other pollutants, like particulate matter fewer than 10 μm in diameter (PM_{10}) create environmental ambient air quality concerns when they are released into the atmosphere. There is a need to know the amount of these pollutants being emitted from pig production buildings for simple regulatory purposes, but more importantly to determine how these airborne pollutants can be reduced to levels that will meet the regulatory limits, lower the impact of odours for neighbors, and minimise the risk to workers and pigs.

Materials and methods

A mechanically ventilated (with wall and pit fans) 2400-head deep-pit pig-fattening barn, located in southern Minnesota, was monitored over two pig growth cycles. Several fixed pit ventilation rates; 0, 7, 17, and 34 m^3/hr-pig (0, 1, 2, and 4 pit fans) were used for two-hour periods during the six month study. The project measured concentrations of H_2S, NH_3, CO_2, and PM_{10} using an instrument trailer that was capable of semi-continuously monitoring these airborne contaminants at multiple locations, including wall and pit fans exhaust, an interior room location, and an ambient (background) site.

Results and conclusions

The results from this project were grouped into the concentration and emission for a particular airborne contaminant into winter, spring, and summer periods. The winter data was collected from 26 January to 4 March, 2006. The barn had the pigs removed and new pigs added (or 'turned') in mid March of 2006. The spring data was gathered from 14 April to 15 May that represented when the barn was transitioning from the winter (ceiling inlets) to summer (side wall curtain inlets). The summer data was collected from 17 May to 30 June. The mean ammonia concentrations and emissions are displayed for the four pit ventilation rates with attached error bars that represent a standard deviation in Figure 1. NH_3 concentrations in the center of the room were similar for each season for all four pit ventilation rates including the case when no pit fans (0 m^3/hr-pig) were operating. Ammonia wall fan emissions show a steady decrease as the pit ventilation rates increase while the pit fan emissions show the opposite trend. Although not shown, similar trends were found for hydrogen sulphide concentrations and emissions for each season for all four pit ventilation rates. Also not shown, CO_2 levels and corresponding airflow rates through the barn were quite similar for the 0, 7, 17 m^3/hr-pig pit ventilation rates categories. PM_{10} concentrations did not vary much in this study between the pit and wall exhaust streams except during the winter when there are very low dust concentrations

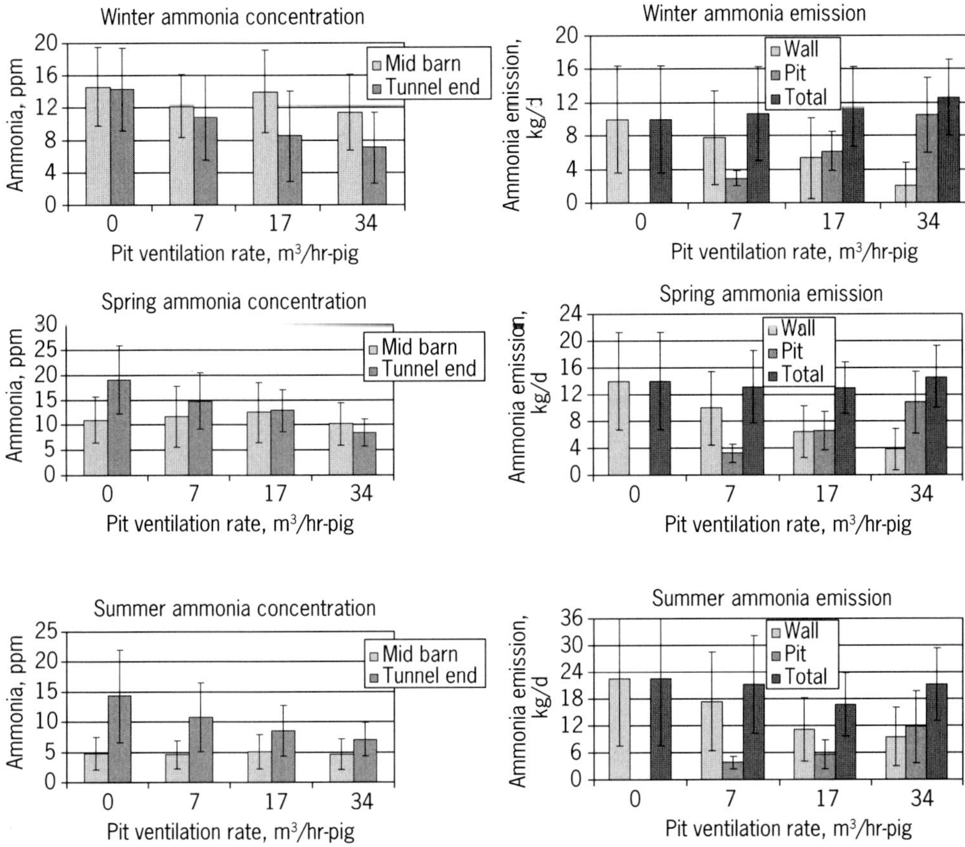

Figure 1. Ammonia concentrations and emissions for the winter, spring, and summer periods.

in the pit exhaust air. Odour concentrations and emissions collected in this study were not as conclusive as NH_3, H_2S, and PM_{10} since odour was only measured intermittingly (monthly). Still pit exhaust did have higher odour levels (as measured by dilution threshold) and emission than wall exhaust. The study shows limited value or benefit for the use of pit fans in a deep-pit pig fattening barn, since similar NH_3 and H_2S concentrations were measured in the center of the barn for all four pit ventilation cases. Thus it seems that exhausting a portion of the barn's ventilation air through the pit has little effect on the room's indoor air quality. The NH_3 and H_2S emissions determined in this study show that a disproportionate amount or mass of these two gases are exhausted from the barn through pit fans if they are operating. This fact should be noted if a producer wants or needs to reduce a pig barn's NH_3 and H_2S emissions, since there would be a considerable benefit to treating only the pit fan exhaust air with an air emission control technology (like biofilters) rather than all of the exhaust air (wall and pit).

Effects on odour and ammonia emissions of different width openings in slatted floors for fattening pigs

C. Fabbri[1]*, N.Labartino*[1]*, G.Bonazzi*[1] *and E.Poletti*[2]
[1]*CRPA spa, Reggio Emilia, Italy*
[2]*CRA, Modena, Italy*

Introduction

Directive 2001/88/EC on pig welfare establishes a maximum width of 18 mm for openings in slatted flooring used for groups of rearing pigs. In the case of finishing pigs over 110 kg, a maximum opening width of 20-22 mm is usually adopted to speed up manure discharge and reduce fouling of both the floor and animal skin. Farmers are concerned that a reduction in opening width may lead to higher degrees of floor fouling, a consequent increase in ammonia emissions and a drop in the animals' general welfare.

Material and methods

The emissions of one batch of fattening pigs were measured in three specially equipped, separate compartments, with slat openings of 18, 20 and 22 mm, in a pig house with ventilation fans extracting air from both the wall and the manure pit. The ventilation rate was continuously monitored plus ammonia and odour concentration levels were recorded using a photoacustic multigas monitor (Bruel&Kjaer, mod 1302) and an olfactometer (Ecoma mod. TO7), respectively. The odour concentration measures were taken during four days distributed along the pig fattening period at three different hours of the day (12 samples per thesis). The manure fouling levels on slatted floor and on the pig's skin were also accurately recorded. Each experimental compartment contained two pens with 15 pigs each from 90 to 160 kg live weight and a vacuum system for removing slurry from the pit.

Results

Slat opening widths significantly influenced the degree of fouling on both the floors and animal skin. The moving animal eliminated less manure in the compartment with the most narrow (18 mm) slat openings than in the other two compartments (20 and 22 mm slat widths) under examination. Manure fouling of the slatted floor and the pig's skin was lower in the other two compartments (Table 1). With an ammonia emission factor of 3.03 kg/pig place/year with in the 18 mm slats room, the 20 mm and 22 mm compartments respectively revealed 3.33 kg/pig place/year and 3.29 kg/ pig place/year. Daily ammonia emission trends (Figure 2) were not perfectly aligned with the ventilation rate (Figure 1) due to the interference of emission peaks, present during the most intense phases of animal movement (liquid feeding times). Odour emissions (Table 2) were on average: 17,603 OU/h/head for the 18 mm compartment, 13,913 OU/h/ head for the 20 mm compartment and 8,031 OU/h/ head for the 22 mm compartment. Odour emissions were on average, respectively 21% and 54% lower in the 20 mm and 22 mm compartments compared to the 18 mm room. Emissions varied with the different gaseous under consideration: there were no statistical differences as far as ammonia was concerned between the three monitored compartments, whilst there were significantly differences in odour emissions. The basically similar ammonia emissions from the 18 mm, 20 and 22 mm

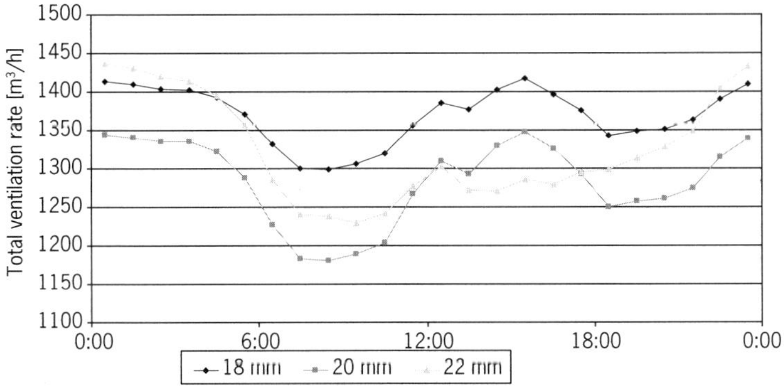

Figure 1. Total ventilation rate of the different thesis during the average day.

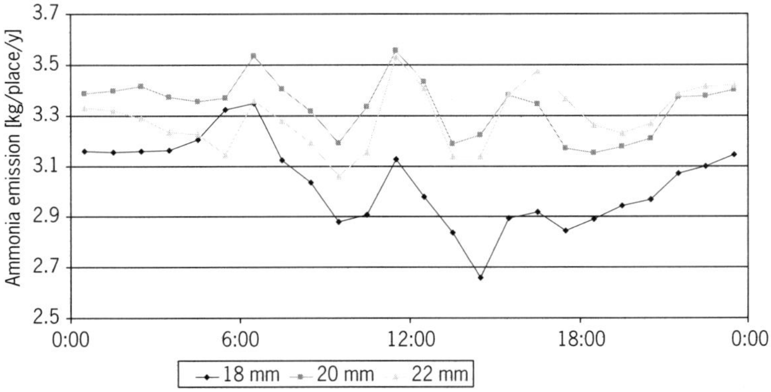

Figure 2. Ammonia emissions measured during the average day.

compartments, despite the increased degree of fouling, was due to the higher ventilation rate in the under floor pit of the latter compartments. In fact, greater slat width promoted the flux of ventilated air under the slat increasing, thus, exchanges between the slurry surfaces and air. On the contrary, the 18 mm compartment produced the highest odour emissions, followed in turn by the 20 mm and 22 mm ones. This was caused by the greater quantities of solid manure, the main cause of odorous compost emissions (indole, skatole), on the slatted surface and the pig bodies in the compartments with smaller slat openings. No significant difference was instead found, in animal performances, morphological anatomical assessments or animal behaviour.

Table 1. Reduction of three fouling parameters of two thesis (20 and 22 mm opening slat) compared to Directive standard thesis (18 mm opening slat).

Opening slat 20 mm			Opening slat 22 mm		
Fouling floor	**Slat wetting element**	**Skin fouling**	**Fouling floor**	**Slat wetting element**	**Skin fouling**
21%	5%	19%	31%	22%	36%

Table 2. Odour emission rate (OU/h/head) measured during the experiment.

	Opening Slat		
	18 mm	**20 mm**	**22 mm**
Average values	17,603	13,913	9,177
Reduction cf. standard (18 mm)		21%	54%

Conclusion

Results show greater fouling of pig bodies and floors in compartments with reduced slat floor openings, but this does not affect ammonia emissions and no significant difference was found between the three experimental situations. On the contrary, higher odour emissions are evident in the compartment with reduced openings, due to the higher fouling by faeces of both floor and animal skins. Other experiences have been planned to investigate these aspects.

Reduction of the number of slots for concrete slatted floor in fattening buildings: consequences for pigs and environment

N. Guingand and V. Courboulay
IFIP, French Institute for pig and pork industry, Domaine de la Motte au Vicomte, BP 35104, 35651 LE RHEU Cedex, France

Introduction

Recommendations concerning welfare indicate the necessity for pigs to have a laying area with solid floor. This study was integrated in a global program concerning the conception of new pen in order to improve pig welfare in respect to European regulations and to protect environment.

Material and methods

The study was carried out at the IFIP Experimental Farm at Villefranche-de-Rouergue (Aveyron, France) on fattening pigs from 30 to 115 kg live weight. The study covered two fattening periods with 144 animals per round. Two identical rooms excepted for the floor were compared, each divided into 4 pens of 18 pigs. In the control room, the whole surface of the pen floor was made with classic concrete slats with 15% open area (8 cm wide slats and 1.8 cm wide gaps). For the second room (noted mixed room), the floor was made with the same concrete slatted floor on the last third of the pen (back). For the rest of the surface, the number and the length of slots were reduced to attempt only 5% open area in order to create a resting area (Figure 1).
Fresh air entered via a ceiling of perforated sheeting and air exhaust was under-floor extraction with chimney. Climate characteristic imposed per batch were identical for both rooms during the whole study. Dust collections were achieved at 1.5 m above the floor in the middle of the corridor. Ammonia concentrations were measured in the ambient and in the exhaust air. In the ambiance, diffuse passive tubes were used in three locations (front, middle and back) in two pens per room and at two heights (0.3 and 1 m above the floor). In the exhaust air, ammonia concentration was measured using the bubbling method in acid solution. Odorous air samples from the exhaust duct were analysed by the CERTECH Laboratory (Seneffe- Belgium) by olfactometric method for the determination of the odour threshold. Equipment and procedures

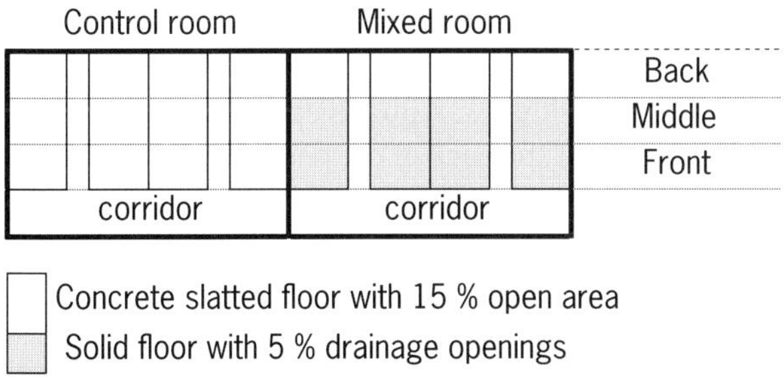

Figure 1. Characteristics of rooms implicated in the study.

are conformed to current recommendations (NF EN 13725). Ambient temperature and exhaust ventilation rate were recorded every 15 minutes during the whole study. For both rooms, pen dirtiness was noted every week and pig dirtiness every 3 weeks. Data concerning pig behaviour were analysed from video recording every 3 weeks.

Results and discussion

Table 1. Dust, ammonia and odours in relation with season and treatment.

	Control room		Mixed room	
	Hot period	Cold period	Hot period	Cold period
Dust ($mg.m^{-3}$)	0.76 ± 0.07	2.01 ± 0.21	1.18 ± 0.32	2.37 ± 0.76
	1.51 ± 0.70[a]		1.89 ± 0.86[a]	
Ammonia emission per pig ($g.d^{-1}$)	17.0 ± 1.4	13.0 ± 3.7	17.6 ± 1.1	17.4 ± 2.9
	15.0 ± 3.3[a]		17.5 ± 1.9[a]	
Odour emission per pig (10^6 u.o. d^{-1})	1.46 ± 0.77	0.88 ± 0.4	1.31 ± 0.28	0.79 ± 0.27
	1.17 ± 0.63[a]		1.05 ± 0.38[a]	

[a]Average value for the two rounds without season effect.

Values obtained in this study in the control room for dust, ammonia and odour are in accord with those already published in similar conditions (Guingand, 2003). Analyses of data concerning pig behaviour didn't show any significant difference between treatments. Pig activity was similar for both rooms for the whole period; only at the beginning of the fattening period, the difference between the resting area and the excretion location was clear for the mixed room. For the whole period, the volume of slurry stored under the pigs was more important at the back of the pens of the mixed room (60% of the excreted dry matter). This area was dirtier with a difficulty for the floor to evacuate the dejections towards the pit. Pen and pig dirtiness contributed to increase ammonia in the ambiance (particularly at the back of the pen) and in the extracted air. The reduction of open area on the two third of the pen leaded to increase the air flow rate at the back of the pen with higher open area and then to increase exchanges between slurry and air. Once more, no relation was established between ammonia and odour. Although pen and pig dirtiness were affected by the reduction of open area, no effect was observed on odour emission.

Conclusions

In our study, the reduction of slots had only effect on pig and pen dirtiness and on ammonia emission. No effect on pig behaviour was observed. This new conception of pen could be ameliorate by the creation of gap at the back of the pen to facilitate the evacuation of dejections or by using metal slatted floor.

References

Guingand, N, 2003. Air quality in pig facilities in relation with physiological stages. In: Proceedings of the Annual Meeting of the EAAP, Roma (Italy), 31 August-3 September 2003, pp. 195.

Reduction of stocking density in pig units: effects on odour, ammonia and dust

N. Guingand
IFIP – French Institute for pig and pork industry, Domaine de la Motte au Vicomte, BP 35104, 35651 LE RHEU Cedex, France

Introduction

The evolution of animal welfare in European regulations would lead to an increase of surface per pig. The low percentage of new piggeries could oblige pig breeders to reduce the number of pigs per pen in order to respect the animal welfare regulation. The aim of this study was to determine the effect of reducing the stocking density by only reducing the number of pigs without any modification on the configuration of rooms on gaseous emissions.

Material and methods

The study was carried out at the IFIP Experimental Farm at Romillé (35 – France) on fattening pigs from 30 to 115 kg live weight. The study covered two fattening periods with 102 animals per batch. Two identical rooms were compared, each divided into 6 pens. In the control room (noted D 1.43), 10 pigs were allocated per pen leading to a stocking density of 1.43 pigs per m^2. In the second room (D1), the stocking density is 1 pig per m^2 (7 pigs per pen). In both rooms, pens had totally concrete slatted floor with slurry storage underneath during the whole period of fattening (pit depth 1.2 m). Fresh air entered via a ceiling of perforated sheeting and air exhaust was under-floor extraction with chimney. Climate characteristic imposed per batch were identical for both rooms during the whole study. For both rooms, measurements were achieved each two weeks from the beginning of the study until the day of first slaughter. Dust concentration was measured using the gravimetric method. Dust collections were achieved at 1.5 m above the floor in the middle of the corridor. Ammonia concentrations were measured in the ambient and in the exhaust air. In the ambiance, diffuse passive tubes were used in three pens (front, middle and back) and at two heights (0.3 and 1 m above the floor). In the exhaust air, ammonia concentration was measured using the bubbling method in acid solution. Odorous air samples from the exhaust duct were analysed by the CERTECH Laboratory (Seneffe- Belgium) by olfactometric method for the determination of the odour threshold. Equipment and procedures are conformed to current recommendations (NF EN 13725). Ambient temperature and exhaust ventilation rate were recorded every 15 minutes during the whole study.

Results and discussion

Ventilation rate per pig were 32.9±14.8 $m^3.h^{-1}$ and 24.9±7.5 $m^3.h^{-1}$ respectively for D1 and D1,43 rooms. Results of dust, ammonia and odours are given in the table 1. Values obtained in room D1,43 for dust, ammonia and odour are in accord with values already published in the literature for fattening pigs (Guingand, 2003; CORPEN, 2003). In both rooms, dust and ammonia concentrations were lower in hot period than in cold period in relation with the effect of ventilation rate linked to ambient temperature. The effect of stocking density was only observed on dust and ammonia concentrations. In accord with the animal and feed origins of dust, the reduction of pig number in D1 leaded to a reduction of dust concentration in the ambient air. The increase of ammonia volatilisation in room D1 could be explained by the reduction of slurry stored in the pit. Migration of ammonium ions could be increased by the lower height of slurry in room D1. With the reduction of pig number, contact area between slurry stored in the pit and ambient air was increased, explaining the increase of ammonia volatilisation in the ambient and in the exhaust air. Effect of reduction of stocking density on ammonia emissions was more important during the hot period linked to the effect of temperature on ammonia volatilisation and to higher ventilation rate per pig in room D1. No effect of stocking density on odour emission was observed in this study. Once more, no relation was established between ammonia and odour emitted by pig units. Measurements of pen dirtiness were achieved in a parallel study (Courboulay, 2005) and no difference was noticed between treatments participating to explain the absence of effect on odour emission.

Table 1. Dust, ammonia and odours in relation with season and stocking density.

	Room D1.43		Room D1	
	Hot period	Cold period	Hot period	Cold period
Dust concentration (mg.m^{-3})	1.07 ± 0.38	2.06 ± 0.82	0.89 ± 0.31	1.45 ± 0.49
	1.61 ± 0.82[1]		1.22 ± 0.50[1]	
Ambient ammonia concentration (ppm)	9.0 ± 5.4	21.0 ± 11.4	13.9 ± 10.3	22.0 ± 9.4
	15.6 ± 11.0[1]		18.3 ± 10.6[1]	
Ammonia emission per pig (g.d^{-1})	7.8 ± 1.6	12.3 ± 2.9	14.6 ± 2.8	13.4 ± 3.8
	10.3 ± 3.3[1]		13.9 ± 3.3[1]	
Ammonia emitted per pig (g)	0.91	1.22	1.44	1.31
	1.07[1]		1.37[1]	
Odour concentration (o.u.m^{-3})	2,675 ± 170	3,720 ± 1,688	2,300 ± 141	3,400 ± 700
	3,255 ± 1319[1]		2,771 ± 720[1]	
Odour emission per pig (10^{6}o.u.d^{-1})	1.68 ± 0.2	1.72 ± 0.6	1.88 ± 0.9	1.63 ± 0.3
	1.71 ± 0.4[1]		1.77 ± 0.2[1]	

[1]Average value for the two batches without season effect.

Conclusions

If the reduction of pig number per pen could appear as a solution to respect the European regulation, consequences on gaseous emissions and particularly on ammonia can not permit to propose it to pig breeders without technical modifications on buildings.

References

CORPEN, 2003. Estimation des rejets d'azote-phosphore-potassium-cuivre et zinc des porcs.Influence de la conduite alimentaire et du mode de logement des animaux sur la nature et la gestion des dejections produites. 41 pp.

Guingand, N., 2003. Air quality in pig facilities in relation with physiological stages. In: Proceedings of the Annual Meeting of the EAAP, Roma (Italy), 31 August-3 September 2003, pp. 195.

Courboulay, V., 2005. Conséquences d'une augmentation de la surface par animal sur les performances, les lésions et le comportement du porc à l'engrais. Journées Rech. Porcine 37: 465-470.

Air exchange through slatted floors in houses for fattening pigs

K.-H. Jeppsson and G. Gustafsson
Swedish University of Agricultural Sciences, Dep of Agricultural Biosystems and Technology, P.O. Box 86, 230 53 Alnarp, Sweden

Introduction

The design of the ventilation system affects the air movements in the pig fattening house and the natural air exchange through the slatted floors created by the air movements. The air below the slatted floors is a source of air with high ammonia concentration. From a partially slatted pig fattening house (25% slatted floor area), 20% of the ammonia release originates from the manure channel (Aarnink *et al.*, 1995). Between 28 and 56% of the ammonia release originates from the manure channel in a fully slatted house (Rom, 1995). The aim of this project was to develop a method determining the natural air exchange through slatted floors with tracer gas technique and to examine how the ventilation system and ventilation rate affect the air exchange and the ammonia release.

Material and methods

The measurements were done in a research building for fattening pigs during one growing period. The pig fattening house had 8 partly slatted pens (40% slatted floor of 9 m^2 total pen area) with 6 pigs in each pen. The manure channel was 1.2 m deep and 1.1 m wide. The manure was removed from the channel each morning with mechanical scrapers. During the investigation the air exchange through the slatted floor was examined at two ventilation rates (80 and 150 $m^3h^{-1}pig^{-1}$), with two types of air inlets (slotted inlets above the pen front and porous ceiling) and two types of air outlets (high air outlet and low air outlet). The ventilation rate was continuously measured with impellers. Two fans placed in the wall (1.5 m above floor level) and in connection with the manure channel created high air outlet (HE) and low air outlet (LE). The air temperature in the house was 17 °C during the measurement period. The tracer gas (SF_6) was pumped into the manure channel and injected in five places, one under each pen. The amount of tracer gas was constant and regulated using a gas flow meter. The concentration of tracer gas (infrared spectrophotometer) and ammonia (detection tubes) was measured in the room, in the exhaust air and at five places in the manure channel.

Results and discussion

To examine the air exchange through slatted floors by use of tracer gas technique, two assumptions must be fulfilled. The concentration of the tracer gas must be constant and the tracer gas must be able to mix with the air. The tracer gas used (SF_6) has 5 times higher density than air and this could have affected the results especially with LE. The air exchange through the slatted floor showed a large variation due to the air movements in the fattening house and in the manure channel (Table 1). The air exchange was about 10 times higher with HE than with LE. With LE the fan creates a forced air flow from the fattening room to the manure channel and the air exchange is only 1% of the ventilation rate. The air exchange also increased with increased ventilation rate. Difference in air exchange between slotted inlets and porous ceiling was not found. The total ammonia release corresponds with other measurements in fattening houses by Aarnink *et al.* (1995) and Rom and Dahl (1996). The total ammonia release from the fattening building and from the manure channel was higher with LE than with HE (Table 2) especially at high air rate. With LE the ammonia concentration in the fattening room was reduced. The ammonia release increased with increased ventilation rate. The part of the ammonia release that originates from the manure channel with HE was high compared with 20 % from a partially slatted fattening house according to Aarnink *et al.* (1995). This could probably be explained by the larger slatted floor area and a higher ventilation rate. The even higher part of ammonia release from the manure channel with LE was expected.

Table 1. Ventilation rate, air exchange through the slatted floor and air exchange rate.

	n	Ventilation rate $m^3h^{-1}pig^{-1}$		Air exchange $m^3h^{-1}pig^{-1}$		Air exchange rate
		Mean	Std	Mean	Std	
HE, low air rate	6	82.5	5.9	11.5	9.9	0.13
HE, high air rate	5	152.8	2.6	19.8	13.3	0.13
LE, low air rate	12	82.6	5.7	0.8	0.4	0.01
LE, high air rate	8	153.3	7.5	1.5	1.2	0.01

Table 2. The total ammonia release and the release from the fattening room and manure channel.

	n	Ammonia release							
		Total g $pig^{-1}day^{-1}$		Fattening room g $pig^{-1}day^{-1}$			Manure channel g $pig^{-1}day^{-1}$		
		Mean	Std	Mean	Std	%	Mean	Std	%
HE, low air rate	4	4.2	0.7	2.6	0.6	62	1.6	0.7	38
HE, high air rate	5	9.3	1.4	3.9	1.7	44	5.4	3.0	56
LE, low air rate	11	6.6	0.7	2.8	0.3	44	3.7	0.7	56
LE, high air rate	6	18.0	5.7	6.6	2.7	37	11.4	4.0	63

Conclusions

The conclusions of the investigation are; tracer gas technique is an interesting method to determine the air exchange and the ammonia release through the slatted floor with different ventilation systems; the air exchange through the slatted floor was 13% of the ventilation rate with high air outlet (HE); the air exchange through the slatted floor was only 1% of the ventilation rate with low air outlet (LE).

References

Aarnink, A.J.A., A. Keen, J.H.M. Metz, L. Speelman and M.W.A. Verstegen, 1995. Ammonia emission patterns during the growing periods of pigs housed on partially slatted floors. J. Agric. Eng. Res. 62: 105-116.

Rom, H.B., 1995. Ammonia emission from pig confinement buildings - System analysis and measuring methods. Ph.D. Thesis, Royal Veterinary and Agricultural University, Copenhagen, Denmark.

Rom, H.B. and P.J. Dahl, 1996. A system approach to describe the dynamics of ammonia emission from pig confinement buildings. Paper 96E-010, AgEng`96, Madrid, Italy.

The reduction in ammonia emission by cooling the surface of liquid manure in slurry channels

R. Myczko, J.L. Jugowar and A. Myczko
Institute for Building, Mechanization and Electrification of Agriculture in Warsaw, Branch in Poznan, 67 Biskupińska Str., 60-463 Poznan, Poland

Introduction

Agriculture is the major source of ammonia emission to the environment. Emissions occur especially from animal houses and from manure utilisation. High concentrations of ammonia in the air negitively affect its quality, especially in animal hosues. Aarnink (1997) showed that exceeding the level of 7 ppm causes negative influence on stockmen health and over 11 ppm reduces growth rates and increases pig's diseases. For these reasons, research has increased into new technology for the reduction in gaseous emission from agriculture in EU countries, including Poland. BAT Reference Document (2004) includes the technology of cooling the slurry surface by floating heat exchangers in order to reduce the ammonia release. Groenestein and Huis in't Veld (1996) had tested a first installation, as one of the solutions for improving the microclimate in pig house. The experiment was conducted in laboratory conditions. The major objective was to describe the technical parameters for an installation built from flat panel heat exchangers in order to recover the heat from slurry stored in ducts under slatted floor. The influence of heat recovery on the reduction in ammonia emission was also examined. The paper presents some chosen factors of technical parameters in multifactorial experiment.

Material and methods

The research objectives were related to the installation for cooling the slurry surface. The installation was made of flat polyurethane panels combined in series. A Thin-walled panel was build out of 11 cube-shaped tubes and 2 stub pipes enabling the constant laminar flow of cooling water in closed cycle. The research was planned as multifactorial experiment and conducted in laboratory set up with closed storage (fermenting chambers) imitating the slurry channels. They were filled with slurry from a pig farm with typical management. In each series of experiments, one of chambers was the control test chamber, where the chosen parameters of natural processes, occur during the slurry storage, had already been registered priviously. The results from this test chamber were the reference values for the results from the remaining experimental chambers. In experimental chambers, the panel installation of heat exchanger was placed in various configurations relative to the slurry surface. The circulation of water inside the panels was pumped by an electronically controlled pump; the water tank with a spirally scrolled pipe immersed in the water was the heat receiver. The air inside the chambers was changed 4 times per hour at a constant flowrate. All technical elements of the research set up were equipped with electronic drivers and meters. Ammonia concentration measurements were done by photoacoustic spectroscope Multigas Monitor 1312 according to Hintz and Schröder (2001) guidance. In the study of Myczko (2006), the detailed methodology of the research conducted is shown. Constant parameters during the experiments were: ventilation air temperature of 21 °C, inlet cooling agent temperature of 5.6 °C and cooling agent flow rate of exactly 25 l/h. The registered parameters were the following: the temperature of air leaving the fermentation chambers, slurry temperature directly under its surface and at half depth, the temperature of cooling water in the outlet of the heat exchanger, and the concentration of ammonia and water vapour in the air leaving the chambers. In the experiment, two cases which a differce in the flow direction of the cooling water within the panels was studied. This enabled to evaluate the amount of heat recovered from slurry and its influence on the ammonia emission.

Results and discussion

The amount of heat collected by the panel heat exchanger, received from fermentation chamber with evaporation surface of 1 m^2, was 19 MJ when the circulated water was used to cool the air above the slurry surface. The opposite direction of circulation resulted in 18 MJ of heat generated from chamber. The total emission in experimental chambers was calculated as the quotient of volumetric air velocity and difference between ammonia concentrations in ventilation air. The obtained results were: 72.6 mg NH_3/h in the control test; 24.3 mg NH_3/h with water circulation from air to slurry, and 30.7 mg NH_3/h with opposite circulation. The conducted research indicates that the slurry surface of 1 m^2 and a temperature of 22 °C emits 41.82 g NH_3/24h. The application of panels witch capacity of 6.5 resulted in a reduction of the ammonia emission by 66.48% when water circulating inside the panels first cooled the air above the surface. The flow of cooling water in opposite direction decreased the emission by 57.7% in comparison to control test. After conversion, an emission reduction from 15.27 kg NH_3 per year to 5.12 kg NH_3 per year was achieved. Comparing the results with those from the research of Groenestein and Huis in't Veld (1996) is difficult because they didn't specify the technical parameters of the installation. Also, they studied other additional factors that would influence the ammonia emission from livestock buildings. The floating elements of polypropylene in that experiment were cooled with groundwater with a temperature of 10 to 11°C. In their research, one series of experiments were conducted with cooling swiched on and off every two weeks. In that situation, the effect was estimated taking the influence of the section into account. The effect of cooling was an emission reduction with 28%, but the emitting surface and the capacity of slurry pit were not reported. In case of the current laboratory experiments without the presence of animals and with constant temperature conditions, the lower temperature of water circulating within the installation resulted in a larger reduction of the ammonia emission.

References

Aarnink, A.J.A., 1997. Ammonia emission from houses for growing pigs as affected by pen design, indoor climate and behaviour. Rapport 97-03, Wageningen, 173 pp.

BREF, 2002. Reference Document on Best Available Techniques for Intensive Rearing of Poultry and Pigs. IPPC Seville 2002, 383 pp.

Groenestein, C.M. and J.W.H. Huis in't Veld, 1996. Meat pig stable with cooling of the manure surface in the cellar. W: Practice research to the ammonia- emission of stables XXVII. IMAG-DLO, Wageningen.

Hintz, T. and O. Schröder, 2001. Optoakustisches-IR-Spectrometer. W: Massmetoden für Ammoniak-Emissionen aus der Tierhaltung. KTBL Darmstadt, pp. 73-77.

Myczko, R., 2006. Different ways for reduction in gaseous emission in slated floor systems of animal housing. Probl.Inż.Rol. IMBER, PTIR Warszawa 1(51): 135-142.

Animal welfare, ammonia emission and economical values

C.M. Groenestein
Animal Sciences Group, Wageningen-UR, The Netherlands

Introduction

In 2003, the blue print of the Comfort Class housing system for fattening pigs was completed (Projectgroep Diergericht Ontwerpen, 2003). Its design is intended to meet the biological needs of pigs (adapted from Bracke *et al.*, 1999). The challenge of sustainable development of animal housing is to reconcile the three P's of Profit, Planet and People. The methodological approach of the design procedure (Van den Kroonenberg and Siers, 1999) made it possible to evaluate the designers' brief of requirements in terms of aspects other than welfare, such as environment or economy. This way conflicts of interest become clear between animal welfare on the one hand and environmental and economical values on the other, and solutions to overcome dilemmas are comprehensible.

Materials and methods

The brief consisted of 58 welfare requirements (Groenestein and Schouten, 2003). Four experts on welfare scored these requirements according to their contribution to welfare: the better the requirement served the welfare, the higher its score on a scale between 0 to100. A low score for welfare is still beneficial for the pig's well-being, because, as a starting point all requirements were based on the needs of the pig. Two environmental experts and two economical experts also scored the requirements in terms of their likely impact on ammonia emission and economical impact respectively: a negative score was given if implementation of the requirement would increase ammonia emission or costs, and a positive score if it would diminish it. The higher the score on a scale between -100 to +100, the more beneficial the requirement would be for the environment or the economy. Based on the scores the requirements were then put together in a diagram and divided into five groups: low scores for welfare, high scores for ammonia or economy (OK); high for welfare and high for ammonia or economy (WIN-WIN); low for welfare and low for ammonia or economy (COMPROMISE); high for welfare and low for ammonia or economy (DILEMMA); and low and high for welfare and around zero for ammonia or economy (NEUTRAL).

Results

Figure 1 presents the result of the environmental and economical assessment of the 58 welfare requirements. Each diamond in Figure 1 represents a requirement. For details on all the requirements, see Groenestein *et al.* (2003) and Groenestein (2006). Implementing the requirements in the NEUTRAL area into the design of a pig house will have little or no effect on ammonia emission or costs. Of the other requirements 12 will reduce ammonia emission and 2 will reduce costs (OK and WIN-WIN). The remaining requirements, however, (COMPROMISE and DILEMMA) show a conflict of interest: good for welfare, less good or bad for the environment or the economy. Most inconvenient for reconciliation are the dilemma's, implementation of welfare requirements have not enough basis if environmental and economical needs are not met. It appeared that 5 out of 8 requirements in the dilemma quadrant of ammonia and welfare concerned the surface area of the pen. Implementing these requirements in the design of a pig house does not in itself enhance ammonia emission, but has theoretically the potential to do so

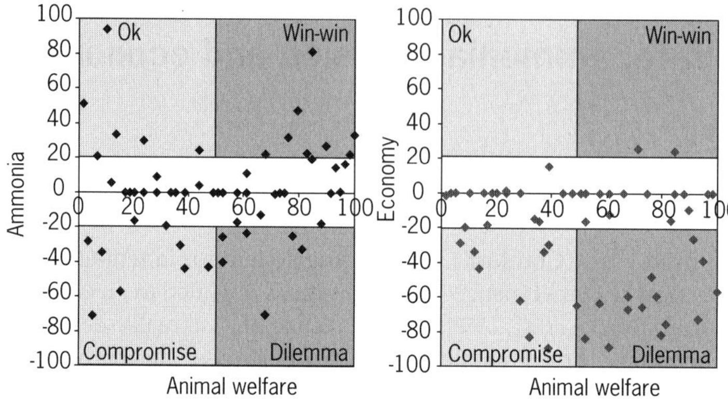

Figure 1. Welfare requirements formulated for the design of a pig house in which animal needs are met, scored on their impact on welfare and their effect on ammonia emission (left) and economical values (right).

according to the expert judgement: if fouled with slurry, the floors can become emitting areas and the larger an emitting surface is, the greater the ammonia emission will be (Elzing and Monteny, 1997). Figure 1 shows that the economy conflicts more with animal welfare than the environment. The dilemma quadrant of economy contains 14 requirements. They concern the surface area, the availability of substrate, ad libitum feed and water and a minimum number of feeding places. Insight in these dilemma's can help researchers to find solutions or give public authorities directions to stimulate developments or discourage unwanted situations.

Conclusions

With a methodological approach welfare requirements of a housing system for pigs can be assessed in view of environmental and economical goals. Requirements with common interest and with conflict of interest can be identified which gives directions to solutions where welfare, environmental and economical needs can be met, to reconcile the three P's.

References

Bracke, M.B.M., B.M. Spruijt and J.H.M. Metz, 1999. Overall animal welfare reviewed. Part 3: Welfare assessment based on needs and supported by expert opinion. Neth. J. Agr. Sci. 47: 307-322.

Elzing, A. and G.J. Monteny, 1997. Modeling and experimental determination of ammonia emission rates from a scale model dairy-cow house. Transactions of the ASAE 40: 721-726.

Groenestein, C.M. and W.G.P. Schouten, 2003. Diergericht ontwerpen: Van programma van eisen tot voorbeeldontwerp van een welzijnsvriendelijke vleesvarkenshouderij. IMAG-rapport 2003-07, Wageningen, The Netherlands, 16 pp.

Groenestein, C.M., W.G.P. Schouten, K.H. De Greef, R.G. Ten Hoope and M. De Jong, 2003. Diergericht ontwerpen: Methodisch in plaats van intuïtief naar een ontwerp voor een diervriendelijk houderijsysteem. Wageningen, IMAG rapport 2003-06, 27 pp.

Groenestein, C.M., 2006. Environmental aspects of improving sow welfare with group housing and straw bedding. PhD thesis, Wageningen University, The Netherlands, 146 pp.

Projectgroep Diergericht Ontwerpen, 2003. Welzwijn in de toekomst, over varkenswensen voor varkensstallen. Wageningen-UR, The Netherlands.

Van den Kroonenberg, H.H. and F.J. Siers, 1999. Methodisch ontwerpen. Ontwerpmethoden, voorbeelden, cases oefeningen. Educatieve Partners Nederland BV, Houten, The Netherlands.

NH_3 emissions in French straw-based pig-on-litter breeding systems

M. Hassouna[1,2]*, P. Robin*[1] *and C. Texier*[3]
[1]*INRA, Agrocampus Rennes, UMR 1069, Sols, Agronomie, Spatialisation, 35000 Rennes, France*
[2]*INRA, UMR1079 Systèmes d'Elevage Nutrition Animale et Humaine, 35590 Saint Gilles, France*
[3]*IFIP, Institut de la filière porcine, La motte au Vicomte, BP3, 35651 Le Rheu, France*

Introduction

The pig-on-litter system is a promising solution regarding the animal welfare and the odours issues. Studies on these systems revealed a weak variability in total nitrogen volatilisation from the building (around 60% of the excreted nitrogen) but a high variability in the partition between the different forms of nitrogen (NH_3 and N_2O or N_2) (Robin *et al.*, 2004). Very few data are available on the emission factors of straw-based pig-on-litter systems. The aim of this study was to quantify the NH_3 emissions from such a system and to enrich the current database.

Material and methods

NH_3 emissions were observed in 3 buildings noted ITP, B1 and B2 (Table 1) with management options that are currently practised in France. During the winter of 2002, we measured NH_3 emissions from 3 identical rooms containing 40 or 30 pigs reared on straw litter (ITP). The main differences between these 3 rooms were the amount of litter per pig and the available surface of litter per pig (1.4 and 1.0 m^2/pig). The ventilation was estimated using a simplified method based on the total heat production in the room. From June 2003 to February 2004, we successively monitored the gas emissions in two commercial pig farms with the available surface of litter per pig 1.4 (B1) and 2.6 m^2/pig (B2). The ventilation was estimated using a reference method based on the tracer gas technique and the above simplified method. During both measurement campaigns, we used a photoacoustic multi-gas monitor (INNOVA, 1312) to monitor gas concentrations (NH_3, N_2O, CO_2, CH_4, SF_6, H_2O). The water, nitrogen, phosphorus and potassium mass balances for each experiment were calculated to check the validity of the results.

Results and discussion

In the commercial buildings (B1 and B2), the tracer gas method confirmed the results obtained with the simplified method provided the heat production of the litter and the animal activity (particularly during daytime) were taken into account. The ranges of mean hourly emissions for each livestock building presented in Table 1 show that the variability of ammonia emissions can be as important during a rearing period as between the different experiments. In the livestock buildings with a normal stocking density (between 1 and 1.4m^2/pig), regular straw supplies and thermal insulation of the building, NH_3 emissions did not clearly depend on the indoor climatic conditions, the rearing conditions and the litter management. Our results were in good agreement with those observed in Denmark and in the United Kingdom (100 à 400 mg NH_3/h.pig) (Groot Koerkamp *et al.*, 1998). Thus, we propose for systems with a normal pig density an emission factor for ammonia in the range 15-25% of the excreted nitrogen for pigs reared at least 10 weeks with a litter input in the range 50-80 kg/pig. For systems with a low pig density,

Table 1. Characteristics of buildings, climates, pigs, food, litter and emissions.

Livestock building		**ITP**	**ITP**	**ITP**	**B1**	**B2**	**B2**	**B2**
Room ventilation (N: natural; D: dynamic)		**N+D**	**N+D**	**N+D**	**N**	**N**	**N**	**N**
Room insulation (N: non insulated; Y: insulated)		**Y**	**Y**	**Y**	**Y**	**N**	**N**	**N**
Food biphase (kg N/pig)					**5.4**	**4.5**	**4.5**	**4.5**
Period		**Winter Summer 2002**	**Winter Summer 2002**	**Winter Summer 2002**	**Winter Summer 2003**	**Winter 2001-Summer 2002**	**Winter 2001-Summer 2002**	**Summer 2003**
Duration (weeks)	Winter	18	14	20	14	13	13	13
	Summer	21	21	16	14			
Number of pigs	Winter	30	40	30	103	3x23	3x23	3x23
	Summer		35	40	106			
Initial weiht (kg)	Winter	35	35	32	35	55	55	55
	Summer	25	24	24	35			
Final weight (kg)	Winter	146	117	146	115	115	115	115
	Summer	143	144	110	105			
Measurement technique		Simplified method			Tracer gas	Simplified method		Tracer gas
Initial straw quantity (kg/pig)	Winter	16	58	16	30	10	10	10
	Summer	16	16	16	30			
Straw supply (kg/day.pig)	Winter	0.8	0	0.5	≈0.6	≈0.6	≈0.6	≈0.6
	Summer	0.5	0.7	0.6	≈0.6			
Litter surface (m^2/pig)	Winter	1.4	1.0	1.4	1.2	2.6	2.6	2.6
	Summer	1.4	1.2	1.0	1.2			
NH_3 emission (mg N-NH_3/h.pig)	Winter	Between 200 and 1300			Between 200 and 1500	Between 30 and 200		
	Summer					Between 100 and 600		

we noticed a lower ammonia volatilisation particularly during winter. We propose for systems with a litter surface/pig more than $2m^2$ an ammonia emission factor in the range 5-15% of the excreted nitrogen.

Conclusion

This surveys led to a first set of emission factors for NH_3 for French pig-on-litter-systems but also emission factors for N_2O and CH_4. It also constituted a preliminary phase in the development of a simplified method to estimate gaseous losses from animal buildings and during storage of animal effluents. Factors that influence ammonia emissions like pig density or the litter management were identified. Basing on these knowledge and on these new emission factors, we are implementing a model to estimate NH_3 and N_2O emissions from pig-on-litter systems.

References

Groot Koerkamp, P.W.G, J.H.M. Metz, G.H. Uenk, V.R. Philips, M.R. Holden, R.W. Sneath, J.L. Short, R.P. White, J. White, J. Seedorf, M. Schröder, K.H. Linkert, S. Pedersen, H. Takai, J.O. Johnsen and C.M. Wathes, 1998. Concentrations and emissions of ammonia in livestock buildings in Northern Europe. J. Agr. Eng. Res. 70: 79-95.

Robin, P., M. Hassouna and C. Texier, 2004. Emission d'ammoniac et de protoxyde d'azote des porcs engraissés sur litière de paille. Journées de la recherche porcine en France 36: 63-70.

NH_3 and GHG emissions from a straw flow system for fattening pigs

B. Amon[1], V. Kryvoruchko[1], M. Fröhlich[1], T. Amon[1], A. Pöllinger[2], I. Mösenbacher[2] and A. Hausleitner[2]
[1]University of Natural Resources and Applied Life Sciences, Department of Sustainable Agricultural Systems, Division of Agricultural Engineering (DAE), Peter Jordan-Strasse 82, 1190 Vienna, Austria
[2]Federal Research Institute for Agriculture in Alpine Regions, A-8952 Irdning, Austria

Introduction

The straw flow system is an animal friendly housing system for fattening pigs. It can be operated economically efficiently on commercial farms. The straw flow system distinguishes a lying and an excretion area. Only a small part of the pen is soiled with excreta. Additionally, excreta may be frequently removed from the pig house by a scraper. The small emitting surface and the frequent manure removal may contribute to a reduction in emissions. It was to be investigated, if the straw flow system emitted less NH_3 and greenhouse gases (GHG) than a conventional fully slatted floor system. Emissions from covered and uncovered stores were followed as well.

Materials and methods

Emissions of NH_3, N_2O, and CH_4 were continuously measured at a commercial farm in Upper Austria from June 2003 to April 2004. The animal house consisted of three fully separated compartments. Each compartment was forced ventilated by a central exhaust fan. Excreta were collected in a dung channel in the rear of the pen (= dung channel system). In two compartments the dung channel was additionally equipped with a scraper that was operated twice a day and moved the slurry to an outside store (= scraper system). Each compartment was separated into 16 pens that held 10–12 pigs each. Gas concentrations were measured with high resolution FTIR spectrometry. The ventilation rate was continuously recorded in the central exhaust fan. From May 2004 to June 2005 emissions were followed from storage of pig slurry received from a straw flow system. The measurements compared emissions from covered and uncovered slurry stores.

Results

Emissions of CH_4, N_2O, and NH_3 from the straw flow system were always lower than default values for forced ventilated fully slatted floor systems. From the straw flow system CH_4 emissions of 1.24 (dung channel system) respectively 0.54 (daily manure removal system) kg CH_4 per pig place and year were lost. The default value for fully slatted floor systems is 4.0 kg CH_4 per pig place and year (UBA, 2001). N_2O emissions from the straw flow system amounted to 39.95 (dung channel system) respectively 24.54 (daily manure removal system) g N_2O per pig place and year. Fully slatted floor systems are estimated to emit 100 g N_2O per pig place and year (UBA, 2001). NH_3 emissions from the straw flow system were 2.10 (dung channel system) respectively 1.91 (daily manure removal system) kg NH_3 per pig place and year. The default value for fully slatted floor systems is higher: 3.0 kg NH_3 per pig place and year (Döhler *et al.*, 2002; UBA, 2001). During storage, a solid cover was an efficient means to reduce NH_3 and GHG emissions. It is recommended to store pig slurry in covered stores. When stored under cool climatic conditions, pig slurry emitted less NH_3 and GHG than when stored under warm climatic conditions (Table 1). It is recommended to set up the national emission inventory by applying separate emission factors for slurry storage in the cooler and in the warmer half of the year.

Table 1. Emissions from covered and uncovered pig slurry stores under warm and under cool climatic conditions.

Treatment	slurry temperature [°C]	CH_4 [kg m^{-3}]	N_2O [g m^{-3}]	NH_3 [g m^{-3}]
Slurry, warm				
covered, after 50 days	16.4	0.17	30	66
uncovered, after 50 days	17.9	0.25	23	118
covered, after 200 days	15.8	1.42	114	273
uncovered, after 200 days	16.5	4.66	119	380
Slurry, cool				
covered, after 50 days	12.9	0.10	18	19
uncovered, after 50 days	11.6	0.16	36	22

Conclusions

NH_3 and GHG emissions from the straw flow system were lower than default values that are currently applied to estimate emissions from a conventional fully slatted floor system. When the dung channel system was additionally equipped with a scraper and the pig manure was removed to the outside storage on a daily basis, emissions from the straw flow system were further reduced. Stores for pig slurry should be equipped with a solid cover. It is recommended to set up the national emission inventory by applying separate emission factors for slurry storage in the cooler and in the warmer half of the year. The straw flow system combines recommendations of animal welfare and environmental protection.

Acknowledgements

The project was supported by the Austrian Federal Ministry for Agriculture and Forestry, Environment and Water Management.

References

Döhler, H., U. Dämmgen, W. Berg, A. Bergschmidt, R. Brunsch, B. Eurich-Menden, M. Lüttich and B. Osterburg, 2002. Anpassung der deutschen Methodik zur rechnerischen Emissionsermittlung an internationale Richtlinien und Prognose der Ammoniak-Emissionen der deutschen Landwirtschaft und Szenarien zu deren Minderung bis zum Jahre 2010. Final Report für das Bundesministerium für Verbraucherschutz, Ernährung und Landwirtschaft and Umweltbundesamt.

Umweltbundesamt der BRD, 2001. Entwurf: Erstellung eines Gutachtens für einen deutschen Beitrag zur Vollzugsvorbereitung zur Umsetzung der IVU-Richtlinie für den Bereich Intensivtierhaltung, UBA Vorhaben FKZ 360 08 001, Stand November 2001, KTBL Darmstadt, Umweltbundesamt Berlin; In: UBA-Text 75/02 „Beste verfügbare Techniken in der Intensivtierhaltung".

Monitoring of ammonia emissions from two contrasting flooring systems for growing and finishing pigs: fully slatted vs. straw based housing

N. J. Penlington[1], T.G.M. Demmers[2] and B.P. Gill[1]
[1]Meat and Livestock Commission, PO Box 44, Winterhill House, Snowdon Drive, Milton Keynes, MK6 1AX, United Kingdom
[2]Silsoe Research Institute, Wrest Park, Silsoe, Bedford, MK45 4HS, United Kingdom

Introduction

The objective of this Defra/MLC funded research was to evaluate the effects of two contrasting flooring systems (straw based vs. fully slatted flooring) on the growth performance, health and welfare of pigs, meat quality, the microbial loading of the housing and feeding environment with respect to gut health and food safety, and environmental impact in terms of dust and ammonia emissions and effluent N loading. A focal point of the research was the development of liquid feeding technology.

Methodology

Detailed methodology is provided in the individual trial reports (e.g. MLC, 2004). Four trials were completed with a total of 4,352 pigs, target start weight of 35 kg, and slaughter weight of 102 kg. Two houses were purpose built, one fully slatted with shallow pits, the other with solid floors and straw bedding, and pigs were fed using a state-of-the-art automated liquid feeding system. Each house contained 4 rooms, with 4 pens per room. Buildings were designed according to Best Available Technique (BAT) with respect to implementation of the EU Integrated Pollution Prevention and Control (IPPC) Directive in Great Britain. Computer controlled side inlet vents and roof fans ensured room temperature and humidity stayed within the parameters of the pre-set standard curves. Feeding treatments evaluated were: Trial 1: liquid vs. dry feed, Trial 2: true phase vs. single diet feeding, Trial 3: controlled cereal fermentation, and Trial 4: reducing the protein content of diets using amino acids. Pigs were fed *ad libitum* and feed intake was monitored automatically by the liquid feeding system. Pigs were weighed at 2 weekly intervals to determine daily gain and feed conversion ratio from entry to slaughter. Environmental monitoring was carried out during Trials 1, 2 and 4. Ventilation rate was continuously measured for each room using a fan-wheel anemometer installed in the exhaust duct (Demmers *et al.*, 1999). Fan wheel rotation rate and opening angle of the inlet damper were continuously monitored electronically, to give an accurate log of the instantaneous ventilation rate of each room. Each fan-wheel anemometer was calibrated before the start of experimental monitoring (Moulsley and Randall, 1990). Ammonia concentration was measured using a chemiluminescence nitric oxide analyser (Demmers *et al.*, 1999). Measurements were taken at 12 locations, at entry to the exhaust fan within each room and at 4 placed immediately outside the buildings, to correct for ambient ammonia entering the buildings. The analyser was calibrated twice weekly using certified standard gas mixtures. Net ammonia emission rate for each room was estimated as a function of hourly ammonia concentration and hourly ventilation rate, corrected for any incoming concentration of ammonia. The ammonia emission was normalised to the live weight of pigs in each room. The ammonia emission factor (g NH_3-N per LU unit per hour) was calculated from the cumulative emission, where one livestock unit (LU) corresponds to 500 kg of live weight. Effluent output was monitored and samples were determined for dry matter, total N and ammonium N throughout the trials. Dust concentration (Total dust) was measured using an IOM open face sampler at entry to the exhaust fan in each room, and also at one air inlet in one room to provide correction. Sampling took place over three-day periods, once per fortnight when the buildings were full. The average ventilation rate over the period of exposure of each set of dust concentration samplers was multiplied by the average dust concentration (corrected for any incoming dust), to give average Total dust emission rate for each room.

Results

There were no effects of housing system on ammonia and dust emissions and the pooled means for the 3 trials are presented in Table 1. In Trial 4, reducing dietary protein content from 20 to 16% using synthetic amino acids reduced effluent total N and ammonium N and ammonia emission (Table 2), but pigs fed the reduce protein diet had poorer daily gain (851 vs. 821 g/d; P=0.013) and FCR (2.49 vs. 2.68; P=0.000). There were no other effects of feeding treatment or feeding by housing system interaction on environmental measurements.

Table 1. Ammonia and dust emission and dust concentration according to housing system.

Housing system	Fully slatted	Straw based
Ammonia emission (g NH_3-N/LU. hour)	1.04	0.92
Dust concentration (mg/m^3)	1.4	1.1
Dust emission (g/LU. hour)	0.41	0.37

Table 2. Pig effluent production and composition, and ammonia emission for Trial 4.

Feeding treatment	Control	Low protein	SED	P - value
Effluent output (L/d)	6.98	7.68	0.62	0.095
Effluent composition				
DM, %	7.39	8.16	1.50	0.659
Ammonium N (g/kg)	4.15	3.53	0.12	0.038
Total Kjeldahl N (g/kg)	5.71	5.21	0.20	0.049
Ammonia emission (g NH_3-N/LU hour)	1.11	0.73	0.112	0.001

Conclusions

In this major programme of research, we established no differences in the environmental impact from ammonia and Total dust of fully slatted and straw based housing systems for growing and finishing pigs designed and operated under BAT applicable to UK pig production. N loading from effluent and ammonia emissions can be significantly reduced using low protein amino acid supplemented diets under liquid feeding. However, there are negative effects on growth performance from adopting this dietary strategy and this needs to be considered against favourable reductions in N emissions.

References

Demmers, T.G.M., L.R. Burgess, J.L. Short, V.R. Philips, J.A. Clark and C.M. Wathes, 1999. Ammonia emissions from two mechanically ventilated UK livestock buildings. Atmos. Env. 33: 217-227.

MLC. 2004. Finishing pigs: Systems research. Production trial 1 dry versus liquid feeding in two contrasting finishing systems (fully slatted versus straw based housing). Meat and Livestock Commission, Milton Keynes, pp. 59.

Moulsley, L.J. and J.M. Randall, 1990. Propeller fans for ventilating livestock buildings: 1. Measurement of performance. J.Agr. Eng. Res. 47: 89-99.

Poultry housing

Management options for reducing ammonia emissions from poultry litter

P.A. Moore, Jr.[1]*, D. Miles*[2]*, R. Burns*[3]*, D. Pote*[4] *and K. Berg*[5]
[1]*USDA-ARS, Plant Sciences 115, Fayetteville, AR, 72701, USA*
[2]*USDA-ARS, 810 Hwy 12 East, MS State, MS 39762-5367, USA*
[3]*Iowa State University, 3224 NSRIC, Ames, Iowa 50011, USA*
[4]*USDA-ARS, Highway 23, Booneville, AR, 72927, USA*
[5]*Cyngenta Seed, 2225 Grant Blvd., Lebanon, IN 46052, USA*

Introduction

Ammonia (NH_3) emissions from poultry manure can result in high levels of NH_3 in poultry rearing facilities and atmospheric pollution. High NH_3 levels in animal rearing facilities have been shown to cause lower weight gains, poorer feed conversion, reduced egg production, immunosuppression, damage to the respiratory tract and retinal damage (Carlile, 1984). Ammonia contamination of the atmosphere can also result in increased nitrogen (N) loading into aquatic systems and acid precipitation. Hence, a need exists to develop management options for reducing NH_3 emissions from animal manure, such as poultry litter. One management practice that has been used by growers in the U.S. for almost a decade is treatment of manure with aluminum sulfate (alum). Alum-treatment of broiler litter has been shown to reduce NH_3 emissions, as well as reduce runoff of phosphorus and heavy metals (Moore *et al.*, 1995, 1996, 1998, 1999, 2000). Lower NH_3 in poultry houses treated with alum results in better weight gains, feed conversion, and lower energy bills due to reduced ventilation requirements in winter (Moore *et al.*, 1999, 2000). The objectives of this study were to: (1) measure NH_3 volatilisation from poultry litter in broiler houses and following land application, (2) evaluate the factors that affect NH_3 losses from litter, and (3) determine the impact of BMPs on NH_3 volatilisation. In order to complete these objectives, a series of experiments were conducted at a commercial broiler farm in NW Arkansas where NH_3 emissions were being investigated.

Materials and methods

Four commercial broiler houses were equipped with NH_3 sensors, anemometers, and data-loggers which continuously record NH_3 concentrations and ventilation. Two practices for reducing NH_3 losses from the houses were evaluated; (a) alum treatment of litter, and (b) utilising an NH_3 scrubber. Four alum treatments were evaluated; (1) control, (2) 0.09 kg alum/bird as dry alum, (3) 0.045 kg alum/bird as dry alum, and (4) 0.045 kg alum/bird as liquid alum. The NH_3 scrubber was a large box (similar to an evaporative cooler), which was mounted on the exhaust fans. A dilute solution of alum was pumped from a reservoir in the bottom of the scrubber to the top where it was released, causing air that ventilated from the house to pass through a water curtain of dilute alum. Litter incorporation into pastureland (Pote *et al.*, 2003), a third BMP for reducing NH_3 loss, was tested during manure application. Wind tunnels were utilised to compare NH_3 losses from Bermuda grass plots fertilised with litter using conventional surface application and by incorporation.

Results and discussion

Average NH_3 emissions were 15.5 kg/house/day during the time when the birds were in the barns. This is equivalent to 27.4 g NH_3/bird. Ammonia losses between flocks were equivalent to 6.8 g/bird. Following land application there was an additional 7.3 g NH_3/bird lost, making the total amount lost equal to 41.5 g NH_3/bird. Ammonia emissions from poultry litter treated with alum were significantly lower than normal litter. Cumulative NH_3 losses were107, 173, 269, and 406 g/m^2, respectively, for the high rate of dry alum, low rate of dry alum, low rate of liquid alum and control. Although liquid alum resulted in better control early during the flock, dry alum prolonged the NH_3 control. These data mimic observations made by growers. The NH_3 scrubber trapped from 0.7 to 4.8 kg N/day. The evaporate mineral ammonio-alunite [$NH_4Al_3(SO_4)_2(OH)_6$], which is normally found in nature near geysers, was formed in the scrubber. For soils having high soluble phosphorus (P) levels, applications of the scrubber solution at a rate to achieve 100 kg N/ha reduced soluble P levels by 90%, hence, this BMP may allow N to be recovered, while decreasing non-point source P runoff. Ammonia losses from poultry litter following land application totalled 34 kg N/ha (15% of the total N applied), when the litter was broadcast applied onto pastures. However, when litter was incorporated, ammonia losses were virtually zero.

Conclusions

The results from this research indicate that there are viable management options that can be utilised by poultry producers to reduce NH_3 losses.

References

Carlile, F.S., 1984. Ammonia in poultry houses: A literature review. World's Poultry Science J. 40: 99-113.

Moore, P.A., Jr., T.C. Daniel, D.R. Edwards and D.M. Miller, 1995. Effect of chemical amendments on ammonia volatilization from poultry litter. J. Environ. Qual. 24:293-300.

Moore, P.A., Jr., T.C. Daniel, D.R. Edwards and D.M. Miller, 1996. Evaluation of chemical amendments to inhibit ammonia volatilization from poultry litter. Poultry Sci. 75: 315-320.

Moore, P.A., Jr., T.C. Daniel, J.T. Gilmour, B.R. Shreve, D.R. Edwards and B.H. Wood, 1998. Decreasing metal runoff from poultry litter with aluminum sulfate. J. Environ. Qual. 27: 92-99.

Moore, P.A., Jr., T.C. Daniel and D.R. Edwards, 1999. Reducing phosphorus runoff and improving poultry production with alum. Poultry Sci. 78: 692-698.

Moore, P.A., Jr., T.C. Daniel and D.R. Edwards, 2000. Reducing phosphorus runoff and inhibiting ammonia loss from poultry manure with aluminum sulfate. J. Environ. Qual. 29: 37-49.

Pote, D.H., W.L. Kingery, G.E. Aiken, F.X. Han, P.A. Moore, Jr. and K. Buddington, 2003. Water-quality effects of incorporating poultry litter into perennial grassland soils. J. Environ. Qual. 32: 2392-2398.

Experimental results of ammonia emissions from a fattening rabbit farm in Spain

S. Calvet, F. Estellés, B. Hermida, M. Cambra-Lopez and A.G. Torres
Universidad Politécnica de Valencia. Institute of Animal Science and Technology. Camino de Vera s/n 46022. Valencia, Spain

Introduction

Rabbit production is relatively important in several countries such as Spain, being more than 100,000 metric tons of meat in 2004 (FAO, 2006). However, ammonia emissions related to this activity have not been measured under local conditions yet, and therefore have not been included in the National Emissions Inventory. There is also very few information about these emissions in other countries. Italy, Portugal and Mexico consider ammonia emissions from rabbit production in their national inventories, but emission factors are adapted from those corresponding to other species. Experimental values on ammonia emissions are very scarce in literature (see Table 1). A set of experiments was carried out to obtain measured ammonia emission rates in a fattening rabbit farm in summer conditions, as a previous step to propose emission factors for national inventories. A nitrogen balance in the rearing period was also performed.

Table 1. Data on ammonia emissions from rabbit production (kg NH_3/place and year).

	Summer	Autumn – Winter	Reference
Does[1]	0.85 – 0.79	0.78 – 0.68	Hol *et al.* (2004)
Fattening rabbits	0.13 – 0.17	0.08 – 0.11	Hol *et al.* (2004)
Fattening rabbits	0.16	0.03	Calvet *et al.* (2005, 2006)

[1]This concept includes reproductive does and their young rabbits until weaning.

Materials and methods

Two sets of measurements were made in a commercial, mechanically ventilated, breeding rabbit farm in Valencia (Spain) in summer conditions during 2005 and 2006. The farm had about 1,500 animals held in cages (about 9 animals per cage), with deep pit manure collection system, which is typical for rabbit production in Spain. Ammonia emissions were calculated by means of a mass balance in a whole fattening period (5 weeks, from 0.7 kg to 1.9 kg of body weight), taking into account air flow and ammonia concentration. Air flow was calculated measuring the extraction capacity and the operation time of each fan, whereas ammonia concentration was measured using both a photoacoustic gas monitor equiped with a gas multiplexer and an electrochemical sensor. Uncertainty analyses were performed using the Monte Carlo software RiskAmp. A nitrogen balance was also performed in the two cycles by controlling all inputs (feedstuff consumption), outputs (urine and faeces) and accumulation (body retention) in eight cages. The balance was performed weekly during the whole cycle.

Results

Mean ammonia concentrations varied between 1.50 and 2.06 mg/m^3 having a clear sinusoidal daily pattern, and ventilation rate was maximum in both experiments (20,200 m^3/hour), taking into account that average outside temperature was over 25 °C. Emissions resulted to be 15.3 grams of ammonia per animal and fattening cycle (161 ± 51 grams of ammonia per place and year), and 11.63 grams of ammonia per animal and fattening cycle (121 ± 11 grams of ammonia per place and year) in 2005 and 2006, respectively. The results of the final nitrogen balance (yearly values for summer conditions) in both years are showed in Figure 1:

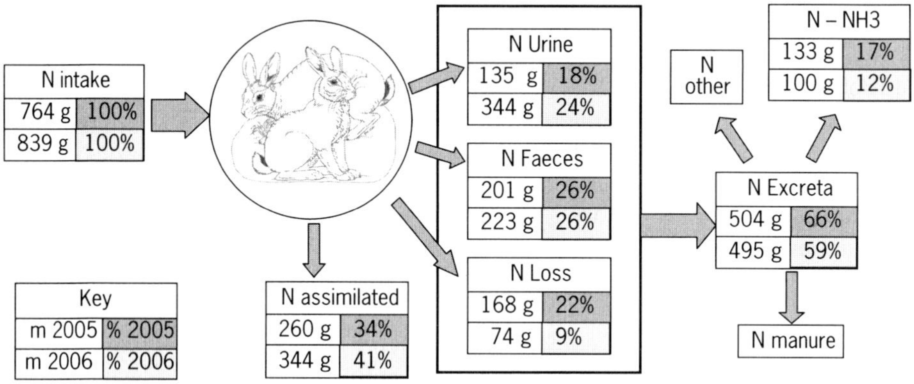

Figure 1. Results of the nitrogen balance in 2005 and 2006 as total N mass and percentage.

Conclusions

Results are consistent with previous reported data, and were also very similar between the two experiments. Differences in uncertainty between both experiments were due to the different ammonia measurement system. A daily pattern in emissions was observed, which could be related to animal activity patterns, which is different in rabbits to other animals. This experiment is now being completed with a set of measurements in winter conditions.

Acknowledgements

This work has been supported by the Spanish Ministry of Education and Science (Project GASFARM AGL2005-07297) and using equipment supported by Generalitat Valenciana, Conselleria de Empresa, Universidad y Ciencia, co-found by FEDER European funds.

References

Calvet, S., F. Carbonell, F. Estellés, P. Ferrer, and A.G. Torres, 2005. Estudio preliminar de las emisiones de amoniaco en granjas de conejos (in Spanish). Book of Abstracts of the III National Congress of Agricultural Engineering, pp. 563-564.

Calvet, S., O. Blumetto and A.G. Torres, 2006. Nitrogen Balance and Ammonia emissions in a rabbit farm in Mediterranean conditions. Book of Abstracts of the XVI CIGR World Congress – Agricultural Engineering for a Better World, pp. 515-516

FAO, 2006. Statistics on animal production. Food and Agriculture Organization of the United Nations. http://faostat.fao.org/. Accessed 3th January 2007.

Hol, J.M.G., A. Scheer and N.W.M. Ogink, 2004. Onderzoek naar de ammoniak- en geuremissie van stallen; Stal voor voedsters en vleeskonijnen (in Dutch). Agrotechnology & Food innovations B.V. Wageningen, the Netherlands. Report Nr. 219, 64 pp.

Experimental results of ammonia emissions from broiler production in Spain

S. Calvet, Y. Úbeda, M. Cambra-Lopez, V. Blanes-Vidal and A.G. Torres
Universidad Politécnica de Valencia. Institute of Animal Science and Technology. Camino de Vera s/n 46022. Valencia, Spain

Introduction

Ammonia emissions in Spain are currently being calculated adapting accepted emission factors developed in other countries (UPV, 2006) but local emission factors should be obtained in order to improve the quality of inventories. Emissions from broiler farms are widely reported in the literature; however, there is poor agreement between methodologies and reporting units. Kosch (2006) summarised emission rates from literature, ranging from 20 to 200 g NH_3 per animal place and year. There is also poor information about the data quality: statistical analyses or uncertainty information. In this work, ammonia emissions from fattening poultry production were measured in a commercial broiler farm in two experiments developed in summer in 2005 and 2006. The experiment is part of a project in which emissions in winter conditions will also be measured.

Materials and methods

The farm is located in Villarreal, in the east part of Spain. The management of the farm was typical for most of the farms in Spain, having cross mechanical ventilation and using rice hulls as bedding material. Animals were reared until 49 days of life (average weight about 2.5 kg), which is normal in Spanish farms. Emissions were measured during two whole fattening periods using a mass balance method. Air flow was calculated by the time of operation of each of the 16 fans, and ammonia concentrations were measured by using both an electrochemical sensor in 2005 (Calvet *et al.*, 2006), and a photoacoustic gas monitor equipped with a gas multiplexer system in 2006. A nitrogen balance was also performed in the whole farm in 2005 by controlling feedstuff consumption, meat production and manure at the end of the cycle. Uncertainties in the results were estimated using a Monte Carlo software (RiskAmp for Excel version 2.08), by means of the general procedure for quantifying uncertainty (Ellison *et al.*, 2000)

Results

NH_3 concentration was always below 15 ppm. NH_3 emissions in 2005 were 29.5±1.4 grams of ammonia per animal and cycle (170±7 grams of ammonia per place and year), and in 2006 were 27.5±0.6 grams of ammonia per animal and cycle (159±3 grams of ammonia per place and year). All confidence intervals are expressed with a 95% confidence level). Figure 1 shows the evolution of the ammonia emission rate in the whole cycle in summer 2006. Results of the nitrogen balance showed that about 44% of the N was assimilated by animals in tissues, whereas N excretion was 578 grams of N (56% of N intake). Ammonia emissions accounted for a 24.3% of the excreted nitrogen.

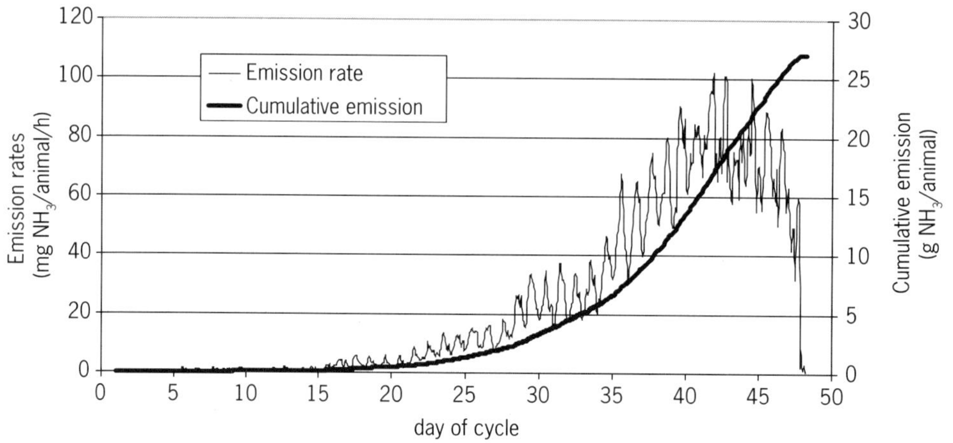

Figure 1. Ammonia emission rates and cumulative emissions in summer 2006.

Conclusions

Results are consistent with reported data in other countries, due to the intensive environmental control in these facilities. Both experiments showed very similar ammonia emission results. Ammonia concentrations are below reported values because of the high ventilation rate required in summer to provide the animals the proper environment. These results are not adequate for winter conditions, because ventilation rates and litter properties are different. Uncertainty analysis using a Monte Carlo procedure can be a very useful tool to improve the quality of reported emission rates and eventually emission factors. Accurate information of uncertainty sources is however required.

Acknowledgements

This work has been supported by the Spanish Ministry of Education and Science (Project GASFARM AGL2005-07297) and using equipment provided by Generalitat Valenciana, Conselleria de Empresa, Universidad y Ciencia, co-found by FEDER European funds.

References

Calvet, S., O. Blumetto and A.G. Torres, 2006. Ammonia emissions and nitrogen balance in a broiler farm in Spain in summer conditions. Book of Abstracts of the XVI CIGR World Congress – Agricultural Engineering for a Better World.

Ellison, S.L.R., M. Rosslein and A. Williams (Eds.), 2000. Quantifying Uncertainty in Analytical Measurement. Second Edition. EURACHEM / CITAC Guide CG 4. 126 pp.

Kosch, R., 2006. Minderungsmassnahmen in der Geflügelmast (in German). In: Emissionen der Tierhaltung. Editors Eurich, B., Grimm, E., Döhler, H. and Dörfler, R. KTBL Schrift Nr. 449, pp. 228-239.

UPV, 2006. Methodology for the estimation of atmospheric emissions from the agrarian sector for the national emission inventory. In press.

Spatial variability of ammonia production in poultry house: biological, chemical, and physical effects at microscale

N. Lovanh[1], K. Cook[1], M. Rothrock[1], D. Miles[2] and K. Sistani[1]
[1]USDA-ARS, AWMRU, Bowling Green, KY 42104, USA
[2]USDA-ARS, WM&FRU, MS State, MS 39762, USA

Introduction

Poultry litter is a valuable fertiliser source for crop production. However, its value as a fertiliser is reduced over time due to the significant losses of nitrogen attributed to the volatilisation of ammonia (Lauer *et al.*, 1976; Pain *et al.*, 1998; Hartung and Phillips, 1994). Ammonia emission and subsequent deposition can be a major source of pollution, causing nitrogen enrichment, acidification of soils and surface waters, and aerosol formation. In the poultry house, ammonia emissions can also adversely affect the health, performance, and welfare of both animals and human operators (Donham, 1990; Donham *et al.*, 1997; Donham and Gustafason, 1982). Understanding factors that affect ammonia emissions in poultry houses is a necessary first step in deploying potential remediation options. In this study, we examined various factors that potentially affect ammonia emissions in a poultry house.

Materials and methods

Studies were undertaken at a poultry house on a Mississippi farm to examine the various factors (i.e. biological, physical, and chemical factors) that potentially affect ammonia emissions in a poultry house. Litter samples were taken in a 36-point grid pattern at 5 m across and 12 m down a 146 m by 12.8 m chicken house plus 8 additional samples from the water/feeder areas. At each sample point, ammonia flux estimates were made, litter moisture and pH were determined, temperature and humidity were recorded, trace minerals and nutrients were analysed, and activities of *urease* enzyme were determined using real time polymerase chain reaction (RT-PCR). Principal component analysis (PCA) of these physical, chemical and biological factors (ca. 22 factors or variables) was carried out to determine the most important factors affecting the ammonia emission.

Results and discussion

PCA module in STATISTICA (Statsoft, Tulsa, OK) was utilised to examine the effect of twenty two physicochemical and biological parameters (i.e. litter temperature, relative humidity, moisture, carbon, nitrogen, pH, *urease* enzyme activity, etc.) for twenty eight sampling locations inside a poultry house out of forty four samples. Ammonia fluxes (ranging from 97 to 2136 $mg/m^2/hr$) from the 28 sampling points were used as the dependent variable. Based on the eigenvalues scree plot (not shown), the original 22 physical parameters were reduced to 10 main factors from the levelling-off point(s) in the scree plot as suggested by Cattell (1966). The factor corresponding to the largest eigenvalue (6.2) accounts for approximately 28.2% of the total variance. The second factor corresponding to the second eigenvalue (5.4) accounts for approximately 24.7% of the total variance. The third factor with eigenvalue of 3.9 accounts for 17.6% of the total variance. The fourth to the tenth factor account for less than 10% of the total variance each. The remaining 12 factors have eigenvalues of less than unity. Further analysis of factor loadings showed that relative humidity, litter temperature, Ca, Al, urease activity, Fe, Pb, P, and Na are the major factors affecting the ammonia flux from poultry litter (Table 1). Multiple analysis of variance using the Hotelling's T-square statistic showed that ammonia flux was also dependent on the locations when taking into account all the biochemical factors (Figure 1).

Table 1. Variable values and their importance in factor loading (influencing the 10 principal components).

Variable	Value Range	Power	Importance
Relative Humidity (%)	79 – 91	0.9918	1
Litter Temperature (°C)	29.5 – 35.0	0.9909	2
Ca (mg/kg)	21,627 – 44,257	0.9868	3
Al (mg/kg)	454 – 8966	0.9866	4
Urease (x 10^7)	0.63 – 21.9	0.9846	5
Fe (mg/kg)	277 – 5586	0.9788	6
Pb (mg/kg)	0.51 – 2.29	0.9742	7
P (mg/kg)	15,807 – 20,957	0.9731	8
Na (mg/kg)	10,828 – 16,818	0.9727	9
B (mg/kg)	67 – 93	0.9702	10

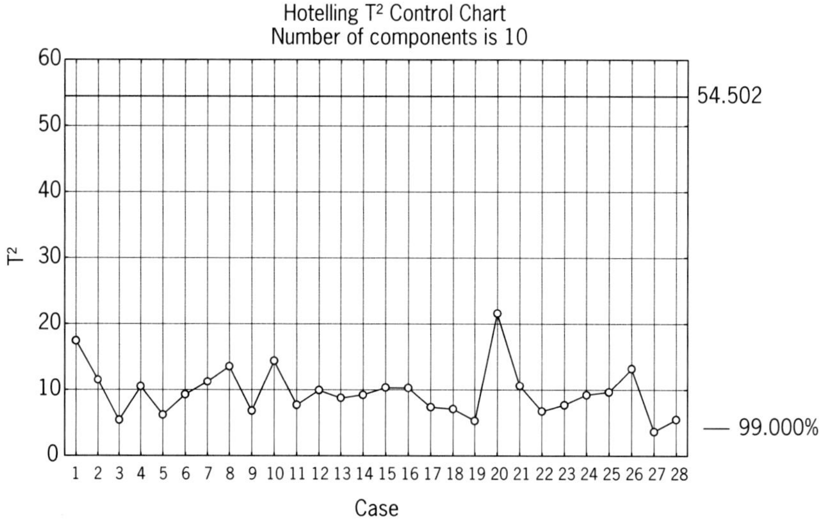

Figure 1. Results from multiple analysis of variance (T^2 = 54.5 at 99% confidence interval). Peaks and valleys show the fluctuations in ammonia emissions at various sampling points.

Conclusions

Ammonia emissions in a poultry house were, to a large degree, affected by sampling location, relative humidity, litter temperature, concentrations of *urease* enzyme, and trace minerals and nutrients (i.e. Fe, Pb, Ca, P and N). Based on these data, biological as well as chemical and physical factors play a major role in ammonia emissions. Thus, understanding the contributing factors in ammonia emissions may provide a rational basis for improving the design and optimising the remediation options for ammonia reduction.

References

Cattell, R.B., 1966. The scree test for the number of factors. Mult. Behav. Res. 1: 245-276.

Donham, K.J., 1990. Relationships of air quality and productivity in intensive swine housing. J. Agric. Practice 10: 15-18.

Donham, K.J. and K.E. Gustafason, 1982. Human occupational hazards from swine confinement. Ann. Am. Conf. Gov. Hyg. 2: 137-142.

Donham, K.J., M.J. Rubino, T.D. Thedell and J. Kammermeyer, 1997. Potential health hazards to agricultural workers in swine confinement buildings. J. Occup. Med. 19: 383-387.

Hartung, J. and V.R. Phillips, 1994. Control of gaseous emissions from livestock buildings and manure stores. J. Agric. Eng. Res. 57: 173-189.

Lauer, D.A., D.R. Bouldin and S.D. Klaususner, 1976. Ammonia volatization from dairy manure spread on the soil surface. J. Env. Qual. 5: 131-141.

Pain, B.F., T.J. Van der Weeden, B.J. Chambers, V.T. Phillips and S.C. Jarvis, 1998. A new inventory for ammonia emissions from U.K. agriculture. Atmos. Environ. 32: 309-313.

Potential means to reduce ammonia emission from laying hen facilities

H. Xin, H. Li and R.T. Burns
Agricultural and Biosystems Engineering Dept., Iowa State University, Ames, Iowa 50011, USA

Introduction

Seeking practical means to reduce ammonia emissions from poultry feeding operations is a priority issue for the U.S. poultry industry. We at Iowa State University have been conducting emission mitigation as well as quantification studies through both laboratory and field-scale tests (Liang *et al.*, 2005; Li, 2006). The potential mitigation strategies we have been examining include (a) dietary manipulation, (b) physical configuration of manure storage stacks, and (c) topical application of some commercially available mineral/chemical agents, including zeolite (grade 14×40), alum of both liquid and powder forms [$Al_2(SO_4)_3 \cdot 14H_2O$], Ferix-3 [$Fe_2(SO4)_3$], and PLT [$NaHSO_4$]. The purpose of this brief paper is to highlight the research findings concerning the efficacy of selected pre-excretion (dietary manipulation) and post-excretion (manure handling and treatment) means on mitigation of NHB_3 emissions for laying hen facilities.

Potential ammonia emission reduction strategies evaluated

Housing and manure handling schemes: Consistent with the literature report, our recent research (Liang *et al.*, 2005) revealed that the *house-level* ammonia (NH_3) emission is much higher for high-rise (HR) laying hen houses (featuring in-house manure storage for about a year) than that for manure-belt (MB) hen houses (featuring frequent removal of manure from the houses). In the case of HR and MB houses in Iowa and Pennsylvania, the MB hen houses emit less than 10% of NH_3 than the HR counterparts. Moreover, the frequency of manure removal affects house-level emissions. It should be noted that house-level emissions are only part of the total emissions at the *farm level* since manure in separate storage facilities, as may exist in some cases, also generates emissions. This is why we need to quantify and mitigate air emissions from manure storage as well as from the houses, as described below.

Stacking profile of laying hen manure in storage: The effects of hen manure stacking profile on NHB_3 emission were evaluated through five surface-area-to-volume ratios (SVR, cm^{-1}) of 5, 10, 20, 40, or 80 with the same base area of 2.8 m^2, all at 25 °C air temperature and 20 air changes per hour over a 40-day storage period. The difference in NH_3 emission per unit manure weight over the 40-day storage period between the 5-cm stack and the 80-cm stack was more than 6 fold. This difference arose from the fact that it is the top sub-layer of the manure stack that is primarily responsible for NH_3 emission. The crust formed near the surface of the manure stack was speculated to provide a physical barrier to NH_3 escape from the stack. Hence, the results indicate when stocking manure, reducing surface to volume ratio will lead to reduction in NH_3 emission. Li (2006) provided a detailed description of the procedures and results.

Dietary manipulation: Ammonia emission from high-rise layer houses was reduced by 10% (298 vs. 268 g NH_3/d-AU) with a nutritionally balanced 1% lower CP diet (Liang *et al.*, 2005). In a lab-scale study using emission vessels, it was shown that NH_3 emissions from manure from hens fed the control (standard) diet was reduced by about 41% over a 14-day storage period, as compared with that from manure of hens fed an experimental Ecocal diet (0.29 $g \cdot kg^{-1}$ d^{-1} vs.

0.17 g·kg^{-1} d^{-1}) (P<0.01) (Li, 2006). In another experiment (Roberts *et al.*, 2007), it was shown that addition of fibre as soy hulls, wheat middlings, or DDGS to the diets of laying hens led to NH_3 emission reduction by up to 50%. The reduction was realised partly through a reduction in the amount of manure uric acid and partly through a lowered manure pH. Egg production and egg mass were not affected by the dietary fibre additions, although feed consumption increased by 2%.

Topical application of manure treatment agents: Topical application of zeolite on laying hen manure was shown to reduce NH_3 emission and the magnitude of emission reduction was generally proportional to the application rate. Adsorption of NH_3 seemed to take effect right after the application, resulting in the largest emission reduction on day 1, with 66%, 91% and 96% reductions at the application rates of 2.5%, 5% and 10%, respectively. Ammonia emission from each of the three application rates (denoted as low, medium and high) was significantly lower than that of the control (P<0.001). However, there was no additional emission reduction between the medium application rate and the high application rate in all cases. The cumulative 7-d NH_3 emission reductions were 64-93% for liquid alum, 81-94% for powder alum, 82-87% for Ferix-3, and 74-92% for PLT. Results of the manure properties showed that manure samples receiving the higher application rates had lower pH and lower ammoniac N content. The practicality and economic feasibility of field-scale application of the treatments remain to be assessed under commercial production conditions.

Ongoing and future work

- Conduct field evaluation and verification tests on dietary manipulation that has undergone lab-scale tests with regards to hen production performance and costs as well as NH_3 emission reduction.
- Explore and assess practicality of field application of lab-tested manure treatment agents.

Acknowledgements

Funding for the studies was provided by the Iowa Egg Council, the U.S. Poultry and Egg Association, and Midwest Poultry Research Program, Iowa State University College of Agriculture, and a USDA Special Project for Air Quality Research.

References

Li, H., 2006. Ammonia emission from manure-belt layer houses and manure storage. PhD dissertation, Iowa State University Parks Library, Ames, Iowa 50011, USA.

Liang, Y., H. Xin, E.F. Wheeler, R.S. Gates, H. Li, J.S. Zajaczkowski, P.A. Topper, K.D. Casey, B.R. Behrends, D.J. Burnham and F.J. Zajaczkowski, 2005. Ammonia emissions from U.S. laying houses in Iowa and Pennsylvania. Transactions of the ASAE 48: 1927-1941.

Roberts, S.A., H. Xin, B.J. Kerr, J.R. Russell and K. Bregendahl, 2007. Effects of dietary fiber and low crude protein on ammonia emission from laying-hen manure. Poultry Science (in press).

Ammonia emissions from laying hen and broiler houses in the U.S.A

E.F. Wheeler[1]*, R.S. Gates*[2]*, H. Xin*[3]*, K. Casey*[4]*, P. Topper*[1] *and Y. Liang*[5]
[1]*Pennsylvania State University, Agricultural Engineering Building, University Park, PA, USA*
[2]*University of Kentucky, Lexington, KY, USA*
[3]*Iowa State University, Ames, IA, USA*
[4]*Texas Agricultural Experiment Station, TX, USA*
[5]*U.S. Department of Agriculture, Agriculture Research Service, Pendleton, OR, USA*

Introduction

The objective of this USDA-supported study was to collect high-quality, representative ammonia emission data from typical broiler and laying hen houses in the United States using a cost-effective strategy. Ammonia emissions were monitored from 10 laying hen and 12 broiler houses on commercial farms in three U.S. regions (Iowa [IA]; Kentucky [KY]; Pennsylvania [PA]) for at least thirteen 48-hour periods over one year. Liang *et al.* (2005) and Wheeler *et al.* (2006) provide complete farm characterisation and data from the study sites.

Methods

Paired repetitions of houses on commercial farms represented current construction with variety in manure management (built-up or new litter each broiler flock; high-rise or manure-belt layer houses), diet treatment (typical vs. reduced-protein diet in HR layer in IA) and climate conditions (cold [PA and IA] or mixed-humid [KY]). In order to economically obtain data from as many houses as possible over the year, the monitoring instruments were taken to one set of houses the first week and another set of houses a second week, following proper bio-security protocols. The interval between 48-hour collection periods was typically three weeks in PA and two weeks in KY and IA. Portable Monitoring Units (PMUs) were designed to monitor ammonia and carbon dioxide (CO_2) concentrations and static pressure difference between interior and exterior conditions. Detailed information is available in Xin *et al.* (2003) and Gates *et al.* (2005). Briefly, the PMU contained two identical electrochemical gas monitors for redundant measurement of ammonia concentration (0-200 ±3 ppm, volume basis; PAC III, Dräger Safety, Inc, Pittsburgh, PA) with plumbing and controls for cycling fresh, outside air (14 or 24 minutes) and poultry house exhaust air (6 minutes) past the sensors. All ammonia sensors were calibrated immediately prior to each study field trip and checked for calibration upon return from the field. Ventilation rate was determined in broiler housing via in-situ measurement of fan capacity (Wheeler *et al.*, 2002), fan on-off times, and house static pressure difference or carbon dioxide balance for layer houses (Li *et al.* 2004). The *in-situ* measurement was conducted with a traversing anemometer array, the Fan Assessment Numeration System (FANS) unit (Gates *et al.* 2004). In short, the FANS consisted of five vane anemometers positioned on a bar that traversed the entire airflow entry area to each fan. The FANS was used to develop performance curves for each individual fan in each broiler house over a range of six typical building static pressure differences (0 to 50 Pa).

Results

Emission rate (ER) for broiler chickens at a given bird age can be relatively uniform from flock to flock throughout the seasons despite the large variations in seasonal house exhaust ammonia concentration and ventilation rate (VR). It is recommended that at least three monitoring periods be used during a flock cycle to determine emission trends. Highest ER was measured during the warmest weather, especially in the mixed-humid climate houses where a very high VR was used to provide convective cooling of birds during hot weather (tunnel ventilation) with VR exceeding that needed for simple heat removal. An estimate of daily NH_3 ER per bird (±standard error) from all data of all four broiler farms was best represented by:

$$ER_b = 0.03(\pm 0.0011) \bullet x$$

where x = flock age (d), if used (built-up) litter
= 0, if new litter and flock age < 7 d
= flock age – 6 (d), if new litter and flock age ≥ 7 d
ER_b = emissions rate, g NH_3 bird^{-1} d^{-1}

For the 10 layer hen houses on three farms, the annual mean emission rates in g NH_3 hen^{-1} d^{-1} ± standard error were 0.90±0.027 for high-rise IA houses with standard diet, 0.81±0.02 for high-rise houses (in IA) with a nutritionally balanced 1% lower crude protein diet, 0.83±0.070 for high-rise PA houses with standard diet, 0.054±0.0035 for manure-belt houses with daily manure removal (IA), and 0.094±0.006 for manure-belt houses with twice a week manure removal (PA).

References

Gates, R.S., H. Xin, K.D. Casey, Y. Liang and E.F. Wheeler, 2005. Method for measuring ammonia emissions from poultry houses. J Appl Poult Res. 14: 622-634.

Gates, R.S., K.D. Casey, H. Xin and E.F. Wheeler, 2004. Fan Assessment Numeration System (FANS) design and calibration specifications. Transactions of the ASAE. 47: 1709-1715.

Li, H., H. Xin, Y. Liang, R.S. Gates, E.F. Wheeler and A.J. Heber, 2004. Comparison of direct vs. indirect ventilation rate determinations in layer barns using manure belts. Transactions of the ASAE 48: 367-372,

Liang, Y., H. Xin, E.F. Wheeler, R.S. Gates, J.S. Zajaczkowski, P. Topper, H. Li and K.D. Casey, 2005. Ammonia emissions from U.S. laying hen houses in Iowa and Pennsylvania. Transactions of the ASAE 48: 1927-1941.

Wheeler, E.F., R.S. Gates, H. Xin, J.S. Zajaczkowski, P.A. Topper and K.D. Casey, 2002. Field estimation of ventilation capacity using FANS. Paper No. 02-4125. Meeting of American Society of Agricultural Engineers. St. Joseph, MI. 11 pp.

Wheeler, E.F., K.D. Casey, R.S. Gates, H. Xin, J.L. Zajaczkowski, P.A. Topper, Y. Liang and A.J. Pescatore, 2006. Ammonia emissions from twelve USA broiler chicken houses. Transactions of ASABE 49: 1495-1512.

Extraction of ammonia by ventilation in an aviary system for laying hens

G. Gustafsson and E. von Wachenfelt
Swedish University of Agricultural Sciences, Dep. of Agricultural Biosystems and Technology, P.O. Box 86, S-230 53 Alnarp, Sweden

Introduction

Raising laying hens in loose-housing systems instead of in cages will mean that more manure will be stored in the buildings. This will cause an increased release and concentration of ammonia. It is therefore important to remove as much as possible of the exhaust air in connection with exposed manure surfaces. The amount of ammonia removed in this way is determined by the ability to capture the gas at the extraction point. To obtain the best possible function in exhausting ammonia is it necessary to achieve the best possible range and uniformity when extracting exhaust air from manure surfaces in long buildings. Uniformity of air flow from the extraction ducts is determined by the distribution of pressure in the ducts. Further, the static pressure drop in the duct must not be so large that it will affect the dimensioning of the fan system. The objective of this study was, thus, to obtain data on the dimensioning of extraction ducts.

Materials and methods

The following ventilation properties were studied in a test duct during extraction of air:

- Uniformity of flow in the exhaust vents along the duct.
- The change in static pressure along the duct.
- Air velocity at different distances from the vents in the test duct.

Results

Uniformity of flow in the exhaust vents: The variations in the air flow along the duct were studied with regard to:

- Absolute change in air velocity along the duct.
- Relative change of air velocity v_{min}/v_{max} in the vents along the duct.
- Variation in exhaust distance along the duct.

These properties were studied both at different area ratios between total area of the vents and the cross section area of the duct, A_v/A_d, and at different ventilation rates. Analyses were made of how the relative variation of air velocity, v_{min}/v_{max}, in the exhaust vents was influenced by the area ratio, as shown in Figure 1. It is clearly seen that the area ratio A_v/A_d must be less than about 1 to maintain uniform exhaust ventilation along the duct. The studies also showed that the relative variation in air velocity v_{min}/v_{max} is independent of the level of the ventilation rate. How the relative variation in the length of the exhaust distance, Z_{min}/Z_{max}, is affected by the area ratio A_v/A_d, was also studied, see Figure 2. The uniformity of the exhaust distance is influenced in about the same way by the area ratio as the air velocity in the exhaust vents. Thus, also here is it important that the area ratio is not too high if a good exhaust function should be guaranteed from manure-covered surfaces. The studies also demonstrated that the uniformity of the exhaust distance, Z_{min}/Z_{max} is independent of the ventilation flow. The length of the

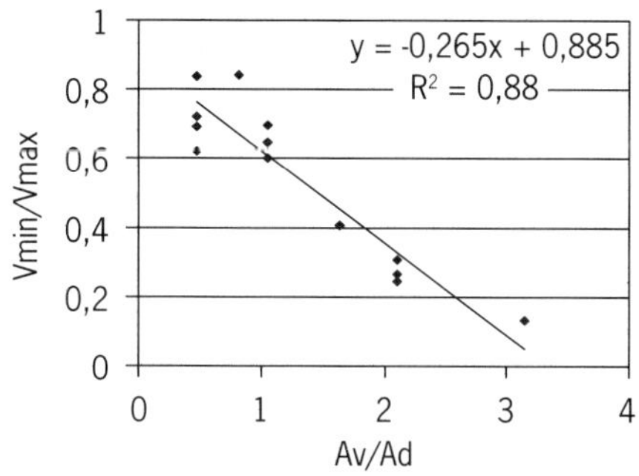

Figure 1. Relation between minimum and maximum air velocities, v_{min}/v_{max}, in the exhaust vents as a function of the area ratio, A_v/A_d.

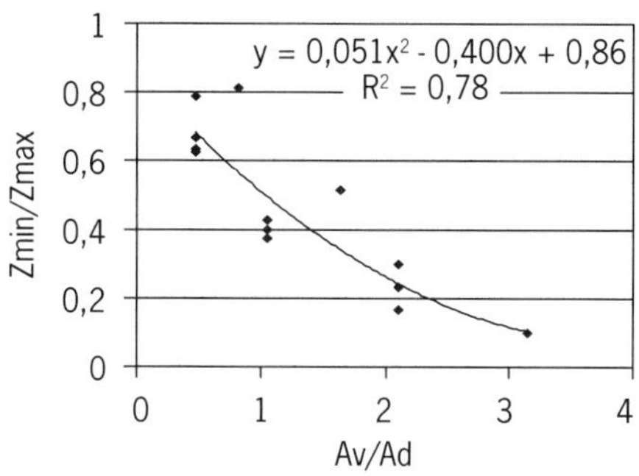

Figure 2. Relation between minimum and maximum lengths of exhaustion, Z_{min}/Z_{max}, as a function of the area ratio, A_v/A_d.

exhaust range is affected by the air velocity and thus also by the ventilation rate. The exhaust range is affected linearly by the level of the ventilation rate. However, it should be observed that the exhaust ventilation has a limited range, maximally 0.3 m from the vents.

Static pressure: The pressure drop can be described by the equation:

$$\Delta P = \frac{a \rho v^2}{2}$$

Where

- ΔP = static pressure, Pa
- a = the total friction coefficient for a duct.
- ρ = air density, kg/m^3
- v = air velocity, m/s

The friction coefficient was, in principle, remained unaffected by the area ratio with an average value of 0.80.

Conclusions

The ratio between the total area of vents and the cross section area of the duct should be less than 1 to maintain uniform exhaust ventilation and exhaust distance along extraction ducts.

Influence of manure storage on ammonia release in a floor housing system for laying hens

E. von Wachenfelt and G. Gustafsson
Swedish University of Agricultural Sciences, Dep. of Agricultural Biosystems and Technology, P.O. Box 86, S-230 53 Alnarp, Sweden

Introduction

The hygienic threshold limit values for ammonia (25 ppm) are often exceeded in floor housing systems for laying hens with long time storage of manure in bins below draining floors. The major reason for high ammonia concentrations is the large amounts of stored and exposed manure. The possibility to reduce ammonia release by reducing the amount of stored manure in bins in floor housing systems for laying hens has therefore been investigated.

Materials and methods

Investigations were carried out in a climate chamber equipped with a floor housing system, Figure 1. The housing system contained a bedding area, a manure bin area with manure conveyors below a draining floor and laying nests that were placed close to an end wall. The conditions when manure is stored in bins below draining floors were simulated by storing manure on the conveyors for several days at constant ventilation rates and temperatures (20-21 °C). Ammonia release was determined from measured concentrations (infra red spectrophotometer Miran 203, Foxboro Analytical) in exhaust air and ventilation rates (hot wire anemometer Alnor). The influence of long time storing of manure in a bin was simulated by storing manure on the conveyors. The influence of increasing storage of manure in the bedding area was investigated during two periods with daily manure removal from the conveyors. Measured ammonia release was then mainly from the bedding.

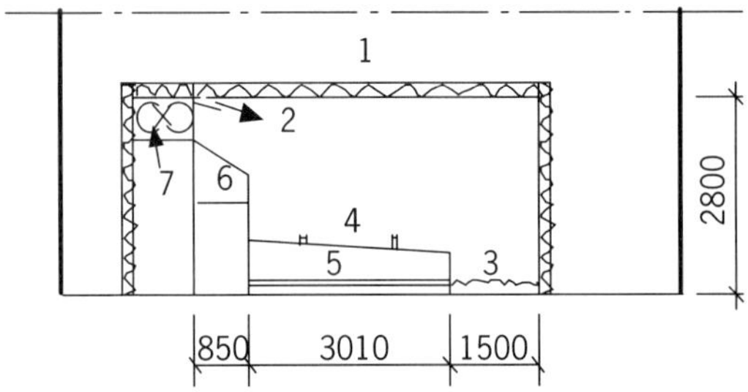

Figure 1. The climate chamber equipped with a floor housing system.

Results

The investigations clearly showed that long time storage of manure in a manure bin will cause a rapid increase in ammonia concentrations. The ammonia concentration exceeded the hygienic threshold limit value of 25 ppm after about 7 days storage of manure in the bin. The lowest concentrations occurred in the exhaust air. At ventilation rates in the range of 0.95–1.6 m^3 hen^{-1} h^{-1} the increase in ammonia release was 0.032–0.037 mg hen^{-1} h^{-1} for each day of accumulation of manure in a bedding of gravel. The release of ammonia from the bedding was also compared

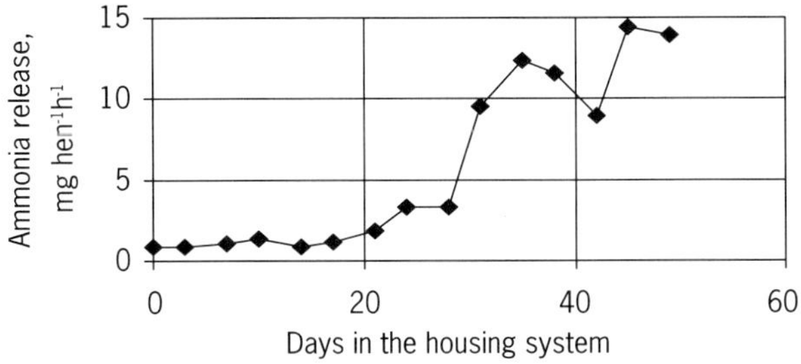

Figure 2. Increase in ammonia release the first part of a housing period at a ventilation rate of 1.01 m^3 hen^{-1} h^{-1} at high exhaust ventilation, daily removal of manure from conveyors and with a bedding of gravel.

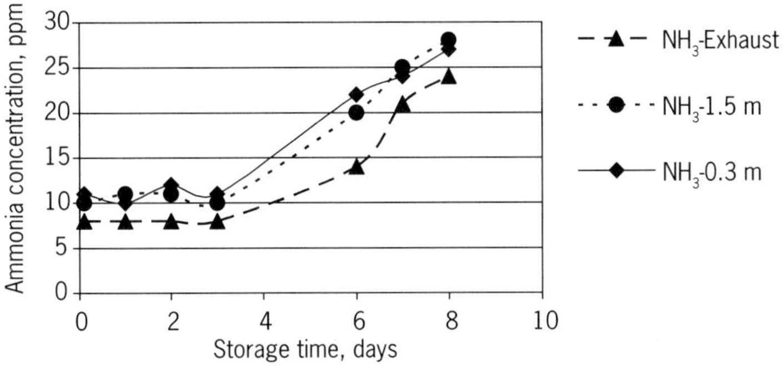

Figure 3. Increase in ammonia concentrations in exhaust air; 1.5 m above the floor and 0.3 m above the floor when manure was stored in a bin at a ventilation rate of 2.2 m^3 hen^{-1} h^{-1}.

with the release from manure stored in the manure bin. The increase in release of ammonia from the bedding was considerably lower than from manure stored in the manure bin. The major reason why the release increased more rapidly from manure stored in the bin was probably that the major part of the manure was left on the elevated draining floor above the bin.

Conclusions

It was possible to maintain the ammonia concentration below the hygienic threshold limit values when manure was removed daily with conveyors. Floor housing systems for laying hens with elevated draining floors should therefore be equipped with manure removal systems that enable frequent removal of manure in the bins.

Costs for housing systems with low ammonia emission in poultry

H.H. Ellen and I. Vermeij
Wageningen UR, Animal Sciences Group, P.O. Box 65, 8200 AB Lelystad, The Netherlands

Introduction

Poultry farmers in The Netherlands will be required to apply housing systems with lower ammonia emissions. This because The Netherlands has to comply with the EU legislation requirement that total ammonia emissions should be lowered to 128 kton by 2010. The available techniques for poultry houses will demand extra investments and costs compared to traditional housing systems. To get a notion of the yearly costs of the applied systems the Animal Sciences Group of Wageningen UR did a study on this topic.

Methods

In the project we looked at the housing systems for laying hens, broiler breeders and broilers. For each species a description was made of the traditional housing. For laying hens this was not only cage housing but also deep litter systems. For broiler breeders and broilers traditional housing is deep litter. Also descriptions were made for the systems that are applied on farms to reduce ammonia emissions. To obtain cost information for all these systems the descriptions were send to contractors and suppliers of the systems. Also information about the use of (extra) energy was collected. Based on the gathered information calculations were made to obtain the additional costs per year for these systems.

Results

In Figure 1 the extra investments and costs are given for laying hen systems that reduce ammonia emissions. For layers in cages the traditional system is a cage with forced air drying on the manure belts with 0.5 m^3 air per hen per hour (0.042 kg NH_3/place/year). Compared with this a cage with improved forced drying (0.7 m^3 air per hen per hour, 0.012 kg NH_3/place/year) has nearly no extra investment or costs. The extra investment in a chemical air scrubber with 90% reduction is about € 3.40 and the total extra costs per year about € 0.80 per place. For alternative housing of laying hens the traditional housing system is the deep litter system with an emission of 0.315 kg NH_3/place/year. Some techniques can be used in this type of housing system, but a poultry farmer can also choose for an aviary system. This because there is no difference on the egg market between these two housing systems. Therefore aviary is compared with traditional deep litter. As seen in figure 1 the aviary systems have a lower investment and also lower yearly costs compared to traditional deep litter. With 0.090 or 0.055 kg NH_3/place/year the emission is much lower than from a deep litter system. The costs of a deep litter system in two stories with manure belts under the slatted floor (0.068 kg NH_3/place/year) are equal to a traditional system. This system is rather popular in The Netherlands. With total costs of € 0.85 to € 1.05 per place per year the chemical air scrubbers are much more expensive.
For broiler breeders the results of the study are given in Figure 2. Compared to the deep litter system as traditional housing (with 0.580 kg NH_3/place/year) the energy costs for the system with forced air drying from above (0.250 kg NH_3/place/year) are very high. This is the result of

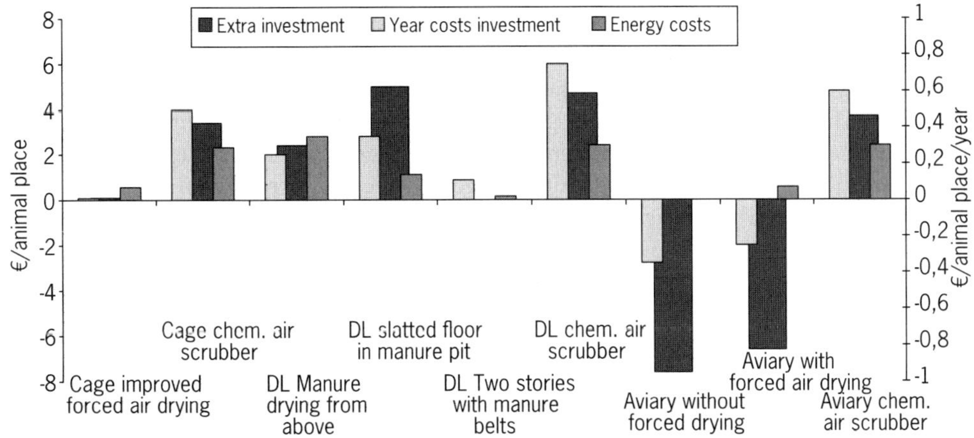

Figure 1. Extra investment and costs for housing systems for layers with reduced ammonia emission.

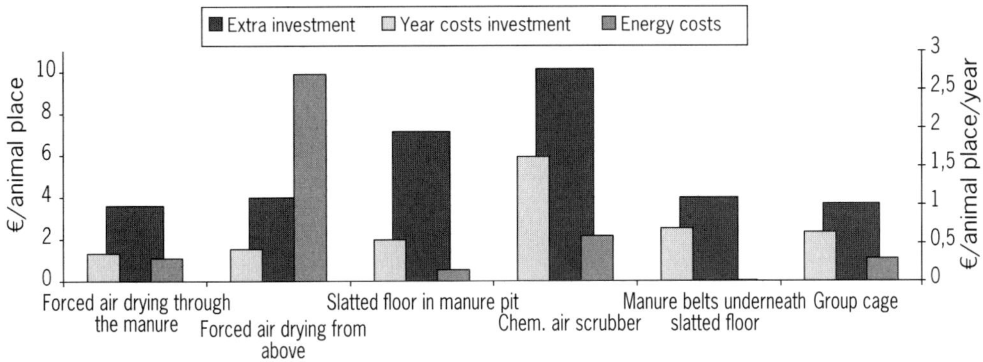

Figure 2. Extra investment and costs for housing systems for broiler breeders with reduced ammonia emission.

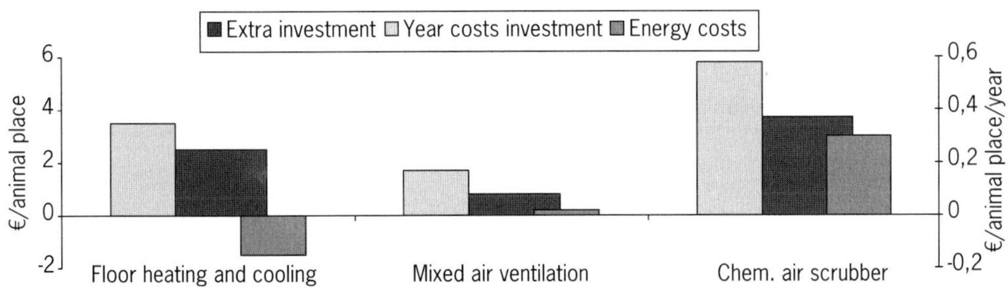

Figure 3. Extra investment and costs for housing systems for broilers with reduced ammonia emission.

the requirements of this system that the air should be at least 24 °C. With an amount of 2.5 m^3/animal/hour this costs lots of gas for heating. The extra costs for manure belts underneath the slatted floor (0.245 kg NH_3/place/year), the forced air through the manure (0.435 kg NH_3/place/year) and the slatted floor in the manure pit (0.230 kg NH_3/place/year) are almost the same.
For broilers there are not so many systems available. Some developed systems appeared not to be practical so the are no longer built. In Figure 3 the costs are shown for systems that are currently used in The Netherlands. The system with floor heating and cooling (0.045 kg NH_3/place/year) has a reduction on energy costs because of using storage of warmth in the soil during the end of one growing period and using it at the beginning of the next. The mixed air ventilation system (0.037 kg NH_3/place/year) has a low investment and also low costs.

Improving air quality in low density floor systems for laying hens

H. von Wachenfelt
Swedish University of Agricultural Sciences, Dep. of Agricultural Biosystems and Technology, P.O.Box 86, S-230 53 Alnarp, Sweden

Introduction

A field investigation of laying hens on floors was carried out consisting of 18 randomly chosen hen-houses in southern Sweden. The object was to assess knowledge of existing laying hen floor systems in regard to climate and indoor environment.

Material and methods

The study was conducted through interviews and answering sheets at 18 producers. The answering sheets consisted of questions regarding animal housing system, insulation level and heat balance, manure- and ventilation systems. Studied climate factors were ammonia, carbon dioxide and relative humidity. A simplified method (Gustafsson *et al.*, 2000) was used to evaluate the climate factors. Measurements were conducted during 5 working days in March to obtain values during a period with low ventilation rate. A relative humidity-sampler and carbon dioxide and ammonia diffusion tubes (8 h measuring period per tube) were placed in a representative place in the middle of the hen-houses. The degree of air pollutants were measured as the difference in concentration in regard to the carbon dioxide difference between in- and outdoor air according to:

$$CR = \frac{(C_2\text{-}C_1)}{(C_{CO2,2} - C_{CO2,1})} \quad (\text{ppm } NH_3/\text{ppm } CO_2)$$

where $(C_2 - C_1)$ is the difference in indoor and outdoor air pollutant concentration and $(C_{co2,2} - C_{co2,1})$ is the difference in carbon dioxide concentration in the indoor and outdoor air. The ratio CR gives a characteristic value for the air pollution in regard to the animal density in the stable independent of the ventilation rate as well as a comparable value to the hygienic threshold limits for air pollutants in poultry houses with bedding set by Swedish Board of Agriculture and National Board of Occupational Safety and Health. These hygienic threshold limits are for ammonia 25 ppm and for carbon dioxide 3,000 respectively 5,000 ppm. The maximum values of the CR- ratio at 25 ppm ammonia is 0.0038 ppm NH_3/ppm CO_2 at the designed outdoor temperature in winter time and with +20 °C indoor air temperature.

Results

The ammonia concentrations in the stables during winter time are shown in Figure 1. Through the design of the diagram the 18 stables can be compared reluctant to a difference in time of measurement. The results indicate that with regular manure removal the levels of NH_3-concentrations fell below 30 ppm at the designed outdoor temperature in winter time. The majority of stables that have permanent manure storage within the stable had NH_3-concentrations exceeding 50 ppm. Identified factors that had an influence on ammonia concentration were for example manure removal system which is shown in Table 1. A statistical analyse was conducted

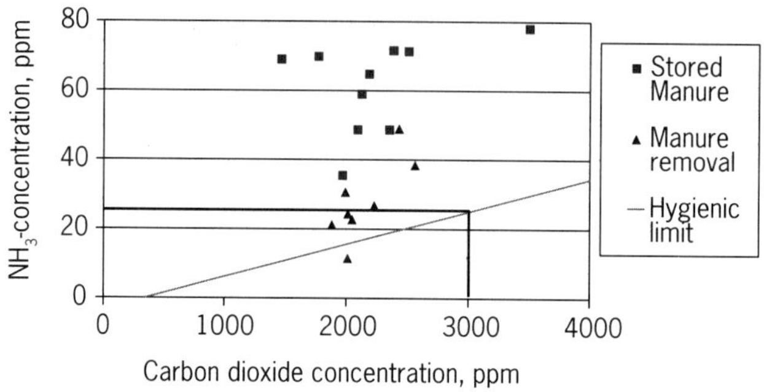

Figure 1. Ammonia concentrations as function of carbon dioxide concentrations at regular manure removal and long time storing of manure. The solid line expresses the hygienic threshold limit.

Table 1. Characteristic relative ratios of CR in regard to different manure handling systems in studied hen-houses. Two houses with manure removal are excluded because of irregular manure removal.

Manure handling system	Number of stables	CR	
		Mean value	Coefficient of variation
Manure storing in house	10	0.0345	0.36
Daily manure removal	6	0.0135	0.28

with ammonia as response factor including surrounding environmental factors that could have an influence on the stable climate factors. The influence from fodder was excluded. The R^2-value of the regression model was 0.95. The results showed a significance level of P>1% for indoor temperature, relative humidity and degree of insulation but a P> 0.1% for daily manure removal.

Conclusion

The regression analysis of ammonia emission from hen houses showed that the larger amount of manure in the stable the greater ammonia concentration. For the indoor climate factors, an increase in indoor temperature and higher level of insulation decreased NH_3 while an increase of relative humidity increased the NH_3. Older stables often had a lower degree of insulation and with that lower temperature and higher relative humidity plus permanent manure storage inside the house. Daily manure removal was the sole factor that had greatest impact on the ammonia concentration in the stables.

References

Gustafsson, G., H. von Wachenfelt and C. Swensson, 2000. A simplified method for evaluation of ammonia release in animal production systems. Proceedings: AgEng 2000, 00-AP-023. European Association of Agricultural Engineers.

Manure application

Effect of application method on ammonia volatilisation from manure applied to grassland in the Netherlands and Belgium

J.F.M. Huijsmans[1], G.D. Vermeulen[1], J.M.G. Hol[2], H. Cnockaert[3] and P. Demeyer[3]
[1]Plant Research International, P.O. Box 16, 6700 AA Wageningen, The Netherlands
[2]Wageningen UR, Animal Science Group, P.O. Box 65, 8200 AB Lelystad, The Netherlands
[3]Institute for Agriculture and Fishery Research (ILVO), P.O. Box 1, 9820 Merelbeke, Belgium

Introduction

Manure applied on farmland is an important source of ammonia (NH_3) volatilisation. Therefore, reduction of NH_3 volatilisation from this source is an important instrument in the environmental policy in the Netherlands and Belgium. Both in the Netherlands and in Belgium data are available on NH_3 volatilisation after manure application on grassland with different techniques. The aim of the research reported in this paper was to analyse the effects of application technique and other influencing factors on NH_3 volatilisation, using the combined data of the Netherlands and Belgium.

Materials and methods

The available volatilisation data originate from field experiments on grassland in which NH_3 volatilisation was measured on up to five comparable plots. Plots differed in application technique, application rate, type of manure applied or grass height. The experiments were carried out in different periods of the year and on different fields, to cover a large range of soil and weather conditions. In the Netherlands NH_3 volatilisation was measured in 59 experiments (totally 110 plots) in the period 1989-1993 (Huijsmans *et al.*, 2001) and in 27 experiments (totally 89 plots) in the period 1995-2003. In Belgium NH_3 volatilisation was measured in 10 experiments (totally 28 plots) in 2002 and in 8 experiments (totally 24 plots) in 2003. NH_3 volatilisation was measured after application of a known amount of manure on circular plots with a radius varying from 20 to 24 m, both in the Netherlands and in Belgium. The volatilisation of NH_3 following manure application was measured for at least 96 hours using the micrometeorological mass balance method as applied by Huijsmans *et al.* (2001). The effect of the application techniques was analysed by statistical estimation of the total volatilisation (in %TAN applied), being the asymptote of curves fitted through the cumulative volatilisation data for each plot. Only data

Table 1. Number of observations per application techniques (Total) and number of observations suitable for comparison of application techniques (Restricted).

	Netherlands		Belgium	
	Total	Restricted	Total	Restricted
Broadcast spreading	81	41	12	12
Band spreading[1]	29 (shoe)	19 (shoe)	5 (hose)	5 (hose)
Shallow injection	89	25	35[2]	17[2]

[1]Band spreading was carried out by trailing shoe or foot (Netherlands) or trailing hose (Belgium).
[2]Concerns also 6 observations on a very shallow injection method.

from experiments with at least two different application techniques were analysed (so-called Restricted data), see Table 1.

The effect of other factors than application technique (external factors) on NH_3 volatilisation was statistically analysed using all volatilisation rate data of all plots and all measuring intervals after application. External factors analysed were weather (wind speed, air temperature, relative humidity, radiation), soil type (sand, peat, clay), soil moisture content, type of manure (cow, pig), manure characteristics (TAN content, dry matter content, pH), application rate, and grass height. The effect of external factors on NH_3 volatilisation was analysed for each application technique separately as applied by Huijsmans *et al.* (2001).

Results and discussion

The NH_3 volatilisation after band spreading with trailing hoses (Belgium experiments) was not significantly different ($P<0.05$) from broadcast spreading (Table 2). Neither was the application with trailing shoes (Netherland experiments) significantly different from broadcast spreading at the level of $P<0.05$, but it was at $P<0.10$. The NH_3 volatilisation after shallow injection was significantly lower than after broadcast spreading and was also significantly lower than the NH_3 volatilisation after band spreading by trailing hose or shoe.

The effect of external factors on NH_3 volatilisation was not analysed for band spreading with trailing hoses as the number of observations was too limited. The effect of soil type, grass height, and radiation could not be analysed because the effect was fully linked with the country. The NH_3 volatilisation rate after broadcast spreading increased with increasing TAN content, application rate, wind speed and temperature whereas an increase in relative humidity decreased NH_3 volatilisation. The effect of external factors after application with the trailing foot was the same as found for broadcast spreading, but an increase in grass height lowered volatilisation and an increase in dry matter content of the manure increased volatilisation. For shallow injection the ammonia volatilisation increased with increasing application rate, dry matter content of the manure, wind speed and temperature and decreased with an increase of relative humidity.

The NH_3 volatilisation after broadcast spreading was 72% of TAN in both the Netherlands and Belgium. Analyses of the data of the two countries separately showed that NH_3 volatilisation after shallow injection was 13% of TAN in the Netherlands and 28% of TAN in Belgium. The relatively high application rate in Belgium (32 m^3/ha), compared with the Netherlands (22 m^3/ha) and the adjustment of the shallow injector may have caused the higher volatilisation in Belgium. The statistical models that were designed yielded valuable information about the factors that influence NH_3 volatilisation, and about the magnitude of their effects.

Table 2. Effect of application technique on the ammonia volatilisation.

Parameter	Broadcast spreading	Trailing hose	Trailing shoe	Shallow injection
Total volatilisation (% of TAN)	72 (a)[1]	34 (a)	28 (a)	20 (b)

[1]Different letters indicate a significant difference ($P<0.05$)

References

Huijsmans, J.F.M., J.M.G. Hol and M.M.W.B. Hendriks, 2001. Effect of application technique, manure characteristics, weather and field conditions on ammonia volatilization from manure applied to grassland. Netherlands Journal of Agricultural Science 49: 323-342.

An algorithm for estimation of the influence of crop and crop heights on ammonia emission from band spread livestock slurry

M.N. Hansen[1], R.E. Thorman[2], T. Misselbrook[3] and S.G. Sommer[1]
[1]*Dept. of Agricultural Engineering, Danish Institute of Agricultural Sciences, Schüttesvej 17, Horsens, Denmark*
[2]*ADAS Boxworth, Battlegate road, Boxworth, Cambridge, CB23 4NN, United Kingdom*
[3]*IGER North Wyke, Okehampton, Devon, EX20 2SB, United Kingdom*

Introduction

Ammonia (NH_3) volatilisation reduces the fertiliser nitrogen (N) efficiency of livestock slurry applied to agricultural land and increases the degree of uncertainty in predictions of crop available manure N. Uncertainty in slurry fertiliser efficiency may lead the farmer to oversupply the crops with N resulting in increased N pollutant losses. Further, the volatilised NH_3 may when deposited as ammonium-N (NH_4^+-N), contribute to undesired changes of oligotrophic ecosystems. Ammonia volatilisation from land applied slurry is reduced by trailing hose/shoe application (band spreading) (Sommer *et al.*, 1997). However, the NH_3 reduction effect of band spreading is in general higher when used in growing crops (Sommer and Olesen, 2000; Smith *et al.*, 2000), due to lower temperatures and wind speeds in the crop canopy and due to leaf absorption of emitted NH_3 (Sommer *et al.* 1997). Nevertheless, to our knowledge no models on NH_3 emission incorporate the effect of crop and crop height in the calculation of NH_3 emission from trailing hose/shoe applied slurry. Therefore, this study has reviewed studies of the NH_3 reduction effect of band spreading of slurry with the objective of developing algorithms for calculating the reduction efficiency of the band spreading technique. The developed algorithm was used in combination with the ALFAM model, which predicts NH_3 emission from bare soil, to provide monthly emission coefficients for estimating NH_3 emission from slurry applied to bare soil, cereal crops and grassland.

Materials and methods

The reduction potential (*RE*) of the band spreading technology was assessed by relating the emission from slurry applied with these application techniques to cropped and uncropped soil.

$$RE = \frac{F_{NH3,c}}{F_{NH3,b}} \qquad (1)$$

Where $F_{NH3,c}$ is the NH_3 emission from slurry band spread to a crop (cereals or grass ley) and $F_{NH3,b}$ the emission from slurry band applied to bare soil or surface broadcast spread on soil and crop. Data were obtained from field studies of NH_3 emission from livestock slurry applied to cropped and bare soil. (Sommer *et al.*, 1997, Sommer and Olesen, 2000; Misselbrook *et al.*, 2002).

Results and discussion

The NH_3 reduction potential of band spreading in winter wheat was found to be linearly related with cereal height (Figure 1). The NH_3 volatilisation from band spread slurry was reduced more efficiently when slurry was applied to a grass ley compared to application to a cereal crop at a similar height (results not shown), probably due to the denser canopy of the grass ley, which has a larger leaf area index (*LAI*) than cereals. The NH_3 volatilisation from slurry applied to bare soil increased during the spring and summer seasons due to higher temperatures, while volatilisation of NH_3 from livestock slurry applied to a crop was found to decrease as crop heights increased during the growing season (Figure 2).

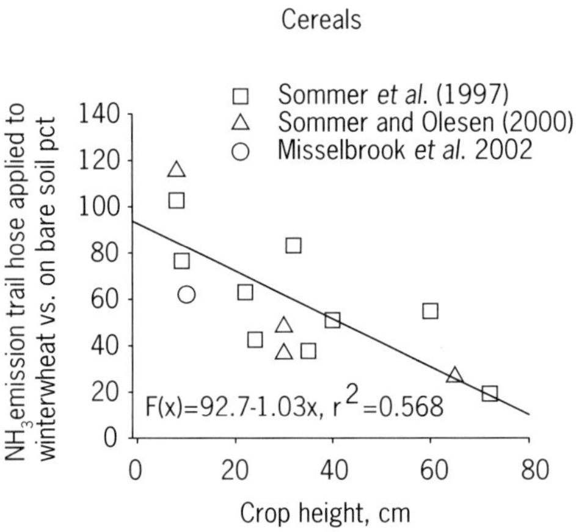

Figure 1. Reduction in ammonia volatilisation following trailing hose application of livestock slurry to winter wheat at increasing heights related to surface broadcast spreading of slurry.

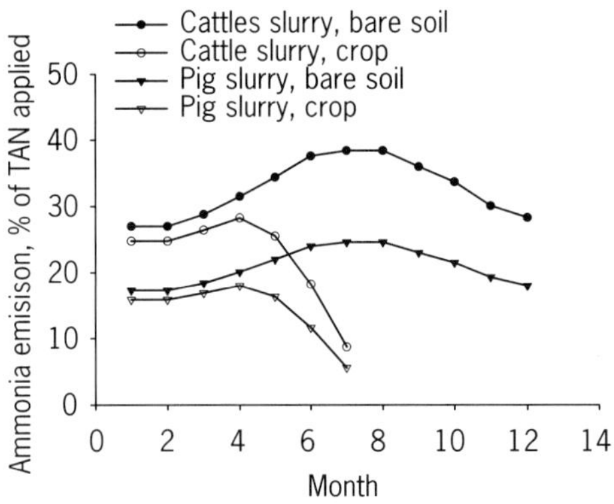

Figure 2. Ammonia volatilisation after trailing hose application of typical Danish and English cattle or pig slurry to bare soil and winter wheat.

Conclusion

The use of an algorithm for estimation of the influence of crop and crop heights on ammonia emission illustrates that crop, and the height of the crop in particular, influence the emission of ammonia. Therefore, allowance for crop and crop heights should be made when calculating the emission of ammonia from band spread livestock slurry.

References

Misselbrook, T.H., K.A. Smith, R.A. Johnson and B.F. Pain, 2002. Slurry application techniques to reduce ammonia emissions: Results of some UK field-scale experiments. Biosys. Eng. 81: 313-321.

Smith, K.A., D.R. Jackson, T.H. Misselbrook, B.F. Pain and R.A. Johnson, 2000. Reduction of ammonia emission by slurry application techniques. J. Agric. Eng. Res. 77: 277-287.

Sommer, S.G., E. Friis, A. Bach and J.K. Schjorring, 1997. Ammonia volatilization from pig slurry applied with trail hoses or broadspread to winter wheat: Effects of crop developmental stage, microclimate, and leaf ammonia absorption. J. Environ. Qual. 26: 1153-1160.

Sommer, S.G. and J.E. Olesen, 2000. Modelling ammonia volatilization from animal slurry applied with trail hoses to cereals. Atmos. Environ. 34: 2361-2372.

Comparison of ammonia emissions when using different techniques after application of slurry in Spain: summary of results 2004-06

M.J. Sanz[1], C. Monter[1], F. Sanz[1], G. Montalvo[2], J.L. Palau[1], P. Illescas[2], C. Piñeiro[3] and M. Bigeriego[4]
[1]Fundación CEAM, Spain
[2]Tragsega, S.A., Grupo Tragsa, Spain
[3]PigCHAMP Pro-Europa, S.A., Grupo Proinserga, Spain
[4]Spanish Ministry of Agriculture, Fisheries and Food, Spain

Introduction

In Europe, the ammonia (NH_3) emission to air from agriculture (animal husbandry) is at the same level as the emission of nitrogen oxides from traffic and stationary combustion plants together (Ferm, 1998). Continuous control of NH_3 emissions will make a positive impact on the air quality. At the same time, NH_3 emissions contribute to the acidification of soils (Moller and Schieferdecker, 1985), eutrophication of terrestrial ecosystems and surface waters (Roelofs, 1986), and constitute a large fraction of the fine particles that can affect human health and the radiation balance. Modern agriculture is the largest source of nitrogen releasing to other ecosystems. The land spreading of animal manure represents approximately one-third of the total NH_3 emissions from agriculture, so there has been much interest in the development of abatement measures in this area. The aim of this study, financed and coordinated by the Spanish Ministry of Agriculture, Fisheries and Food through a broader project carried out by Tragsega S.A., was to evaluate the efficiency for reducing NH_3 loss of different slurry application techniques at field-scale on representative grassland and arable soils in the Central Plateau of Spain (Segovia).

Material and methods

There are three main types of slurry application systems in use in Spain: (1) broadcast spreader with a splash plate to distribute slurry into the surface of the land, that can be incorporated (buried) in the soil several hours after application or not; (2) band spreader that discharges slurry just above ground level in strips or bands; (3) trailing shoe spreader with a shoe added to each hose, allowing the slurry to be deposited onto the soil. A fourth technique is the injection at several centimetres depth, but is less used due to the cost. The application of similar ammonium nitrogen load in mineral fertiliser was used as control treatment. To test the reductions of NH_3 concentrations in the air, several experiments at field scale were performed during 2004-06 in Segovia (Spain). Air ammonia concentration in different points (2 m height) around the application plots (up to 200 m distance from the border) and gradients (from 1.8 to 3 m height) above the application plots were measured with passive samplers. In order to measure background concentration one of the sampling locations was stablished at 1.3 km upwind of the application plots. Samplers were exposed for different periods of time (from 5 to 120 h depending on the experiment) after spreading. With the aim to optimise the experimental setup and select a correct plot orientation in both experiments, a meteorological tower was installed one month before the experiments were started at each of the study sites. Meteorological variables were also measured both before and during the experiments (wind speed, direction, temperature and radiation, at different height levels). The same load of nitrogen was applied in all treatments done in parallel for each experiment, and N-NH_3 concentration of the slurry was determined. To measure NH_3 concentrations in the atmosphere, three types of passive samplers were used (diffusion tube with and without diffusion barrier, as described by Sanz *et al.* (2005), and box sampler type – Ferm type). An approximation of the aerodynamic gradient method (Genermont *et al.*, 1998) was employed to estimate the relative NH_3 emission rates for each plot at the different exposure times.

Results and discussion

In all experiments the NH_3 emissions for the mineral fertiliser application were near 0 $\mu g/m^2 s$, except in one where likely the plot was contaminated with near plot emissions. Whereas the broadcast spreading with splash plates resulted in the highest emissions. A summary of the reductions (range of reductions) observed for all experiments is given in Figure 1.

The slurry application systems evaluated were effective in the reduction of NH_3 loss from grassland and arable soils. In other studies, application techniques for slurry (shallow injection, band spreading and trailing shoe), resulted in reductions of NH_3 emission of 70-95% compared with surface broadcast application (EC, 2003). However, earlier experiences using field-scale equipment and more recent trials with a small-plot applicator suggest that abatement efficiencies under different conditions are not so high, *but instead, more according to our results*' (Sanz *et al.*, 2005).

	Grasslands	Arable after harvest	Arable planted cereal
Splashplate	0	0	0
Splashplate buried		16-42	
Bandspreader	58	25-39	26
Trailing shoe spreader	49		
Injection		38	
Mineral fert.	100	(41)94-100	80
Mineral fert, buried		100	

Figure 1. Percentage of reduction observed respect to Splash plate (0%) for all experiments. White cells = no experiment done, light grey = 1, grey = 2, dark grey = 3 or more. Shaded cells indicate practice not possible.

References

European Comission, 2003. Integrated Pollution prevention and Control. Reference Document on Best Available Techniques for intensive Rearing of Poultry and Pigs (BREF).

Ferm, M., 1998. Atmospheric ammonia and ammonium transport in Europe and critical loads - a review. Nutr. Cycl. Agroecosys. 51: 5-17.

Genermont, S., P. Cellier, D. Flura, T. Morvant and P. Laville. 1998. Measuring Ammonia Fluxes after slurry spreading under actual field conditions. Atm. Env. 32: 279-284.

Moller, D. and H. Schieferdecker, 1985. A relationship between agricultural ammonia emissions and atmospheric sulphur dioxide content over industrial areas. Atm. Env. 19: 695-700.

Roelofs, J.G.M., 1986. The effect of air-borne sulphur and nitrogen deposition on aquatic and terrestrial heathland vegetation. Experimentia 42, 372-377.

Sanz, M.J., C. Monter, C.R. Antequera, G. Montalvo and M. Bigeriego, 2005. Influencia del sistema de aplicación de purines sobre las emisiones de amoniaco a la atmósfera en una explotación agrícola. Porci 87: 55-68.

The effect of slurry application rate on ammonia losses from bandspread and shallow injected applications

E. Sagoo[1], J.R. Williams[1], B.J. Chambers[2], H.Collis[1], T. H. Misselbrook[3] and D.R. Chadwick[3]
[1]ADAS Boxworth, Battlegate Road, Boxworth, Cambridge, CB23 4NN, United Kingdom
[2]ADAS Gleadthorpe, Meden Vale, Mansfield, Nottinghamshire, NG20 9PF, United Kingdom
[3]IGER North Wyke Research Station, Okehampton, Devon, EX20 2SB, United Kingdom

Introduction

Ammonia (NH_3) emissions from UK agriculture were estimated at 218,000 tonnes NH_3-N per year in 2000, with losses following the land spreading of farm manures responsible for c.36% of total emissions. Previous work in the UK has shown that at application rates of 30-35 m^3/ha slurry bandspreading and shallow injection on grassland can reduce ammonia emissions compared with surface broadcasting by c.40 and 70%, respectively. However, the effectiveness of bandspreading and shallow injection techniques in reducing NH_3 emissions can be variable, with factors such as application rate (which can be controlled by the farmer) and the retention of slurry in a band (for bandspreading) or slot (for shallow injection), likely to be important in controlling NH_3 emissions. This paper reports results from a Defra-funded study to evaluate the effects of slurry application rate and method on NH_3 emissions from cattle slurry applied to arable land and grassland.

Materials and methods

Two experiments were carried out on clay textured soils at ADAS Boxworth (Cambridgeshire, England). Cattle slurry was applied to cereal stubble using broadcast and trailing hose techniques in September 2004, and to grassland using broadcast and shallow injection techniques in July 2005. The slurry was applied at five application rates (20, 35, 50, 65 and 80 m^3/ha), with each treatment replicated three times. The plots were 6 m wide and 12 m long, and arranged in a randomised block design. A specialist plot applicator was used to apply the slurry which was representative of commercial practice, with the injection slots 20 cm apart and the trailing hoses 30 cm apart. Ammonia emissions were measured from each plot for seven days following application, using windtunnels.

Results and discussion

In September 2004, the trailing hose application to cereal stubble did not significantly reduce NH_3 emissions compared with the surface broadcast application, even though the high dry matter slurry (7.5% dry matter) remained in a band, covering 24% of the soil surface area at the 20 m^3/ha application rate and 50% of the soil surface at the 80 m^3/ha rate. Ammonia emissions were equivalent to a mean of 20% of the total N applied (range 13-24%) from the broadcast application and a mean of 17% of the total N applied (range 12-22%) from the trailing hose application (Figure 1a). The higher hydraulic loading from the trailing hose slurry application (8-16 mm per unit of surface area occupied by the slurry) compared with the surface broadcast application (2-8 mm), meant that whilst the slurry stayed in a band, infiltration of the high dry matter slurry into the clay soil was restricted. Hence, the NH_3 emission patterns and rates were similar from the two techniques, even though the emitting surface area of the trailing hose

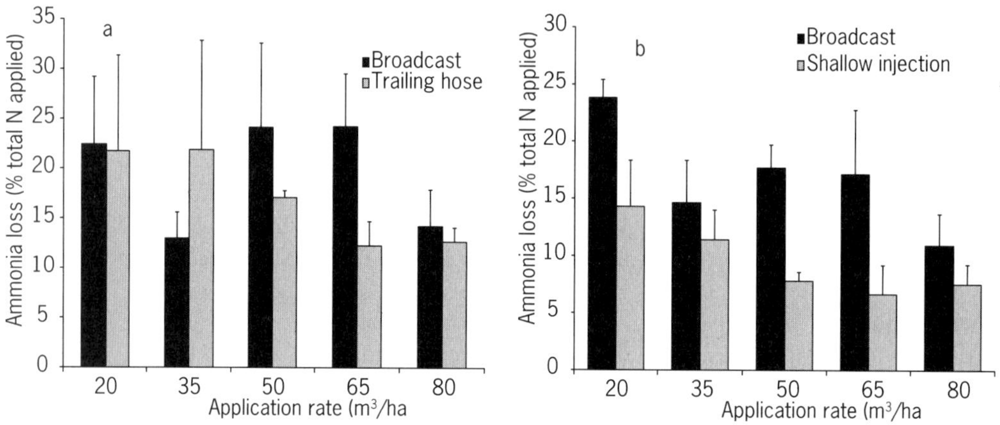

Figure 1. Ammonia losses following broadcast and trailing hose slurry application on cereal stubble (a) and following broadcast and shallow injection slurry application on grassland (b).

application was only 25-50% that of the surface broadcast application. There was no relationship (P>0.05) between broadcast or trailing hose slurry application rates and NH_3 emissions in September 2004.

In July 2005, shallow injection on grassland reduced (P<0.001) NH_3 emissions by an overall mean of 42% compared with surface broadcast application (Figure 1b). The cattle slurry (5.7% dry matter, 2.9 kg/m^3 total N) was effectively retained in the 2.5-3.5 cm deep injection slots. Ammonia emissions were equivalent to a mean of 17% of the total N applied (range 11-24%) from the broadcast application and a mean of 10% of the total N applied (range 7-14%) from the shallow injection application. The lower emissions from the shallow injection treatment were a result of the slurry being retained in the 2-3.5cm deep injection slots (17-31% of the ground surface area was covered by the slurry) and infiltrating rapidly into the dry clay soil. On the shallow injection treatment ammonia emissions decreased as a proportion of the total N applied with increased application rate (P<0.05; r^2 = 81%) reflecting the smaller emitting surface area per volume of slurry applied.

Conclusions

Slurry application to cereal stubble using a trailing hose did not significantly reduce NH_3 emissions compared with surface broadcast application. Although the slurry remained within a band (surface area occupied 18-30%), the combination of a high hydraulic loading rate (8-16 mm) and high slurry dry matter content (7.5%) restricted the rate of slurry infiltration into the heavy clay soil. Shallow injection of slurry into the grassland clay soil (at application rates between 20 and 80 m^3/ha) reduced NH_3 emissions by *c.*40% compared with surface broadcast application, largely as a result of the lower slurry emitting surface area. Also, on the shallow injection treatment, NH_3 emissions decreased as a proportion of the total N applied with increased application rate. These data indicate that for slurry bandspreading and shallow injection techniques to successfully reduce NH_3 emissions, slurry needs to be retained in a band/slot (to give a smaller emitting surface area) *and* to rapidly infiltrate into the soil. If both conditions are not satisfied, the NH_3 reduction benefits of bandspread/shallow injection spreading techniques will not be realised.

The effectiveness of different soil incorporation techniques as a means of ammonia abatement following the application of solid manure

R.E Thorman[1], J. Webb[2], S. Yamulki[3], T.H. Misselbrook[3], G. Bennett[4], S. McMillan[5] and R. Matthews[3]
[1]ADAS Boxworth, Battlegate Road, Boxworth, Cambridge, CB23 4NN, United Kingdom
[2]ADAS Wolverhampton, Wergs Road, Wolverhampton, WV6 8TQ, United Kingdom
[3]IGER North Wyke, Okehampton, Devon, EX20 2SB, United Kingdom
[4]ADAS Gleadthorpe, Meden Vale, Mansfield, Nottingham, NG20 9PF, United Kingdom
[5]ADAS Drayton, Alcester Road, Stratford upon Avon, Warwickshire, CV37 9RQ, United Kingdom

Introduction

Agriculture is the principal source of atmospheric ammonia (NH_3) in the UK, accounting for 265 kt NH_3 per year or c.80% of total emissions (Misselbrook *et al.*, 2000). The land spreading of farm manures is estimated to be responsible for approximately one third of the total UK NH_3 emissions from agriculture, of which c.50% is estimated to come from solid manures. The rapid soil incorporation of manures on arable land and prior to grass re-seeds, therefore, has the potential to significantly reduce NH_3 emissions. Moreover, rapid incorporation has been identified as a cost-effective abatement option (Webb *et al.*, 2006a). This paper will present data from a study to quantify the effectiveness of incorporating solid manures into the soil by plough, disc or tined cultivation as a means of NH_3 abatement.

Materials and methods

Six experiments were carried out at four UK sites: two experiments at site 1 on a loamy sand soil, at ADAS Gleadthorpe, central England; two experiments at site 2 on a clay soil, at ADAS Drayton, central England; one experiment at site 3 on a sandy loam soil, at IGER North Wyke, south west England and one experiment at site 4 on a clay loam soil, at IGER North Wyke. Ammonia emissions were monitored from replicated (x4) plots (3 x 10 m) following an early spring or early autumn application of solid manure. The plots were established on cereal stubble or on bare arable ground. Cattle farm yard manure (FYM), pig FYM, layer manure or broiler litter were spread at a target application rate of 250 kg N ha^{-1} and either left on the soil surface or immediately incorporated by ploughing (c.20 cm deep), discing (c.10-15 cm deep) or tine cultivation (c.10 cm deep). Ammonia emissions were monitored for up to 2 weeks after manure application using a modified wind tunnel technique based on the design of Lockyer (1984). One wind tunnel per plot was set up immediately after the manure had been incorporated (immediately after the manure was spread on the unincorporated plots). The NH_3 abatement efficiency was determined as the mean reduction in emission achieved by the incorporation technique divided by the mean emission measured from the manure left on the soil surface.

Results and discussion

There was a significant ($P<0.001$) effect of incorporation on the efficiency of NH_3 abatement (NH_3 reduction efficiency). Compared with the NH_3 loss from surface applied manure, immediate incorporation by plough reduced NH_3 emissions by 88% and incorporation by disc and tine reduced emissions by 62% and 56%, respectively. These results are comparable to the default reduction efficiencies of 80%, 60% and 50%, respectively, used in the MAVIS model (Webb *et al.*, 2006b), which was developed to identify the optimum method of incorporating solid manures to reduce NH_3 following spreading. In validating the model using four of the six experiments reported here, the authors reported that although incorporation by the more efficient plough was slow, in practice under field conditions, ploughing always reduced NH_3 to a greater extent than the faster, but less efficient disc or tine implements. There was a significant ($P<0.05$) interaction between soil type (heavy/light) and incorporation. On the heavy soils (sites 2 & 4) the NH_3 reduction efficiency was in the order plough > disc > tine, with mean abatements of c.80%, c.65% and c.50%, respectively. Whereas, the reduction efficiency on the lighter soils (sites 1 & 3) was in the order plough > tine > disc, at 95%, 65% and c.60%, respectively. These differences were probably related to a relatively poor performance of the discs on the light soil at site 1 and of the tines at site 4 where the soil was heavier and the ground conditions were hard. There is some indication that incorporation by plough and tine may abate NH_3 emissions to a greater extent on the lighter soils. In the case of ploughing, this is not unexpected, as complete inversion would be expected to be more difficult on heavy soil. There was also a significant ($P<0.01$) interaction between incorporation and manure type. Whereas ploughing reduced NH_3 emissions by c.80-95% across all manure types, incorporation by disc reduced emissions from pig FYM and poultry manures by c.70%, but only c.45% for cattle FYM. Incorporation by tine reduced NH_3 emissions from broiler litter by c.75%, but only c.50% from the FYMs and layer manure. These differences are likely to be because it is more difficult to incorporate straw-based manures (FYM) by non-inversion techniques than poultry manures (particularly broiler litter), due to the bulkier nature of the former. Furthermore, cattle FYM tends to breakdown less readily than pig FYM.

Conclusions

Following the land application of solid manures, incorporation by plough is more effective at abating NH_3 emissions than by disc or tine, although soil type and/or manure type may influence the degree of abatement.

Acknowledgements

Funding of this work by the UK Department for the Environment, Food and Rural Affairs (Defra) is gratefully acknowledged.

References

Lockyer, D.R. 1984. A system for the measurement in the field of losses of ammonia through volatilisation. J. Sci. Food Agr. 35: 837-848.

Misselbrook, T.H., T.J. Van der Weerden, B.F. Pain, S.C. Jarvis, B.J. Chambers, K.A. Smith, V.R. Phillips and T.G.M. Demmers, 2000. Ammonia emission factors for UK agriculture. Atmos. Environ. 34: 871-880.

Webb, J., G.G. Anthony and S. Yamulki, 2006. Validating the MAVIS model for optimising incorporation of litter-based manures to reduce NH_3 emissions. T. ASABE. 49: 1905-1913.

Webb J., M. Ryan, S.G. Anthony, A. Brewer, J. Laws, F.A. Aller and T.H. Misselbrook 2006. Identifying the most cost-effective means of reducing ammonia emissions from UK agriculture using the NARSES model. Atmos. Environ. 40, 7222-7233.

Evaluation of open and closed slot shallow injection of slurry into ley under Swedish conditions

L. Rodhe[1], A. Etana[2], G. Gebresenbet[3], M. Pell[4], T. Rydberg[2] and S. Yamulki[5]
[1]JTI, Swedish Institute of Agricultural and Environmental Engineering, P.O. Box 7033, 750 07 Uppsala, Sweden
[2]Dep. Of Soil Sciences, SLU, Box 7014, 750 07 Uppsala, Sweden
[3]Dep. of Biometry and Engineering, SLU, Box 7032, 750 07 Uppsala, Sweden
[4]Dep. Of Microbiology, SLU, Box 7025, 750 07 Uppsala, Sweden
[5]Environmental and Human Science Division, Forest Research, Alice Holt Lodge Farnham, Surrey, GU10 4LH, United Kingdom

Introduction

Spreading of slurry onto ley can lead to high losses of ammonia. However, incorporation of the slurry, directly after application, effectively reduces ammonia emissions. On ley, there is a need for special devices, which are capable of incorporating the slurry without crop damage. The main objectives were to evaluate open and closed slot injection methods on typical soils in Sweden in different aspects as (1) slurry placement, (2) ammonia emissions, (3) yield and crop nitrogen uptake, and (4) draught requirement. Finally, a full-scale injector with tubulator tines was compared with band spreading regarding both ammonia and nitrous oxide emissions.

Materials and methods

Conventional slurry tankers with different types of injectors were used to shallow-inject (less than 0.05 m) slurry into open slots after the first cut (Rodhe and Etana, 2005). Application methods used were; (a) pressurised injection with closed spot (PI); (b) shallow injection 1 with open slots; V-shaped disc tine (SIO1); (c) shallow injection 2 with open slots; tine consisting of two angled disc coulters (SIO2). In practice, the PI also worked as shallow injection into open slots as the slurry was partly still on the surface. The cattle slurry was applied at a rate of 25 t/ha corresponding to about 100 kg total N or 50 kg NH_4-N/ha. In order to find a tine that shallow-injected the slurry into closed slots on grassland, a tubulator tine was designed and compared to SIO2. The ammonia emissions were measured using the equilibrium concentration method, based on passive diffusion sampling close to the ground (Svensson, 1993) during the following four to five days after spreading. Gas samples for estimating methane and nitrous oxide emissions were also collected from static chambers during 7 weeks following the slurry applied in closed slots or band spread on the surface. In the five different field experiments, the statistical model randomised block design was used.

Results and discussion

Figure 1 presents the measured ammonia emissions. In experiments 1-4, with open slot injectors, only one with SIO2 could place the slurry below the soil surface in all soils tested. Ammonia release was on average 39% of the total ammoniacal nitrogen (TAN) applied with SIO2, about half the level with band spreading (75%) (Rodhe and Etana, 2005). The tubulator tine required significantly lower vertical forces to penetrate the soil than the SIO2 and with draught force similar as the SIO2. The ammonia losses after injection with the tubulator were 1.6% of the TAN

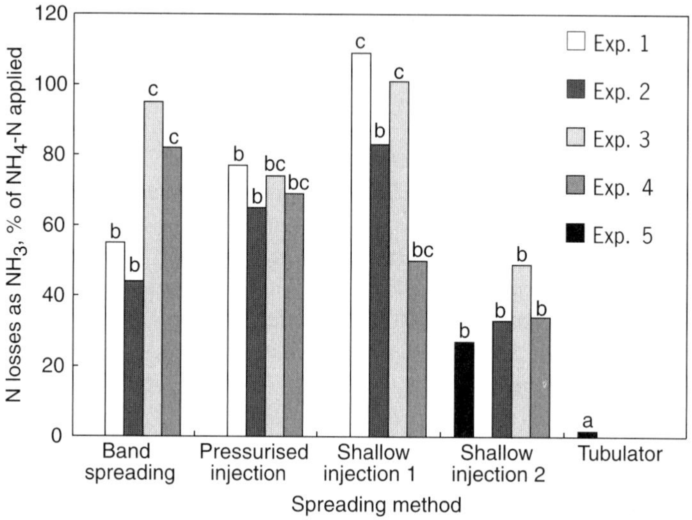

Figure 1. Nitrogen losses as ammonia (% of applied NH_4-N) after spreading of cattle slurry with five different techniques from five different experiments. Means with different letters within each experiment are significantly different ($P<0.05$).

applied and the corresponding loss for SIO2 was 27% (Rodhe *et al.*, 2004). Closed slot injection with tubulator tines prevented ammonia release but gave rise to slightly higher N_2O emissions than bandspreading (Rodhe *et al.*, 2006).

Conclusions

The general conclusion is that with appropriate design of the injector tines, such as the tubulator, slurry could be placed below the soil surface even in hard soil conditions, with minimised ammonia emissions but at an extra energy input, investment cost and slightly higher nitrous oxide emissions.

References

Rodhe, L., T. Rydberg and G. Gebresenbet, 2004. The influence of shallow injector design on ammonia emissions and draught requirement under different soil conditions. Biosys. Eng. 89: 237-251.

Rodhe, L. and A. Etana, 2005. Performance of slurry injectors compared with band spreading on three Swedish soils with ley. Biosys, Eng. 92: 107-118.

Rodhe, L., M. Pell and S. Yamulki, 2006. Nitrous oxide, methane and ammonia emissions following slurry spreading on grassland. Soil Use and Management 22: 229-237.

Svensson, L. 1993. A new dynamic chamber technique for measuring ammonia emissions from land-spread manure and fertilizers. Acta Agric. Scand. B. 44: 33-46.

Quantification of ammonia flux from land application of swine effluent

K.R. Sistani[1]*, A.C. Williams*[2] *and D.A. Mays*[3]
[1]*USDA-ARS, Animal Waste Management Research Unit, 230 Bennett lane, Bowling Green, KY 42104, USA*
[2]*Southern University, Baton Rough, LA, USA*
[3]*Alabama A&M University, Huntsville, AL, USA*

Introduction

Animal feeding operations (AFOs) are one of the major sources of airborne contaminants including ammonia (NH_3), which add to air and water pollution (Mallin, 2000; Walker *et al.*, 2000). According to U.S. Environmental Protection Agency (USEPA, 2002, 2004), livestock (>71%) and fertiliser (15%) contributed to the total national ammonia emissions in 1998.

Material and methods

A field experiment was conducted on previously established tall fescue forage site as a randomised complete block design. The soil at the site is Hartselle fine sandy loam (Typic Hapludult). Plots were circular with a 15 m radius. Swine effluent was collected from an aerobic swine waste lagoon after mechanically agitating the effluent to ensure a uniform mixture and maximising the nutrient concentration of the effluent. The effluent application rate was based on the N requirement of tall fescue (67 kg N/ha per application). There was a minimal variability in nutrient content among the swine effluent used at each application. The NH_3 losses following three swine effluent applications per year was determined using the method recommended by Wood *et al.* (2000). Briefly, a 318 cm rotating mast with wind vane was erected in the center of each plot that received effluent. A separate mast was installed several hundred meters away from the plots to measure the background NH_3 concentration. There were five collection points on each mast at 40, 70, 150, 220, and 300 cm, respectively. The collection tubes were 200 mm long, which were coated with a solution of 3% oxalic acid dissolved in acetone to capture NH_3. The tubes were changed at twelve-hour intervals for the first week then twenty-four hour intervals for the second week. Tubes were then extracted with 2 mL of deionised water for NH_3 determination.

Results

Results are presented in Table 1. The percent NH_3 losses were calculated based on the 67 kg N/ha applied as swine effluent corrected for the background NH_3 volatilisation at each effluent application. The losses for the first week after effluent application were much greater than the second week, consistently for each application. The order of losses for the timing of the effluent application were as follow: August > May> March.

Table 1. Daily total ammonia losses, after effluent application for 2 weeks (3-year average).

Application date	Days	Loss range (kg/ha)	Total loss (kg/ha)	Loss from total N applied (%)
March	1-7	0.61- 1.71	6.3	9.3
March	8-14	0.12- 0.73	2.7	3.3
May	1-7	0.86- 2.12	9.67	15
May	8-14	0.18- 0.79	4	5
August	1-7	0.89- 2.44	11	20.3
August	8-14	0.17- 0.60	3	5.3

References

Mallin, M.A., 2000. Impacts of industrial animal production on rivers and estuaries. American Scientist 88: 26-37.

USEPA, 2002. National air pollutant emission trends, 1990-1998. EPA-454/R-00-002, March. Washington, D.C.:U.S. Environmental Protection Agency.

USEPA, 2004. National emission inventory, Ammonia emissions from animal husbandry. Washington, D.C.:U.S. Environmental Protection Agency.

Walker, J.D., V.P. Aneja and D.A. Dickey, 2000. Atmospheric transport and wet deposition of ammonium in North Carolina. Atmos. Environ. 34: 3407-3418.

Wood, C.W., S.B. Marshal and M.L. Cabrera. 2000. Improved method for field scale measurement of ammonia volatilization. Commun. Soil Sci. Plant Anal. 31: 581-590.

Modelling ammonia volatilisation following urea fertilisation in a winter wheat-maize rotation in China

A. Pacholski[1], G.-X. Cai[2] and M. Roelcke[3]
[1]Institute of Crop Science and Plant Breeding, Christian-Albrechts-University of Kiel, Hermann-Rodewald-Str. 9, 24118 Kiel, Germany
[2]Nanjing Institute of Soil Science, Chinese Academy of Sciences (CAS), Beijing East Road 71, 210008 Nanjing, P.R. China
[3]Institute of Geoecology, Braunschweig Technical University, Langer Kamp 19c, 38106 Braunschweig, Germany

Introduction

In recent years several mechanistic model approaches have been developed for the calculation of ammonia (NH_3) losses following slurry or organic fertiliser application in the field (e.g. Génermont *et al.*, 1997, Wu *et al.*, 2003). However, since the development of the model by Rachhpal-Singh and Nye (1986a) no major attempts have been made to further develop a deterministic model for the simulation of NH_3 volatilisation losses after fertilisation with urea or other synthetic ammonium-based fertilisers with high inherent volatilisation potential from upland soils. With regard to the high proportion of urea fertilisers used in many developing countries including several emerging economies (e.g. China) and the increase of its use in no-till agricultural systems an improvement of the available modelling approaches is desirable.

Material and methods

Fertiliser-N losses by NH_3 volatilisation were measured with an Integrated Horizontal Flux (IHF) Method equipped with pivoted passive flux samplers following urea fertilisation (as urea granules) by different application methods (surface broadcast, incorporated fertiliser (mixed in the plough layer), fertilisation followed by irrigation). Experiments were carried out in winter wheat and irrigated maize at Fengqiu Experimental Station, Henan Province, China. Apart from the volatilisation losses meteorological conditions were monitored and soil cores as well as surface soil samples (0-3 mm) were taken for determining soil pH and soil nitrate, NH_4^+ and urea concentrations. The model of NH_3 volatilisation from applied urea by Rachhpal-Singh and Nye (1986a) was chosen as the basis for the simulations. As this model was designed for the simulation of NH_3 losses from soil without plant cover under constant laboratory conditions the model was expanded with subroutines for the calculation of the effect of variable *in situ* conditions as temperature, soil water content and wind speed. The effect of the fertiliser application method was considered in the simulations. Plant growth and plant cover effects were neglected. Model parameters were calculated by the new subroutines and determined in independent laboratory experiments (e.g. urease activity). Remaining parameters were chosen as given in the paper by Rachhpal-Singh and Nye (1986 a,b).

Results and discussion

Figures 1 and 2 show the curves of simulated and measured cumulated NH_3-N losses in two example experiments with maize (Figure 1) and winter wheat (Figure 2). In both experiments the plants in the field were still very small (5-10 cm). The magnitude of the NH_3 losses was well reflected in the simulated data. Particularly for the treatments with fertiliser mixed into the soil (fertilisation followed by irrigation, incorporated fertiliser) the absolute height of the NH_3 efflux, the dynamics of the process as well as surface soil urea concentrations were matched by the model simulations (Figures 1-4). The model simulations also resulted in a pronounced reduction of NH_3 losses by incorporation of the fertiliser, however, simulation results for surface applied urea were not likewise satisfactory. In both examples presented NH_3 losses were overestimated by the model. In addition, the dynamics of the volatilisation process was not properly reflected in the model curves. These discrepancies could mainly be attributed to the process of nitrification (data not shown) not included in the model and specific processes at the soil surface, such as severe drying out of the soil or increased NH_3 losses from very wet irrigated soils under high evaporation rates (faster dissolution of granules, faster urea hydro-lysis, strong water flux in direction of the soil surface). The neglect of plant-soil interactions was presumably of low importance for the goodness of the simulation of these experiments.

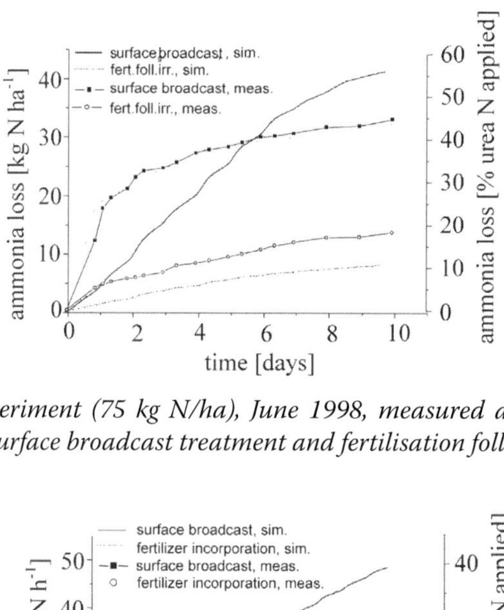

Figure 1. Maize experiment (75 kg N/ha), June 1998, measured and simulated cumulative ammonia losses for surface broadcast treatment and fertilisation followed by irrigation.

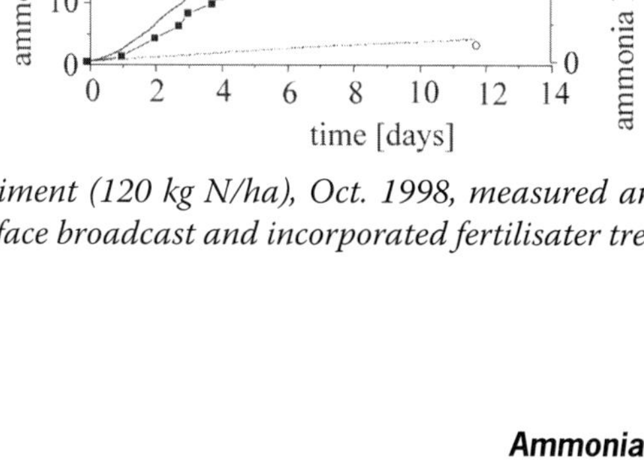

Figure 2. Wheat experiment (120 kg N/ha), Oct. 1998, measured and simulated cumulative ammonia losses for surface broadcast and incorporated fertilisater treatment.

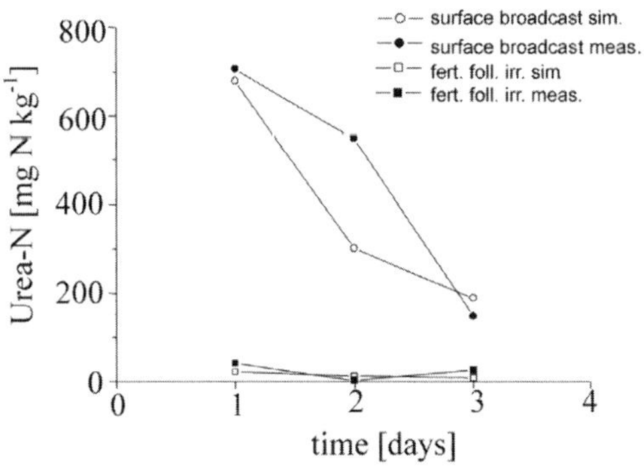

Figure 3. Maize experiment (75 kg N/ha), June 1998, simulated and measured urea contents at the soil surface (0-3 mm), surface broadcast treatment and fertilisation followed by irrigation.

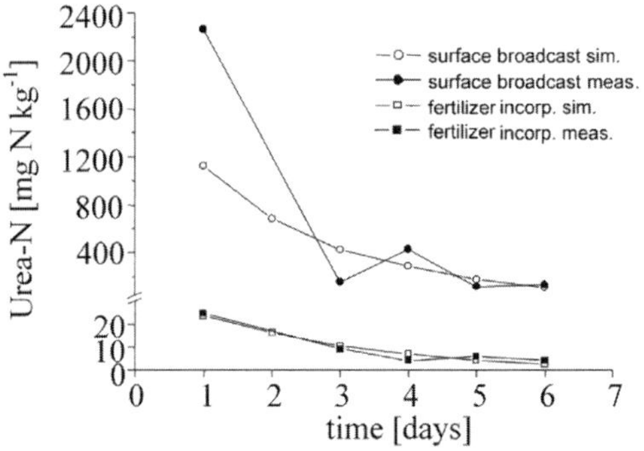

Figure 4. Maize experiment (120 kg N/ha), Oct. 1998, simulated and measured urea contents at the soil surface (0-3 mm), surface broadcast and incorporated fertilisater treatment.

References

Génermont, S. and P. Cellier, 1997. A mechanistic model for estimating ammonia volatilization from slurry applied to bare soil. Agric. Forest Meteorol. 88: 145-167.

Rachhpal-Singh and P.H. Nye, 1986 a. A model of ammonia volatilization from applied urea. I. Development of the model. J. Soil Sci. 37: 9-20.

Rachhpal-Singh and P.H. Nye, 1986 b. A model of ammonia volatilization from applied urea. II. Experimental testing. J. Soil Sci. 37, 21 - 29.

Wu, J., D.L. Nofziger, J.G. Warren and J.A. Hattey, 2003. Modeling ammonia volatilization from surface-applied swine effluent. Soil Sci. Soc. Am. J. 67: 1-11.

Evaluation of ammonia emissions from chemical fertilisers land spreading

E. Dinuccio, F. Gioelli, N. Dorno and P. Balsari
Dipartimento di Economia e Ingegneria Agraria, Forestale e Ambientale, Sezione di Meccanica. Università di Torino, Italy

Introduction

Nitrogen (N) is the mineral fertiliser most applied to agricultural land (Follet, 2001). N also forms some of the most mobile compounds in the soil-plant-atmosphere system and this results not only in financial losses for farmer but also a detrimental impact on the environment. For such reasons there is mounting concern to find solutions to improve the use of mineral fertilisers and to mitigate the N negative impact on the environment. Nitrification inhibitors (NI) are additives added to ammonium fertilisers with the aim to improve the nitrogen efficiency by retarding the conversion of ammonium to nitrate after fertiliser application. Since addition of NI to fertilisers is able to maintain soil N in ammonium form, consequently, nitrous oxide (N_2O) and nitrate (NO_3^-) losses are less likely to occur but ammonia (NH_3) volatilisation may be increased (Menéndez *et al.*, 2006). While the potential of NI to reduce N_2O emissions is well known (Weber *et al.*, 2004), at this moment few data are available on ammonia emissions. The purpose of this study was to analyse NH_3 emissions from four chemical fertilisers: three traditional nitrogen fertilisers (Ammonium Nitrate, Urea, Prilled Urea) and a mineral fertiliser containing the nitrification inhibitor 3,4-dimethylpirazole phosphate (DMPP).

Material and methods

Trials were carried out in summer time (average environmental temperatures were in the range of 20.6 to 23.6 °C) in open field conditions on loamy-sand and loam undisturbed soils both characterised, at trial start, by a soil moisture close to 50% of their field capacity. Before fertiliser spreading, soil samples were collected for the determination of their main chemical-physical characteristics (Table 1). Analysis were carried out by the laboratory of the Department of Exploitation and Protection of the Agricultural and Forestry Resources (DI.VA.P.R.A) of the University of Turin.

The system used for measurement of NH_3 emission was a modified version of the funnel method (Balsari *et al.*, 1994), a dynamic chambers arranged to collect the air from below the captor (0.138 m^2 surface) with a flow rate of approximately 9 l/min. NH_3 emission from all fertilisers was measured with three replicates in a randomised block design. Each fertiliser was applied

Table 1. Main soils chemical and physical characteristics.

Type of soil	Soil characteristics							
	pH	Clay (%)	Silt (%)	Sand (%)	$CaCO_3$ (%)	C (%)	Total N (%)	CEC (Cmol/kg)
Loamy sand	8.50	3.10	10.60	86.30	17.40	0.35	0.11	0.09
Loam	5.40	12.10	50.90	37.00	absent	1.20	0.14	7.32

on the soil area covered by the captors of the funnel system with an application rate of 100 kg N/ha. Trials had a duration of 21 days. Sampling of acid traps were done after 2, 7, 14 and 21 days after fertiliser application. Significant differences in results were investigated using the ANOVA procedure.

Results and discussion

Ammonia nitrogen losses from all fertilisers spread on the loamy-sand soils were significantly higher than those on the loam soils (Figure 1). The lower loss of NH_3 nitrogen was observed with Ammonium Nitrate in both soils. On loam soil, emission from traditional Urea was significantly higher than those generated by Prilled Urea. No differences between the two types of Urea were observed when applied on loamy-sand soil. Fertiliser with DMPP showed a different behaviour according to soil typology: high NH_3 emission were measured when land spread on loamy-sand soil, while contained emissions were measured when applied on loam soil. The use of nitrification inhibitors had significant effect on NH_3 losses reduction when fertiliser was spread on loam soil. Soil chemical and physical characteristics (such as soil pH and CEC) appear to have affected NH_3 emissions.

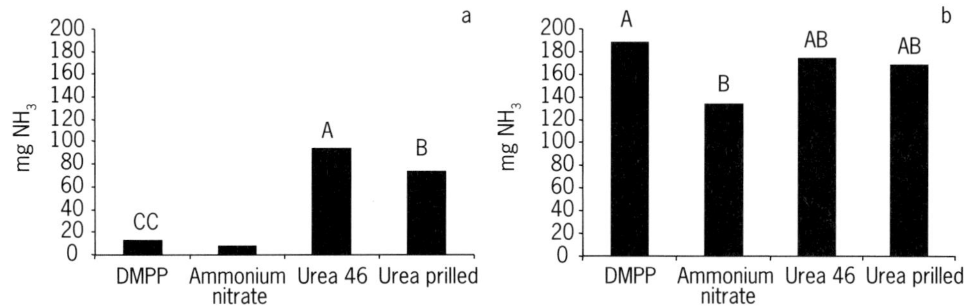

Figure 1. Total ammonia emission recorded during the trial from loam (a) and loamy-sand (b) soils. Means with unequal letters are significantly different (P<0.01).

References

Balsari, P., S. Bechis and F. Girodengo, 1994. First results of tests on ammonia losses along with and after the distribution of slurry in the fields. In: Treatment and re-utilization of farm effluents and sludges. Conte Editore, pp. 507-520.

Follet, R.F., 2001. Nitrogen transformation and transport processes. In: Nitrogen in the Environment: Sources, Problems, and Management. R.F. Follet and J.L. Hatfield (Eds.), Elsevier Science, pp. 17-44.

Menéndez, S., P. Merino, M. Pinto, C. González-Murua and J. M. Estavillo, 2006. 3,4–Dimethylpirazole phosphate effect on nitrous oxide, nitric oxide, ammonia and carbon dioxide emissions from grasslands. J. Env. Qual. 35: 973-981.

Weber, A., R. Gutser, H.-J. Michel, H. Wozniak, G. X. Chen, H.Xu and H.-J. Niclas, 2004. Dicyandiammide and 1H-1,2,4-Triazole – a new effective nitrification inhibitor for reducing nitrous oxide emissions from cultivated land. International conference on Greenhouse Gas Emissions from Agriculture Mitigation Options and Strategies, Leipzig, Germany, pp. 273-275.

Ammonia emissions from sewage sludge management: an underestimated problem?

J. Dach[1] *and F. Nicholson*[2]
[1]*Institute of Agricultural Engineering, Agricultural University of Poznań, Poland*
[2]*ADAS Gleadthorpe, Meden Vale, Mansfield, Nottinghamshire, United Kingdom*

Introduction

The growth of the European population together with increasing concern over human impacts on the environment has led to continuous growth in the quantities of sewage sludge produced by municipal wastewater treatment (EEA, 2005). In the year 2000, the amount of sewage sludge produced in Europe was estimated to be around 17,000,000 tonnes of dry matter (Darees Boucher *et al.*, 1999) and is likely to increase in the future. The total nitrogen (N) content in dry matter is usually 3.5-6%. Thus sewage sludge management introduces about 0.5-1 millions tonnes of total nitrogen each year into the European environment. Depending on the treatment and disposal method, a significant proportion of this nitrogen could be lost to the atmosphere by ammonia (NH_3) volatilisation. Only a small proportion (around 5%) of agricultural land in the EU is currently treated with sewage sludge (EEA, 2005). However, agricultural use of sewage sludge has without doubt many beneficial aspects (Jensen and Jepsen, 2005). The requirements of EU legislation such as the Urban Waste Water Treatment Directive and the Landfill Directive, which limit other disposal options for sewage sludge, may tend to increase application to agricultural land (EEA, 2005). In some countries, 'advanced' treatment processes such as thermal drying or lime stabilisation are being introduced in response to pressures to limit the pathogen content of sludge applied to land, whilst in other countries sewage sludge composting will be the main method of treatment in the future.

Problem identification

At present, sewage sludge management is not considered to be an important source of ammonia emissions. The EMEP/CORINAIR Emission Inventory Guidebook (2006) presents more than 20 treatments or waste disposal routes, including sewage sludge spreading to agricultural land (SNAP code: 091003). The authors estimate the contribution of sewage sludge spreading to the total ammonia emission from 28 European countries to be around 0.1%. They also state that the amount of ammonia produced by sludge spreading is determined by the dry matter content of the sludge and the total amount of ammoniacal nitrogen present. However, there are two factors which could dramatically alter this estimate. New sewage sludge treatment technologies often produce sewage sludge with a much higher dry matter content (composted sludge typically has a dry matter content of around 40%, whilst thermally dried products may contain 95% dry matter) and different NH_4-N contents. The second factor is related to new national and EU regulations which are moving away from the spreading of non-treated sewage sludge on agricultural land and replacing it with 'advanced' treated products such as composted materials. However, in the chapter 'Compost Production from Waste' (SNAP code: 091005) no ammonia emission is reported. This seriously underestimates the potentially very large (on a European scale) quantity of ammonia which could be produced during composting of sewage sludge.

Potential NH_3 emission from sewage sludge

Sewage sludge can be an important source of ammonium from the enzymatic hydrolysis of proteins, urea and uric acid, present in sludge. During decomposition of carbonaceous bulking agents the micro-organisms use part of this NH_4–N for their own metabolism but the remainder can be lost by NH_3 emission (Darees Boucher *et al.*, 1999). This emission is especially important during sewage sludge composting because of air circulation, quick growth of pH and temperature. The theoretically 'easiest' way of decreasing ammonia volatilisation is to increase C/N ratio to 20-30. However, the real-scale practices differ very often from this optimal situation. The addition of carbonaceous amendments (straw, wooden chips, etc.) to achieve a C/N ratio over 20 is expensive, so the C/N ratio is usually between 10 and 12 in composting plants. Thus the low initial C/N ratio is related with high concentrations of NH_4-N throughout the composting process and leads to ammonia volatilisation (Czekała *et al.*, 2006). Also recent experiments carried out at the Agricultural University of Poznań in the framework of 'CleanCompost' UE6 Framework Program have showed the high ammonia losses during composting of sewage sludge in conditions of insufficient carbonaceous amendment.

Conclusion

As the quantities of sludge applied to agricultural land are likely to increase further in the future, it is important to obtain more robust measurements of ammonia emissions following land spreading (research priority mentioned by Sutton *et al.*, 2000) and during sludge treatment prior to land spreading. Present data in national and European Ammonia Emissions Inventories is based on little quantitative information and is not relevant to modern sludge treatment processes and the range of enhanced and conventionally treated sludge products being recycled to farmland (especially composts). With usually low C/N ratio in composted sewage sludge (10-12), the ammonia losses from the real scale treatments are usually higher than the emission from composting of typical agricultural materials like solid manures (where C/N is usually between 18-25). However, lack of complete gaseous emission data in this area underlines the necessity of research concerning ammonia and GHG emissions from sewage sludge management.

References

Czekała, J., J. Dach and A. Wolna-Maruwka, 2006: The usage of bioreactor for modelling of sewage sludge composting. Water-Environmental-Village Areas (in press).

Darees Boucher, V., J.C. Revel, M. Guiresse, M. Kaemmere and J.R. Bailly, 1999 Reducing of ammonia losses by adding $FeCl_3$ during composting of sewage sludge. Water, Air and Soil Pollution 112: 229-239.

EEA, 2005. The European environment - State and outlook 2005. 584 pp.

EMEP/CORINAIR, 2006. Emission Inventory Guidebook. Published by European Environment Agency.

Jensen, J. and S.E. Jepsen, 2005 The production, use and quality of sewage sludge in Denmark. Waste Management 25: 239-247.

Sutton, M.A., U. Dragosits, Y.S. Tang and D. Fowler, 2000. Ammonia emissions from non-agricultural sources in the UK. Atm. Env. 34: 855-869.

Handling and storage

Turnover and losses of NH_3 during manure composting: a critical review

G.L. Szántó[1], H.V.M. Hamelers[1], W.H. Rulkens[1] and A.H.M. Veeken[2]
[1]Sub-department of Environmental Technology, Wageningen University and Research Centre, Bomenweg 2, 6703 HD Wageningen, The Netherlands
[2]Lettinga Associates Foundation, Bomenweg 2, 6703 HD Wageningen, The Netherlands

Introduction

Composting is a well developed treatment for various types of organic wastes, e.g. it is the main waste treatment method for green waste and biowaste. However, because of concerns related to N (nitrogen) losses and NH_3 (ammonia) emissions it is not frequently applied for the treatment of solid manures. In manure composting literature there is a large variation in published values for N losses and NH_3 emissions. Manure composting leads to N losses ranging from 4-60% and from 3-75% of NH_3-$N_{initial}$ for NH_3 emissions. This indicates that insufficient knowledge of NH_3 dynamics is available in order to improve emission control. NH_3 dynamics determine the changes in the NH_3 pool as shown below:

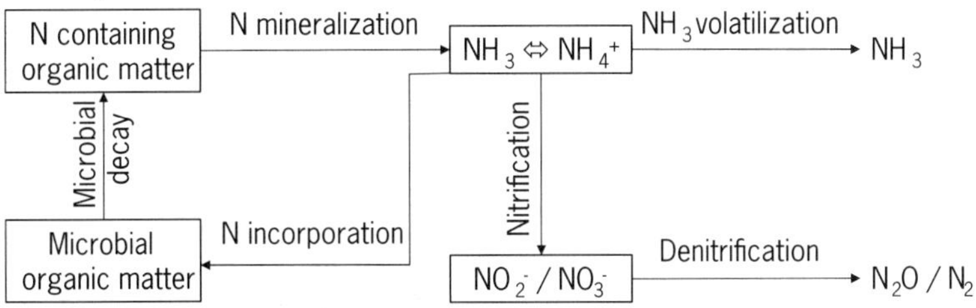

The NH_3 pool in a composting process is supplied by the mineralisation of the N containing organic matter and by the available NH_3 initially present. It may be depleted by incorporation into microbial biomass or conversion to N_2 and/or N_2O through nitrification-denitrification. NH_3 may also be lost to the environment through volatilisation. Microbial decay provides new organic N for mineralisation to NH_3. Analysing the relationships between these processes and main composting parameters is likely to provide a better insight in reducing NH_3 emissions.

Methodology

The reviewed composting designs were classified into three groups: aerobic storage, extensive and intensive composting (Szántó, 2007). Aerobic storage is a process where aeration is led by diffusion and minimal or no control is exercised. This is not surprising as the objective of this 'treatment' is preservation rather than conversion. Agitation (turning and mixing) enhances the natural convection mechanism in extensive composting. Intensive composting methods use forced aeration and exhibit a high level of control. Additionally, intensive composting designs are characterised by the application of closed reactors using gas (NH_3) collection and treatment. The main substrate parameters (C/N ratio, $NH_{3\text{-}initial}$, structure, pH) and process parameters (O_2, temperature, pH, agitation) were correlated to NH_3 emissions and N loss for the three composting designs. Only papers describing an amendment (straw, leaves, etc.) aided manure composting process were reviewed.

Results and discussion

The analysis on the relation of composting parameters and NH_3 dynamics resulted in the following categorisation (more details can be found in Szántó (2007)):

- *N mineralisation:* the initially available NH_3 and the mineralised N from organic matter is in an NH_3-NH_4^+ equilibrium in the compost beds. This equilibrium is directly linked to the pH and temperature parameters of the substrate.
- *NH_3 volatilisation:* The results suggest that next to substrate optimisation, it is the O_2 level and the aeration control (forced or agitated compost bed) of the process that is dominant in NH_3 dynamics. Its influence can only be controlled within limits by the adjustment of parameters such as C/N ratio.
- *Nitrification-denitrification:* Substrate related influence on NH_3 dynamics appeared to depend chiefly on the addition of amendments. Literature showed that next to increasing C/N ratio of the mixture, amendments also improved moisture content and structural parameters (porosity, structural strength) of the compost bed. Agitation appeared to be the dominant process related parameter to improve bed structure. It improved the bed structure resulting in a more porous therefore more aerobic bed suitable for nitrification.
- *N incorporation:* availability of more C from amendments increased the C/N ratio of the compost beds, resulting in a higher degree of N incorporation.

Conclusions

The review showed that the relation between NH_3 dynamics and composting parameters are differing in the three composting groups. C/N ratio optimisation of starting substrates turned out to have more effect on NH_3 losses from storage and extensive composting.

Although in literature the C/N ratio was considered a central parameter in controlling NH_3 emissions, intensive composting studies showed relatively high NH_3 emissions and comparable total N losses with extensive compostings. It is likely that higher aeration rates flushed out initial NH_3 preventing its incorporation. This phenomenon reduced the amount of N for incorporation or nitrification as the NH_3 was collected at these designs.

Increasing the rate of aeration reduced the controlling effect of an optimal C/N ratio on NH_3 retention. Therefore, it is suggested that the concentration and especially the degradation rates of N and C containing solids and concentration of free NH_3 are studied to estimate NH_3 volatilisation instead of the C/N ratio.

As aerobic storage processes showed practically no substrate or process optimisation, the lack of O_2 control led to largely anaerobic beds, resulting in uncontrolled NH_3, but also N_2O and CH_4 emissions. Extensive compostings operating with natural convection offered limited NH_3 volatilisation control (such as turning); therefore reduction of NH_3 emissions could not always be guaranteed. Intensive composting processes using forced aeration did offer gas collection and adequate degradation combined with the possibility of substrate control.

In general, intensive compostings showed higher NH_3 emissions. In extensive composting processes N was lost in the form of N_2 and N_2O due to nitrification-denitrification. The review indicates that process selection supported by substrate and/or process parameter optimisation can considerably influence NH_3 dynamics and total N losses in composting.

References

Szántó, G.L., 2007. NH_3 dynamics in swine manure composting. PhD-thesis, Sub-department of Environmental Technology, Wageningen University, The Netherlands.

Predicting ammonia emissions from carbon and nitrogen biodegradability during animal waste composting

M. Hassouna[1,2], *J.-M. Paillat*[3] *and P. Robin*[1]
[1]*INRA, Agrocampus Rennes, UMR 1069, Sols, Agronomie, Spatialisation, 35000 Rennes, France*
[2]*INRA, UMR1079 Systèmes d'Elevage Nutrition Animale et Humaine, 35590 Saint Gilles, France*
[3]*CIRAD, INRA, Agrocampus Rennes, UMR 1069, 35000 Rennes, France*

Introduction

Composting is a useful technique to facilitate the management of livestock manure, which is one of the major problems in regions of intensive livestock production (Mustin, 1987). However during this process of organic matter transformation, emissions of pollutant gases like NH_3, N_2O, CO_2 and CH_4 (Houghton *et al.*, 2001) can occur. An experimental design was built to quantify the relationship between NH_3 and CO_2 emissions and the nitrogen and carbon forms in the initial composting material.

Material and methods

Eight heaps (1.4 m^3) were constituted with different mixtures of wheat straw, sawdust, pig slurry and manure, sugarbeet molasses, urea and water and placed in isolated enclosures (8 m^3) in a controlled climate building (Paillat *et al.*, 2005). These 8 heaps represented the range of biodegradability of nitrogen and carbon in the livestock manure (Figure 1). We monitored NH_3, N_2O, CO_2, CH_4 and H_2O emissions during the thermophilic composting phase (until 56 days) using multigas analysers (Innova 1312, Brüel et Kjaer 3426 and a infrared CO_2 sensor Gascard II Edinburgh sensors).

Results and discussion

Four parameters were used to describe the NH_3 emission kinetics (Figure 2): (1) the response time to reach maximum intensity, mainly affected by the initial micro-flora; (2) the amplitude of the peak, mainly depended on C biodegradability and also on initial micro-flora; (3) the emission

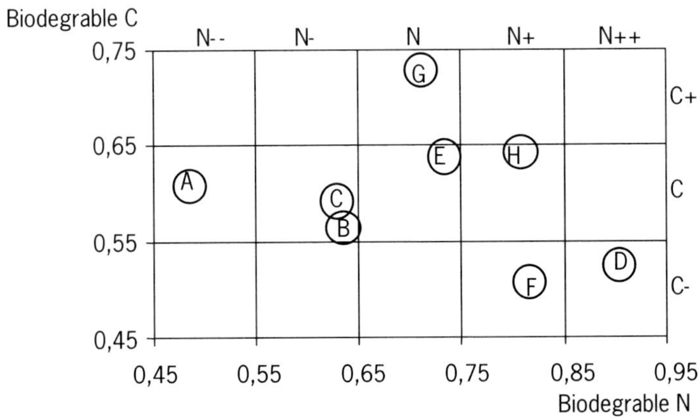

Figure 1. Classification of mixtures from their available N (SN:TN) and biodegradable C.

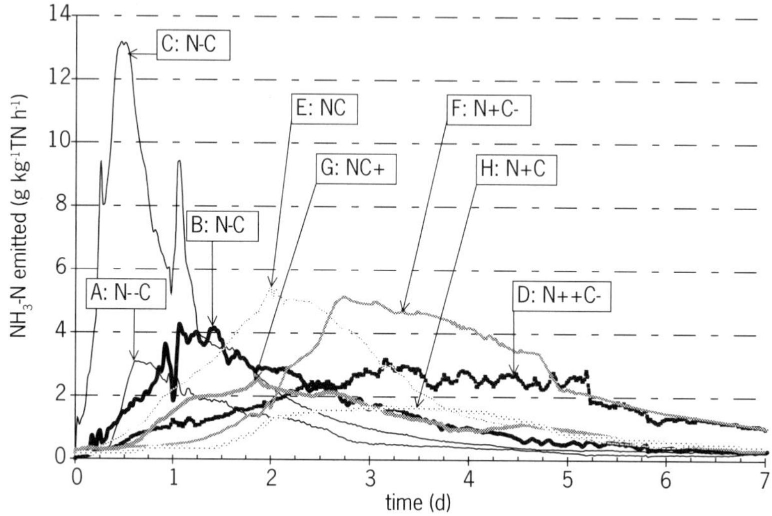

Figure 2. Kinetics of NH_3-N emission for the eight piles (A to H) as influenced by microflora and the N and C biodegradability (+ -) of the initial substrate.

duration, mainly depended on N biodegradability; and (4) the cumulative emission, which varied from 16.5% to 48.9% of the initial nitrogen depending both on C and N biodegradability (Paillat *et al.*, 2005). A predictive model for NH_3 emissions for the thermophilic phase was proposed. The variability in total NH_3 emissions is well explained by the contents of soluble elements and hemicellulose in the dry matter (SH-vs Van Soest fractioning), and soluble nitrogen (SN 12 h extraction at 4 °C in water). It is given by the following equation:

NH_3-N emitted (in g kg^{-1} TN) at 56 days = 16.38 SN –0.903 SH-vs + 643.7
(N=8; P<0.05; R^2=0.82).

References

Houghton, J.T., Y. Ding, D.J. Griggs, M. Noguer, P.J. Van der Linden, X. Dai, K. Maskell and C.A. Johnson, 2001. Climate change 2001: the scientific basis. Report of the IPCC working group I, 96p.

Mustin, M. 1987. Le compost, gestion de la matière organique. Edition François Dubusc, Paris, 954 pp.

Paillat, J.M, P. Robin, M. Hassouna and P. Leterme, 2005. Predicting ammonia and carbon dioxide emissions from carbon and nitrogen biodegradability during animal waste composting. Atmos. Environ. 39: 6833-6842.

Investigations into ammonia emissions during composting

I. Körner
Hamburg University of Technology; Institute of Waste Resource Management, Harburger Schloßstraße 36, 21079 Hamburg, Germany

Introduction

During composting, NH_4^+/NH_3 develops through ammonification mainly from the proteins contained in waste. In the course of the composting process, NH_4^+/NH_3 can be converted into NO_2^-/NO_3^-, immobilised into organic compounds, leached, or stripped as NH_3. The NH_3 emission via the exhaust air is affected by means of aeration or, in a restricted manner, by natural convection. As the waste employed for composting and the process controls are extremely variable, a universally applicable data basis for NH_3 emissions during composting is not yet available. Therefore, they have been examined in a total of 53 composting tests for the most widely differing substrates and composting conditions.

Experimental procedure

The test set-up of the pilot plant with its 100 l-bioreactors, as well as the measurement technique used and the chemical analysis were described in Körner *et al.* (2003). Real waste (biowaste, digestion residues, MSW, chicken excrement, monowaste, sewage sludge) and well defined model waste (consisting of 7–10 of the following components: apples, potatoes, turnips, wheat, peas, meat and bone meal, wood, bark, straw, leaves, grass, sand) was applied. The process control varied as far as aeration rates, temperature profiles, water contents, portions of structural material and pH values are concerned. By means of the investigations, N balances were established in accordance with the method indicated in Körner and Stegmann (2002). In this publication, the NH_3 emissions during the aerated composting phases were evaluated with regard to the cumulative NH_3-N emission per test series and to the average NH_3-N emission rates per measuring phase (period between two substrate samplings).

Results and discussion

NH_3 was quantitatively the most important discharged N compound. In all test series, NH_3-N emissions could be observed in extremely different ranges (0.1–88.2% N_{total}-0 or 0.004–2.5% DM)[1]. The test series could be divided into categories with regard to their total NH_3-N emissions (Table 1). NH_3-N could be detected in the exhaust air during approx. 80% of all measuring phases. In the remaining measuring phases, the NH_3-N emissions were not recorded due to their insignificance. In the 'Moderate' and 'Much' categories, the average emission rates were comparable and significantly higher than in the other two categories. A correlation between the total emission per test series and the maximum emission rates could be ascertained (Table 1). The NH_3-N emissions followed the scheme presented in Figure 1, usually starting immediately with the beginning of composting. In only 21% of the test series, emissions started later. In 96% of the test series, the average NH_3-N emission rates first increased to the maximum which lasted during the entire intensive rotting phase and decreased afterwards. However, the chronological course

[1] The discharge was related to the dry matter and not to the total N in the waste, since it turned out that the type of waste is much more important than the amount of the contained N.

was, in different ways, marked by fluctuations. 38% of the test series showed no fluctuations or only slight fluctuations up to ±0.001% DM/d. In 9% of the test series, the fluctuations were found to be in the range of ±0.005% DM/d and in 25% of the test series in the range of ±0.02% DM/d. Even in an advanced stage of composting, mostly low NH_3-N emissions could be detected. They were in the range between 0.00002 and 0.0005% DM/d (25-75% quantile). A concentration of 0.0005% DM/d corresponded to an average of approx. 2.7 mg NH_3-N/m^3 air. The NH_3-N emissions were determined mainly by the portion of NH_4^+/NH_3-N in the substrate. The latter in turn could be attributed to the additional delivery from the ammonification of the N_{org}. The ammonification rates were determined mainly by the ratio of persistant to easily degradable compounds in the substrate (CL_{sub}). The N_{org} portion in the substrate was less relevant. A very high share of persistant components (CL_{sub}=0-1) caused low N_{org} degradation rates and, presumably, increased immobilisation of NH_4^+/NH_3-N. A well-balanced ratio between easily degradable and persistant components (CL_{sub}=1-2) tended to result in high N_{org} degradation rates under favourable ambient conditions, in connection with a rather slight NH_4^+/NH_3-N immobilisation. A very high portion of easily degradable substances (CL_{sub}=4) in turn caused rather lower N_{org} degradation rates as a result of degradation limitations caused by suboptimal ambient conditions. Furthermore, the parameters temperature, pH value, aeration rate and

Table 1. Cumulative total NH_3-N discharges per test series and average NH_3-N discharge rates of a measuring phase, determined in 53 composting experiments.

Category	Test series		ΣNH_3-N discharge in % DM	NH_3-N discharge rate in % DM/d	
	Number	Portion in %		25-75% quantile	maximum
Not significant	7	13.2	<0.025	0.0000–0.0005	0.0002–0.0013
Significant	23	43.4	0.025–0.74	0.0001–0.0068	0.002–0.08
Moderate	15	28.3	0.75–1.5	0.002–0.0229	0.01–0.07
Much	8	15.1	1.5–2.0	0.001–0.0253	0.03–0.1

DM-0: Dry matter of the initial substrate

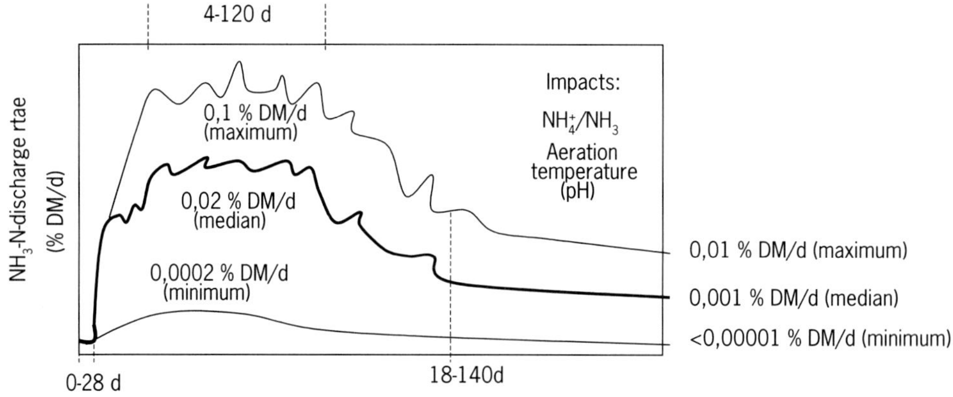

Figure 1. Typical course of NH_3-N discharges during composting.

substrate structure exerted an influence on the NH_3-N emissions. The fluctuations regarding the NH_3-N emission rates during the intensive rotting phase were attributed to the fact that some of these parameters showed alternating scales during composting.

Conclusions

The NH_3-N emission during composting differ significantly in the duration of the emission period and in the maximum emission rates. On the one hand, NH_3 emissions are undesirable, as they are odour-intensive and N is no longer available to the compost as a nutrient. On the other, the NH_3 emissions offer the possibility to recover N in a plant-available form. In either case, the basic data regarding the NH_3-N emission offers the possibility to develop situation-adjusted process controls.

References

Körner, I., J. Braukmeier, J. Heerenklage, K. Leikam, M. Ritzkowski, M. Schlegelmilch and R. Stegmann, 2003. Investigation and optimization of composting processes - Test systems and practical applications. Waste Manag. 23: 17-26.

Körner, I. and R. Stegmann, 2002. N-Dynamics during composting – Overview and experimental results. In: Microbiology of Composting. H. Insam, N. Riddech and S. Klammer (Eds.), Springer-Verlag, Berlin, pp. 143-154.

Emissions of ammonia from a plant for pig slurry processing on sawdust

F. Gioelli, E. Dinuccio, E. Santoro and P. Balsari
Università degli Studi di Torino, DEIAFA, Sez. meccanica, via Leonardo da Vinci 44, 10095 Grugliasco (To), Italy

Introduction

Under the Nitrates Directive (91/676/EC), Nitrate Vulnerable Zones (NVZ) are being established throughout Europe. As other European and Italian regions, some areas of Piemonte are characterised by intensive animal husbandry and often show water contamination due to high nitrates concentration (Regione Piemonte, 2002). Since slurry mainly consists of water, it should be concentrated before transportation to other regions to reduce the resulting transportation costs (Pieters *et al.*, 1999). Furthermore, the odours emission during liquid manure spreading and the reluctance of arable farmers to accept rough animal manures on their lands increases the problem of their correct agronomic reuse. For such reasons, technologies able to transform liquid manure into solid and nutrient rich manures are nowadays meeting the farmers approval and are rapidly spreading all over the national territory. Nevertheless, at present few data are available on the sustainability of such technologies. With the aim to partially cover this gap in knowledge, a plant for slurry processing on sawdust was monitored over two seasons.

Materials and methods

A plant for pig slurry processing in combination with wood sawdust located in the North Western Italy was monitored. The plant is made up of a rectangular platform (10m wide, 60m long, 1.3m high) in reinforced concrete filled with untreated sawdust (~500m^3) and covered with a roof to protect the biomass against rain. A self-propelled machine moving onto rails spreads the slurry from a fattening house over the bed of sawdust, mixes and aerates the biomass twice a day to promote the aerobic degradation of the organic matter. The reaction produces heat allowing the evaporation of the water contained in the slurry so that a solid product (DM >20%) can be obtained. The plant monitoring was repeated in two seasons (winter and summer). The biomass and environmental temperatures were monitored over the two periods of processing and ammonia emissions were measured by means of a set of three wind tunnels (Schmidt and Bicudo, 2002). In order to determine the percentage of nitrogen lost from the biomass as ammonia, the amount and the chemical characteristics of the rough input materials (sawdust and pig slurry) and of the output materials (exhausted material) were determined. A general evaluation of the plant performance in terms of its ability to evaporate water from the slurry was also conducted.

Results

The main chemical characteristics of the input and output materials are shown in Table 1. The temperatures of the biomass increased rapidly during the first two weeks from the activation of the plant (up to 65 °C and 55 °C in summer and winter respectively), then in summer remained steady until the saturation of the sawdust, whereas in winter dropped off to 45 °C after only 40 days of treatment. A big difference in the plant performance over the two seasons was observed.

Greater amounts of slurry per unit of substrate were disposed of under summer conditions (Table 2). As expected, ammonia losses were higher in summer conditions and increased with the progressive saturation of the sawdust bed with the slurry. The highest emission rate was observed after the substrate mixing and aeration phase.

Table 1. Main characteristics of the rough input product and of the output materials.

Season		ST (%)	TKN (%)	NNH_3 (%)	pH (%)
Winter	sawdust	77.2	0.07	-	5.0
	slurry	2.6	0.38	0.23	6.6
	final product	21.1	n.d.	n.d.	7.6
Summer	sawdust	84.6	0.07	-	4.9
	slurry	2.8	0.34	0.18	6.9
	final product	24.8	0.54	0.32	8.4

Table 2. Main results of the monitoring.

	Avg. environment temperature °C	Avg. biomass temperature °C	applied slurry/sawdust		$N\text{-}NH_3$ emission on applied TKN %
			$l\ m^{-3}\ day^{-1}$	$l\ m^{-2} day^{-1}$	
Winter	5.5	41	7.8	8.6	9.4
Summer	24	52	12.5	13.8	23.5

Conclusion

The final product, being characterised by a higher total solids and nutrients content compared to the rough slurry, can be more conveniently transported over long distances. Nevertheless, the investment cost for the plant construction (~180,000€ for a $600m^2$ platform plant) is very high, especially considering that a surface of at least 1 m^2 is required for the disposal of the daily slurry production of a 100kg weight pig. Furthermore, the process determines high ammonia losses to air (especially in summer conditions) due to the mass mixing and the temperature developed during the process. In order to have a whole picture of the plant sustainability, an evaluation of the GHG emissions is currently in progress.

References

Pieters, J.G., G.G.J. Neukermans and M.B.A. Colambeen, 1999. Farm scale filtration of sow slurry. J. Agric. Eng. Res. 73: 403-409.

Regione Piemonte, 2002. Decreto della Giunta Regionale 18 Ottobre 2002, n. 9/R, 55 pp.

Schmidt, D.R. and J.R. Bicudo, 2002. Using a wind tunnel to determine odor and gas fluxes from manure surfaces. Paper No. 024083 presented at XXXII ASAE-CIGR Meeting, Chicago, USA.

Farm-scale evaluation of three cover systems for reducing ammonia emissions from swine manure stores and subsequent land spreading

L. Loyon, F. Guiziou, S. Picard and P. Saint-Cast
Environmental Management and Biological Treatment of Wastes Research Unit, Cemagref, 17 avenue de Cucillé, 35044 Rennes Cedex, France

Introduction

Covering manure stores is a technique to achieve a reduction of NH_3 emissions (Sommer *et al.*, 1993). However few data are available at the farm scale and on the impacts on emissions from the subsequent land spreading. The aim of this study was to assess the efficiency of three cover materials (peat, polystyrene balls and synthetic sheet) to reduce NH_3 emissions during pig slurry storage and subsequent land spreading. The impact of the covers on greenhouse gas emissions (CH_4, CO_2 and N_2O) during the storage was also measured.

Materials and methods

On-farm measurements were carried out for 7-10 weeks over one or two seasons. Experiments with peat and polystyrene balls were carried out at a farm equipped with two below-ground rectangular concrete stores. One store was covered with peat or polystyrene balls (∅:20cm) and the second one was left uncovered as the control. Gaseous emissions (NH_3, N_2O, CH_4, and CO_2) were measured using 2 dynamic chambers and an infrared analyser (Innova 1412). Assessment of the synthetic sheet was carried out at a farm storing slurry in a 1000 m^3 covered store. Gaseous emissions were determined by measuring the concentrations in the headspace under the cover and the air leakage with a tracer gas (CO_2). Ammonia volatilisation following slurry application was measured for three days using a wind tunnel. Slurry was manually applied (50 m^3 ha^{-1}) using a splash-plate spreading technique.

Results and discussion

Slurry characteristics and gaseous measurements from slurry stores and spreading are given Table 1. The results indicated a reduction of about 25% of NH_3 from the peat covered store with an increase of 30% of CH_4 and CO_2 emissions. The spreading step resulted in a reduction of around 8% of NH_3 emissions. However some results (not mentioned in this paper) suggested a higher volatilisation of NH_3 or less CH_4/CO_2 when the peat cover was wet or dry, respectively. Analysis of such variability revealed no significant difference (Student's Test, 0.05) between the uncovered and the peat covered stores. The use of a layer of polystyrene balls reduced ammonia emissions during storage by up to 80% whatever the season. Nevertheless the use of the cover in summer indicated a potential increase in greenhouse gas emission of 20%. No difference was perceived during the spreading operation in winter but a reduction was observed in summer probably due to the high volatile fatty acid level in the covered slurry. The experiment with the synthetic sheet indicates very low emissions factors compared to those of the other experiments. Based on these results a reduction of 97% was achieved compared with emission from the control slurry of the polystyrene ball experiment. These results agreed with other studies (Sneath *et al.*, 2006; Peterson *et al.*, 2004). The spreading of the covered slurry didn't indicate higher NH_3 volatilisation.

Table 1. Slurry characteristics, mean gaseous emissions from uncovered and covered pig slurry stores and ammonia emission following surface-application of covered slurry (given as the % of applied TAN).

	Peat cover (autumn)		Polystyrene balls (winter/ summer)		Synthetic sheet (winter/ summer)
	Control slurry	Covered slurry	Control slurry	Covered slurry	Covered slurry
TAN ($gN\ kg^{-1}$)	2.7	2.3	3.2/2.8	2.9/3.7	2.2/2.4
TKN ($gN\ kg^{-1}$)	4.2	3.3	4.8/4.1	4.3/5.5	2.4/4.7
DM ($g\ kg^{-1}$)	4.9	38	46/45	47/60	8/71
OM ($g\ kg^{-1}$)	35	25	32/31	32/43	3/51
pH	7.6	7.6	7.8/7.8	7.9/7.6	8.2/7.6
Vol. Fat. Acid ($g\ l^{-1}$)	2	0.3	1.1/1.3	0.4/5.1	NM/0.1
NH_3 ($gN\ m^{-2}\ d^{-1}$)	4.1	3.1	3.4/3.5	0.5/0.8	0.013/0.006
N_2O ($gN\ m^{-3}\ d^{-1}$)	ND				
CH_4 ($gC\ m^{-3}\ d^{-1}$)	57	76	67/90	69/136	0.3/1.2
CO_2 ($gC\ m^{-3}\ d^{-1}$)	45	59	35/239	36/114	0.3/1.3
NH_3 from spreading[a]	35	33	29/30	28/8	21/29

[a]Mean of 3 trials, expressed in % applied TAN, ND: no detection, NM: not measured.

References

Peterson, J., J. Agnew, C. Laguë and E. Gaudet, 2004. Greenhouse gas emissions from a NAP-covered earthen manure storage basin. North Central ASAE/CSAE Conference, Annual Intersectional Meeting, September 24 -25, 2004

Sommer, S.G., B.T. Christensen, N.E Nielson and J.K Schjørring, 1993: Ammonia volatilization during storage of cattle and pig slurry: effect of surface cover. J. Ag. Sc. Cambridge 121: 63-71.

Sneath, R.W., F. Beline, MA. Hilhorst and P. Peu, 2006. Monitoring GHG from manure stores on organic and conventional dairy farms. Agr. Ecosystems & Env. 112: 122-128.

Air purification

Combined air scrubbers in livestock production: performances and costs

N.W.M. Ogink and A.J.J. Bosma
Wageningen UR, Animal Sciences Group, P.O. Box 65, 8200 AB Lelystad, The Netherlands

Introduction

Intensive poultry and pig operations concentrated in the south and east of The Netherlands are major contributors to both ammonia, odour and PM_{10} emissions. Current environmental regulatory standards on ammonia, odour and PM_{10} emissions restrict economically preferred scales of operations for many of these farms. An innovation and implementation programme has been set up by the national government to introduce and stimulate the use of a new generation of combined air scrubbers on livestock operations that drastically reduce emissions of both ammonia, odour and dust. Combined air scrubbers consist of two or more scrubbing phases, using different removal principles, to reach high removal efficiencies for all of the mentioned emissions. The aim of this programme is to support environmental policy makers in meeting the national emissions targets, and to provide technical solutions for farmers that enable them to operate at scales of operation demanded by market conditions. Objectives of this paper are to provide:

- Overview of principles applied in combined air scrubber for livestock operations and available removal performances for ammonia, odour and PM_{10}.
- An assessment of investment levels comparing combined air scrubbers, conventional air scrubbers and low emitting housing systems.

Principles applied in combined air scrubbers

Since the nineties, air scrubbers are occasionally implemented on intensive livestock operations to minimise ammonia emissions for the protection of nearby located sensitive ecosystems. Two types of scrubbers are commonly applied: chemical scrubbers and biotrickling filters. Chemical scrubbers are based on the entrapment of ammonia in acid scrubbing liquid and the discharge of ammonium salt solutions. Normally sulphuric acid is applied at pH<4. Melse and Ogink (2005) reported overall average removal efficiency on farm operations of 96%. Removal efficiency for odour was only 27%. In biotrickling filters, ammonia is converted by microbial mass into nitrite and nitrate and regularly discharged from the recirculation liquid. Average ammonia removal efficiency at farm operations amounted 70%, whereas for odour removal an average efficiency of 51% was reported by Melse and Ogink (2005). All these installations are based on single scrubbing wall designs. In the late nineties, combined air scrubbers for livestock operations were initially developed in Germany to ensure a high and sustainable removal of livestock odour. The potentially high odour removal of biofilters was made sustainable by pre-treating incoming air by an acid scrubber thus avoiding high nitrite/nitrate salt formations and subsequent degradation of microbes in the biofilter, avoiding high dust loads that increase pressure differences over the biofilter, and avoiding drying out of the biofilter materials by the moistening of air in the scrubbers. In a later development three phase installations were developed based on three vertical washing walls, where the first wall consists of packing material over which water is re-circulated to mainly remove dust, the second wall operates as an acid scrubber to remove ammonia, and the third wall is designed as an vertical placed biofilter to remove odour. Two-phase setups have been worked out from this concept leaving out the biofilter, where an acid-based treatment is followed by a water-based treatment to remove remaining odours, and where discharge water of the water treatment is used as recirculation liquid in the first acid-based phase. Currently both a two-phase and three-phase combined air scrubber were admitted to the regulatory list of low emission housing systems in The Netherlands, with an assigned ammonia removal efficiency of 85 and 70%, and provisional odour removal efficiencies of 70 and 80% respectively. Removal of PM_{10} fractions have not been reported so far.

Overview of investment levels

The perspectives for a successful implementation of combined air scrubbers in practice depend on the balance between additional technical performances, creating additional scale of operation opportunities for farms, and additional costs. Ogink *et al.* (2007) compared additional investment levels between low emitting housing systems, chemical air scrubbers and combined air scrubbers for main animal categories both for new constructions and modification of existing housing constructions (Table 1). Their results show that for the main pig categories chemical scrubbers for new constructions require lower investments than low emitting housing systems. Combined scrubbers in general require a 25-35% higher investment compared to chemical scrubbers.

Table 1. Additional investment requirements in Euros per animal place for conventional housing systems by implementing best available animal house adaptation for low ammonia emission ('Low NH_3 pen'), single phase chemical scrubber and combined air scrubber; distinguished between animal categories and new constructions or modification of existing constructions.

Animal category	Construction	Low NH_3 pen (€)	Chem. scrubber (€)	Combined scrubber (€)
Sows	New	178	165	220
	Mod.	193	233	288
Fattening pigs	New	38	35	47
	Mod.	30	47	59
Layers	New	0.6	3.8	5.1
	Mod.	0.6	4.0	5.3
Broilers	New	0.8	4.7	6.3
	Mod.	0.8	4.9	6.6

References

Melse, R.W. and N.W.M. Ogink, 2005. Air Scrubbing Techniques for Ammonia and Odor Reduction at Livestock Operations: Review of On-Farm Research in the Netherlands. Transactions of the ASAE. 48: 2303-2313.

Ogink, N.W.M., A.J.J. Bosma, R. Hoste and P.L.M. van Horne, 2007. Prognose investeringsomvang benodigd voor implementatie gecombineerde luchtwassers [Prognosis of investment levels required for implementation of combined air scrubbers in The Netherlands]. Animal Sciences Group of Wageningen UR, report in press.

Trickle bed reactors: optimisation of the washing water processing to increase the ammonia removal efficiency

R. Kosch, M. Olberding, H.F.A. Van den Weghe
University of Göttingen, Research Centre for Animal Production and Technology, Universitätsstr. 7, 49377 Vechta, Germany

Introduction

Air scrubbers are increasingly used for the reduction of ammonia emissions from pig finishing units in Germany in order to maintain the planning and building permission in accordance to the Technical Instructions on Air Quality Control (TA Luft). They can also be considered as a measure to achieve the Guidelines of the UN/ECE-Protocol (2002). The most common scrubber technique is a trickle bed reactor, which can operate either chemically or biologically. Chemical scrubbers show higher ammonia removal efficiency but also cause higher costs due to acid consumption. In order to operate a durable biological scrubber with high removal efficiency, a high elutriation rate is necessary as it is assumed that nitrification and denitrification processes do not take place above a concentration of 3 g N/L (Hahne *et al.*, 2005). At higher concentrations in the washing water and low ammonia concentrations in the waste gas, stripping effects can appear (Kosch, 2005). In this study a trickle bed reactor is being optimised in consideration of three aspects: (1) comparison of biological and chemical operation mode, (2) influence of flood irrigation density, and (3) discharge of ammonia by aerosols, which are not detected by the common measuring techniques (e.g. photoacoustic monitor) with Impinger measurements.

Material and methods

The investigation was carried out with a down stream trickle bed reactor of the company Weda, Germany and was dimensioned for a pig fattening unit with 2000 animals. The assigned carrier material and the drift eliminator of the scrubber had a lattice structure and a volume of 65 m^3. The volume loading in experiments averaged out to 350 and 250 g NH_3 per m^3 carrier material per day. The volume of the washing water was 58 m^3. The biological experiment was conducted in two periods of 11 weeks each, and the washing water was inoculated with activated sludge. The chemical experiment, in which the washing water was adjusted by the use of sulphuric acid to pH 5, ran for two periods of 6 days respectively. The daily elutriation rate of the collection basin amounted to 160 L in all experiments. The gas concentrations in waste and clean gas (NH_3, N_2O, CH_4, CO_2) were taken up with the photoacoustic multigasmonitor 1312 and the multiplexer 1303 (INNOVA, Denmark). NH_4^+, NO_2^- and NO_3^- contents of the washing water were measured intermittently. The discharge of ammonia by aerosols, which are not detected by the photoacoustic monitor, was measured with Impinger bottles filled with acidified water. The absorbed amount of ammonia was detected photometrically.

Results

The averaged ammonia removal efficiency for the periods of the biological mode was 62% and 63%, and 90% in both periods of the chemical mode. Although the averaged removal efficiency of the two biological experiments was very similar, the removal efficiencies of their starting phases were different. A lower efficiency at the beginning of the second experiment can be traced back to the lower added amount of activated sludge, resulting in the biology of the washing water having to be built up in the first three weeks. The formation of N_2O-N amounted 5.9% and 4.4% of the removed NH_3-N in the biological mode, whereas no formation was found in the chemical mode. The ratio of NH_4^+-N:NO_2^--N and NO_3^--N in the washing water in both experiments were similar to each other (1:0.7:0.06). The total nitrogen concentration increased significantly during the first five weeks up to 7.0 g N/L and 5.0 g N/L, and remained at this level until the termination of the experiments. In the chemical mode mainly the ammonium concentration increased, while the nitrite and nitrate concentration remained on a very low level (<0.2 g/L). The averaged acid consumption amounted to 65 kg per day (3.5 kg H_2SO_4/ kg NH_3-N). One remaining question is whether, if the low elutriation rate can be maintained for a longer experiment period without the appearance of stripping effects. The flood irrigation density (Level 1: 10 min. on, 5 min. off; flow rate 45 m^3/h; Level 2: continuous 68 m^3/h, Level 3: continuous 110m^3/h) didn't have a significant effect on the efficiency of ammonia removal. The impingement measurements showed that 7% and 10% of the ammonia nitrogen left the trickle bed reactor in form of aerosols. In consideration of these results the removal efficiency of the biological experiment decreases from 60% to 53% and 52%.

Discussion and conclusions

The ammonia removal efficiency of the chemical experiment was significant higher than in the biological experiment (90% vs. 60 %) and involves no NO_2-Formation. On the other hand, the operating costs in the chemical mode are high due to the high consumption of concentrated sulphuric acid (3.5 kg H_2SO_4/kg NH_3-N). Other than expected, the biological mode did not collapse, and showed a continuous removal process of 60%, despite the low elutriation rate and the high total nitrogen concentration. This was significantly above the maximum concentration of 3 g N/L mentioned by Hahne *et al.* (2005). This mode appears to have the potential to be optimised, provided that the N_2O-formation is low. It was shown that a significant part of the nitrogen left the reactor in form of aerosols, making the use of an appropriate drift eliminator necessary. A significant influence of flood irrigation density on the ammonia removal efficiency could not be observed (ammonia removal efficiencies of 57 to 62%). Thus operating costs can be cut by reducing irrigation at this level of ammonia removal. Impingement measurements verified a substantial discharge of nitrogen in form of aerosols, which are not detectable with common measuring devices. This emphasises the need for utilising an effective drift eliminator.

References

Hahne, J., W. Asendorf and K.-D. Vorlop, 2005. Ammonia separation at waste gas scrubbers (German). In: 7th International Conference: Construction, Engineering and Environment in Livestock Farming 2005. Darmstadt, Landwirtschaftsverlag Münster-Hiltrup, pp. 271-276.

Kosch, R., V. Siemers and H. Van den Weghe, 2005. Efficiency of a Bioscrubber System for the Reduction of Ammonia and Dust Emissions in a Broiler House; ASAE Meeting Presentation; Paper Number: 054164.

Schütze, G., H.-D. Gregor, T. Spranger and H.D. Nagel, 2001. Ökologische Wirkung von Ammoniak. [ecological valuation of ammonia]. In: Emissionen in der Tierhaltung; KTBL-Schrift 406; KTBL, Darmstadt, pp. 9-23.

UN/ECE, 2002. Protocol to the 1979 convention on long-range transboundary air pollution to abate acidification, eutrophication and groundlevel ozone.

Ammonia and odour removal from waste gases of composting processes

M. Schlegelmilch, J. Streese-Kleeberg and R. Stegmann
Hamburg University of Technology, Institute of Waste Resource Management, Harburger Schlossstr. 36, 21079 Hamburg, Germany

Introduction

Ammonia is one of the main issues related to gaseous emissions from composting and anaerobic digestion processes. A major problem is the acidification of the biofilter media caused by ammonia uptake, which may lead to a malfunction of the biofilter, which is well established for the treatment of odorous emissions from waste gases of biological waste treatment processes. To guarantee a reliable operation of the biofilter ammonia should be removed from the waste gas prior to this treatment step. Besides the investigation of odour release during composting and its removal in a bioscrubber/biofilter combination, focus was set on the question to what extend a bioscrubber is able to reduce ammonia emissions.

Materials and methods

Biological waste gas treatment: A bench-scale plant consisting of a bioscrubber/biofilter combination with a modified 1.1 m^3 waste container acting as a composting reactor which produced odorous emissions was used for the experiments. Fresh biowaste was fed into the reactor and turned after 2 and 5 weeks of composting. The bioscrubber/biofilter combination was fed with waste gas from this process at a flow rate of 2 m^3/h. The bioscrubber (2.25 m high, ∅ 80 mm) consisted of two stages (each 0.5 m high) filled with carrier material (HiFlow pall rings, ∅ 15 mm). Water was circulated through the column counter-current to the air flow and regenerated in a 60 L water tank, acting as a bioreactor. The biofilter (1.75 m high, ∅ 0.25 m) was divided into two stages, each filled with about 22 L of screened yard waste compost (>10 mm) as biofilter material (filter load: 45 m^3/m^3h, superficial velocity: 0.01 m/s).
Analyses: The odour concentrations (odour unit per m^3 air [OU/m^3]) were determined according to the DIN EN 13725 (2003) with an olfactometer (TO 6, Mannebeck, Kiel, Germany). Sample bags made of polyterephtalic ester (Nalophan®, K. Nalo, Wiesbaden, Germany) were used. The ammonia concentration was determined with Dräger test tubes, ammonium using photometry (Spectroquant® photometer, Merck).

Results and discussion

In accordance to Krzymien *et al.* (1999), this experiment revealed that critical odour concentrations are released mainly during the first 2-3 weeks of the composting process (Figure 1a. It reaches its maximum after about a week, and slowly starts to decrease to concentrations of about 3,000 OU/m^3, and even below 1,000 OU/m^3 during the last 2 weeks of composting. Figure 1a also shows the odour degradation of the bioscrubber and the biofilter, respectively. While during the first 2 weeks of composting with its high input odour concentrations the degradation efficiency of the whole test plant was about 80%, it decreased to about 30% with lower input concentrations in the curing phase. During that time it was not unusual that the degradation rate of the bioscrubber turned out to be negative, meaning that low input concentrations resulted in higher output concentrations of the bioscrubber. In this case odorous substances are accumulated in the scrubbing liquid, and released with a certain delay. This phenomenon also applies to ammonia (Figure 1b). During this experiment ammonia nearly was completely removed from the waste gas using the bioscrubber/biofilter combination. In the first phase ammonia was mainly accumulated in the bioscrubber and was stripped out and released back into the waste gas flow at a lower concentration level over a longer period of time to be degraded in the biofilter. Comparing the ammonia (Figure 1b) with the odour concentration (Figure 1a) in the exhaust air it can be seen that in this case the ammonia concentration is not a suitable indicator for the level of odorous emissions from composting processes. While the odour concentration had its maximum after 7 days the highest ammonia release from the biowaste was detected after 16 days.

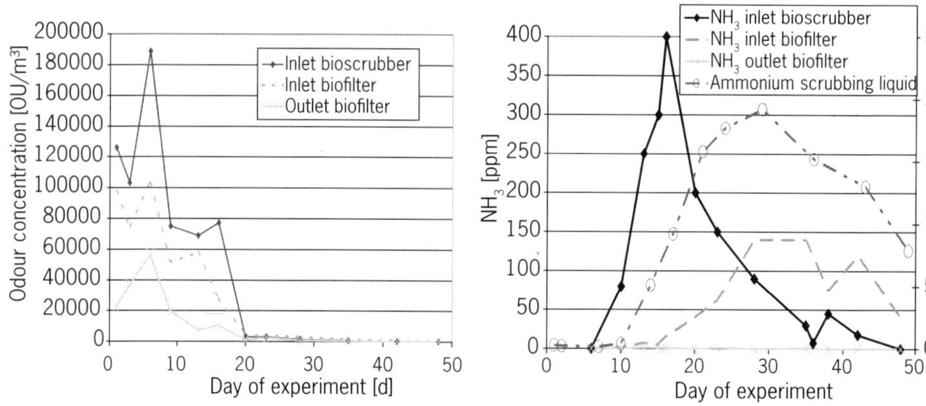

Figure 1. (a) Odour and (b) ammonia (NH_3) concentrations in the waste gas at selected sampling points and (also b) ammonium (NH_4^+) concentrations in the scrubbing liquid during a composting cycle over a period of 7 weeks.

Conclusions

The bioscrubber/biofilter combination has been used effectively to treat odours from waste gas, with the bioscrubber acting as a buffer for odorous substances like e.g. ammonia. However, ammonia could not completely be removed before entering the biofilter. In further studies a H_2SO_4-scrubber is used as a promising alternative, with the additional benefit that the saturated scrubbing liquid might be used as fertiliser if suitably processed.

References

DIN EN 13725, 2003. Luftbeschaffenheit – Bestimmung der Geruchsstoffkonzentration mit dynamischer Olfaktometrie [Air quality – Determination of odour concentration by dynamic olfactometry], Deutsche Fassung EN 13725:2003. Beuth Verlag, Berlin, Germany.

Krzymien, M., M. Day, K. Shaw and L. Zaremba, 1999. An investigation of odors and volatile organic compounds released during composting. J. Air & Waste Man. Ass. 49: 804-813.

Biofiltration with rockwool for removal of ammonia emitted from livestock farming

T. Yasuda, K. Kuroda, Y. Fukumoto, D. Hanajima and K. Suzuki
National Institute of Livestock and Grassland Science (NILGS), 2 Ikenodai, Tsukuba, 305-0901, Japan

Introduction

Composting is the traditional treatment of livestock manure and organic wastes in Japan. However, malodour containing high NH_3 is discharged from the composting process and causes serious environmental problems in surrounding areas. For example, 10-25% of total N was estimated to be lost as NH_3 during composting (Osada *et al.*, 2000). Biofiltration with solid media such as soil, saw dust or rockwool are often used to treat malodour in Japanese livestock farms. In a biofilter, it is considered that nitrifiers oxidise ammonium and nitrite, and then denitrifiers reduce nitrate to N_2. In this study, the performance of a rockwool biofilter was evaluated as a model case.

Materials and methods

Inlet and outlet gases, and solid media were sampled twice (Nov. 2004 and June 2006) from the rockwool biofilter (8 m width×5.8 m depth×3.3 m height) operating for the treatment of the gas from the composting process of cow and swine manure at the National Institute of Livestock and Grassland Science (NILGS) of Japan. Solid media, a mixture mainly consisting of rockwool, urethane and dried chicken faeces, contained approximately 50% moist. The pH of solid media was almost neutral. Concentrations of NH_3, N_2O, and CH_4 were measured using detector tubes, GC with ^{63}Ni-ECD, and FID, respectively. Nitrification and denitrification activities were determined based on the methods of Schmidt and Belser (1994) and acetylene block methods of Tiedje (1994), respectively, with some modifications.

Results and discussion

Inlet NH_3 concentration fluctuated around 0.5–145 ppm because of the turning of composts at three-week intervals. NH_3 was not detected in the outlet of the biofilter at any time. The amounts of ammonium and nitrate contained in the media varied in space within the biofilter. Nitrification activity was estimated to be sufficient to oxidise ammonium which came into the biofilter per day. Denitrification activity, however, was not enough to reduce nitrate produced by nitrification. This suggested that nitrate accumulation occurred because the nitrification activity was higher than the denitrification activity in the biofilter. Inlet and outlet gas concentrations of N_2O and CH_4 were about 4, 3.7, 428, and 302 ppm, respectively, and did not decrease significantly by passing through the biofilter.

Conclusions

Biofiltration with rockwool is effective to remove NH_3 emitted from the composting process. However, further improvement of this technology is required to treat other environmental pollutants such as nitrate, N_2O, and CH_4. Further studies are necessary to understand the microbial activity and community which are responsible for these reactions.

References

Osada, T., K. Kuroda and M. Yonaga, 2000. Determination of nitrous oxide, methane, and ammonia emissions from a swine waste composting process. J. Mater. Cycles Waste Manag. 2: 51-56.

Schmidt, E.L. and L.W. Belser, 1994. Ch. 10 Autotrophic Nitrifying Bacteria. In: Weaver R. W. et al. (editors), Methods of Soil Analysis, Part 2. Microbiological and Biochemical Properties. SSSA Series 5, USA, pp. 159-177.

Tiedje, J.M., 1994. Ch. 14 Denitrifiers. In: R.W. Weaver, S. Angle, P. Bottomley, D. Bezdiecek, S. Smith, A. Tabatabai, A. Wollum, S.H. Mickelson and J.M. Bigham (Eds.), Methods of Soil Analysis, Part 2. Microbiological and Biochemical Properties. SSSA Series 5, USA, pp. 245-267.

Production of aerosols of diameter under 10 µm from stables and its elimination

J. Dolejs[1], O. Toufar[1], M. Dedina[2], J. Nemeckova[1] and J. Knizek[1]*
[1]Research Institute of Animal Production, Pratelstvi 815, Prague 10, Czech Republic
[2]Research Institute of Agricultural Technology, Prague 6, Czech Republic

Introduction

The effect of dust particles on the organism is dependent on their composition, shape and size. The particles of the size above 10 µm in diameter are either totally prohibited from getting into the respiratory system or they are retained in the upper part of the respiratory system. Smaller particles (below 10 µm), however, may reach the lower part of the respiratory system and have negative implications on the self-cleaning mechanisms of the lung. Based on current scientific knowledge (van Leeuwen, 1997), it is not possible to determine a safe threshold concentration for dust aerosol with no ill effect on human health below it. Emissions of dust particles from farm animal facilities are most important for the vicinity of stables. Dust sources in farm animal facilities are above all feeds (fine particles of processed cereals and dried plants), parts of animal skin, urine crystals and solid parts of faeces. However, the concentration of these dust particles is not constant but it varies between years and seasons (Chardon, 1999). In our case air ionisation was chosen for elimination of dust emissions in dairy cows breeding (Dolejs *et al.*, 2006). This work is supposed to provide essential data about dust emissions (fraction PM_{10}) from common farm animal facilities and to show the possibility of the use of ionisation for dust elimination.

Materials and methods

Measuring of PM_{10} emission from farm animal facilities: For measuring of dust fraction emissions 3 facilities were chosen: 2 for cattle (dairy cows and bulls for fattening) and 1 for fattening pigs. Dust concentration was measured in the air outlet from the stable and the air inlet (= concentration in external air) by Dusttrak (method of isokinetic sampling in shaft). The emission of measured fraction is a product of difference in dust concentration (outlet - inlet) and rate of air flow (deduction from air volume ($m^3.h^{-1}$)). The resulting emission was recalculated per animal and day. Three measurements throughout 24 hours were realised at every facility.
Experiment utilising air ionisation under model conditions: Two experiments (A,B) were realised. Dust emission (PM_{10}) and rate of dust sedimentation under conditions of special climatic stable for 4 LU (livestock unit) were measured during the 14 days period. Mentioned measurements were carried out continually from 7th to 14th day of the experiment. The emission of PM_{10} fraction was measured according to methods mentioned before. For the determination of the rate of dust sedimentation various exposure surfaces were installed: asbestos, PVC covering and steel. From the point of view of triboelectric tension of surfaces asbestos (ASB) and PVC covering (PVC) were chosen as extreme and steel (ST) as approximately neutral. The determination of the dust was realised by vacuum-cleaning of the sedimented dust and subsequent gravimetric determination of its weight in 24 h. Air ionisation equipment with electrostatic tension of 7 kV and current of 25 µA was used for air ionisation.

Results and discussion

Measuring of PM_{10} emission from farm animal facilities: Detected values reflect certain dependence of PM_{10} production in cattle on live weight of animals and applied housing technology (Table 1). Net production of PM_{10} dust fraction was 2.3 times higher in dairy cow housed in loose cubicles (643 kg.head^{-1}) than in fattening bulls (362 kg.head^{-1}) housed on slatted floors. In the case of comparison of PM_{10} emission on basis of liveweight of animals, the ratio of weight of pigs (70.8 kg.head^{-1}) and fattening bulls, using the same housing technology (slatted floors), the production of PM_{10} should be 16.3 mg.head^{-1}.day^{-1}. However the detected value was 102.5 mg, i.e. 6.3 times higher than in fattening bulls.

Experiment utilising air ionisation under model conditions: Both experiments (A, B) were realised under temperature conditions between 20.4–20,9 °C, relative humidity 55.3–58.1%, air flow rate 0.34–0.43 m.s^{-1}. The PM_{10} fraction was significantly decreased by 24.1 ($P \leq 0.05$), respectively 28.5% ($P \leq 0.01$) (Table 2). The emissions of PM_{10} was affected by great variability of concentration of emission component of dust fraction (inlet concentration). Average variability during 24 hours was 34.2%, variability of daily averages in a period was 49.7%. Due to air ionisation the rate of dust sedimentation was increased on ASB by 105.1%, resp. 101.0%, on PVC by 90.1%, resp. 94.4 and on ST by 21.4%, resp. 27.9%. The result was again highly significant in sedimentation on surfaces ASB and PVC ($P \leq 0.01$) in both experiments. In the case of sedimentation on ST it was significant only in experiment B ($P \leq 0.05$). The results indicate that air ionisation will serve as one of the ways of dust elimination in stables and elimination of dust emission outside the stables.

Table1. Net production of dust fraction PM_{10} from farm animal facilities.

Species/category	Capacity head	Concentration (µg.m^{-3}) Outlet	Inlet	PM_{10} production/ day g.stable^{-1}	mg.head^{-1}
dairy cows	125	107	98	20.302	191.5
fattening bulls	280	44	31	20.498	83.3
fattening pigs	90	315	116	9.830	102.5

Table 2. The influence of air ionisation on production of PM_{10} fraction and dust sedimentation in dairy cows stable.

Exp.	Period	PM_{10} (µg.m^{-3}) Outlet	Inlet	Production (mg.day^{-1}) Per stable	Per head	Sedimentation (mg.m^{-2}day^{-1}) ASB	PVC	ST
A	ref.	135	106	828.0	207.0	1044.5	565.7	575.5
	exp.	130	106	629.0	157.3*	2141.8**	1075.4**	698.6
	Index				0.759	2.051	1.901	1.214
B	ref.	89	62	711.6	177.9	698.6	379.9	389,4
	exp.	82	61	508.8	127.2**	1404.2**	740.8**	498.4*
	Index				0.715	2.010	1.949	1.279

t-test: difference of exp. and ref. periods; *$P \leq 0.05$, **$P \leq 0.01$.

Conclusion

The relations of species and category of animals, their liveweight and housing technology were detected by measuring of dust emissions (PM_{10} fraction). Air ionisation that decreased emission by almost 30% is one of the ways of dust emission (PM_{10} fraction) elimination. The rate of dust sedimentation was increased too.

Acknowledgements

This paper was elaborated on the basis of data outputs of research projects of MA 0002701402.

References

van Leeuwen, R.F.X., 1997. Update and revision of WHO Air Duality Guidelines for European. European EpiMarker 1: 10-12.

Chardon, W.J., 1999. Emission of fine dust in farming. Report-Plant-Research-International, No.105, 26 pp.

Dolejs, J., O. Masata and O. Toufar, 2006. Elimination of dust production from stables for dairy cows. Czech J. Anim. Sci. 51: 305-310.

Integrated measures

Assessment of pollution swapping risks of EU environmental policies and measures

O. Oenema[1], D. Oudendag[1], G.L. Velthof[1], G.J. Monteny[1] and H. van Grinsven[2]
[1]Wageningen UR, P.O. Box 47, 6700 AA Wageningen, The Netherlands
[2]Netherlands Environmental Assessment Agency MNP, P.O. Box 303, 3720 AH Bilthoven, The Netherlands

Introduction

In response to the environmental side effects of the intensification of agriculture, especially during the period 1960-1990, series of policies and measures have been implemented in the European Union (EU). This relates also to the use of nitrogen (N) in fertiliser and animal manure, as the N losses from agriculture have serious environmental effects. Currently, the use of animal manure and fertilisers are directly and indirectly affected by four categories of EU policies and measures: (1) The reform of the Common Agricultural Policy (CAP), including Cross Compliance measures and Agri-Environmental and Rural Development regulations, (2) Water Framework Directive, including the Nitrates Directive, (3) Air quality Directive and the Thematic Strategy on Air Pollution, and (4) Nature conservation legislation, including the Birds and Habitats Directives. Though there is a tendency towards greater integration of agricultural and environmental policies and also towards integration of single environmental polices, many of the current EU policies and measures still have rather narrow and specific objectives. Because of the large number of policies and measures, and the narrow focus, there is a risk of so called 'pollution swapping.' The study presented here examined the possible 'pollution swapping' mechanisms of 8 EU Directives and Regulations on ammonia (NH_3), nitrous oxide (N_2O) and methane (CH_4) emissions into the atmosphere and N leaching to groundwater and surface waters.

Methodology

Pollution swapping refers to a special side-effect of policies and measures, i.e., the unwanted increase of another pollutant (type 1), and/or the unwanted increase of the emission of the target pollutant elsewhere (type 2). For example, the EU Nitrates Directive aims at decreasing (preventing) the leaching of N to groundwater and surface water, but some of the measures of the Code of Good Agricultural Practice have the potential of increasing NH_3 and CH_4 emissions (Type 1 pollution swapping), whilst others may lead to increased N leaching elsewhere (Type 2), for example when farming activities have to be transferred to other locations, following the designation of Nitrate Vulnerable Zones and the implementation of Action Programs. We used a simple conceptual model and a categorisation of policies and measures in six categories to assess the pollution swapping risks of some of the main environmental policies affecting the use of N in EU agriculture. The 6 categories were: (1) Abatement of single N species emissions, with potential antagonistic effects, (2) Control of N input, with potential synergistic effects, (3) Agri-environmental measures focused on the extensification of agricultural production and environmental protection, with possible synergistic effects, (4) Regulations on animal welfare, with possible antagonistic effects (increase of NH_3, N_2O and CH_4 emissions), (5) Measures aimed at improving the competitiveness of the agricultural sectors, with diverse effects, and (6) Spatial zoning, i.e. delineating areas where measures mentioned under (1), (2) and (3) and perhaps (4) apply, with a risk of type-2 pollution swapping. We also used the simulation model MITERRA-Europe to quantitatively assess the effects of the policies and measures on the emissions of NH_3, N_2O and CH_4 into the atmosphere and N leaching to groundwater and surface waters (Velthof *et al.*, 2007).

Results and discussion

Volatilisation of NH_3 occurs at an early stage in the sequence of processes following the excretion of faeces and urine by animals and or the application of urea and ammonium-based fertilisers. The emission of N_2O and the leaching of N occur at later stages. From this sequence of processes, it is clear that measures that effect the emission of NH_3 will change the total amount of N at an early stage and thereby likely have an effect on the emission of N_2O and the leaching of N too. Conversely, it is unlikely that measures that effect the leaching of NO_3 will *directly* effect the emission of NH_3. The sequence of processes explains to some extent why various NH_3 abatement measures likely have effect on N leaching, and why various N leaching abatement measures have little effect on emission of NH_3, unless the total N input (category (2) measure) is controlled too. Measures of the Nitrate Directive fall in categories (1), (2), and (4), and potentially have both synergistic and antagonistic effects on the emissions of NH_3, N_2O and CH_4 into the atmosphere. Synergistic effects on the emissions of NH_3 and N_2O dominate. Measures aimed at decreasing the emissions of NH_3 according to the Guidelines of the UNECE Convention on Long Range Trans-boundary Air Pollution (CLRTAP) fall in categories (1) and to a lesser extent (2), and potentially have significant antagonistic effects on the emissions of N_2O and CH_4 and also N leaching. The NH_3 emission abatement measures are proposed by the NEC and IPPC Directives. Results of our study also indicate that the potential pollution swapping mechanisms of NH_3 emission abatement measures on N_2O and CH_4 emissions and N leaching can be largely circumvented when the NH_3 emission abatement measures and the Code of Good Agricultural Practice of the Nitrates Directive are integrated and implemented jointly. Clearly, there has to be a much greater emphasis on measures categorised under (2), for example on low-protein feeding and on limits for the amount of plant available N to be applied to crops, and on targets for the fraction of manure N assumed to be plant available. Results of the assessments of the Birds and Habitat Directives (i.e. Natura 2000 management plans and special action plans) suggest that the measures may contribute to decreasing the emissions of NH_3, N_2O and CH_4 and the leaching of N as most of these measures put restrictions on agricultural activities. However, some measures may contribute to type-2 pollution swapping, as some farming activities may have to be transferred from and around the Natura 2000 areas to elsewhere.

Conclusion

In conclusion, pollution swapping is potentially a serious and disturbing side effect of some EU environmental policies. However, this pollution swapping can be largely circumvented by emphasising N input control and joint and integrated implementation of the policies and measures.

References

Velthof, G.L., D. Oudendag, H.P. Witzke, W.A.H. Asman, Z. Klimont and O. Oenema, 2007. Quantification of ammonia emission and nitrate leaching in EU-25 using MITERRA-EUROPE, (this issue).

Prioritisation of ammonia abatement measures, their costs and impacts on nitrate leaching and nitrous oxide emissions using the NARSES model

J. Webb[*], *S.G. Anthony and S. Humphries*
ADAS Wolverhampton, Wergs Rd, Wolverhampton WV6 8TQ, United Kingdom
[*]*Now at, AEA Energy & Environment, Gemini Building, Harwell Business Centre, Didcot, OX11 0QR, United Kingdom*

Introduction

The NARSES model (Webb and Misselbrook, 2004) uses a mass-flow approach to identify the most cost-effective means of reducing ammonia (NH_3) emissions from agriculture and to make unbiased estimates of the total costs. Reducing NH_3 emissions may, by increasing the amount of total ammoniacal-N (TAN) entering the soil, increase subsequent losses of nitrate (NO_3^-) to ground and surface waters and emissions of nitrous oxide (N_2O).

Materials and methods

NARSES output is now linked to the Manner model (Chambers *et al.*, 1999) enabling calculation of the impacts of NH_3 abatement on NO_3^- leaching and N_2O emission at both the national and regional scales. In this paper we are only reporting results at the national scale.

Results and discussion

Ammonia abatement: The most cost-effective measures to reduce NH_3 emissions have been reported in detail by Webb *et al.* (2006). These include measures such as covering poultry manure stores and allowing cattle slurry lagoons to crust which, while simple and inexpensive to adopt, each produce only small ($<1.0 \times 10^3$ t or less) reductions in NH_3 emission. Rapid incorporation of poultry manures and cattle and pig slurries and FYM to arable land are reasonably cost-effective and gave greater reductions (up to c. 3×10^3 t NH_3-N for each measure) over the current implementation of these measures. Other reasonably cost-effective measures were to store all poultry manure prior to spreading and apply pig slurry to grassland by trailing shoe. Hereafter, the unit cost of measures more than doubled. Total NH_3-N abatement with the programmed measures was c. 46×10^3 t out of a total of c. 203×10^3 t NH_3-N from livestock production.
Nitrate leaching: Of the NH_3-N conserved, c. 5.3×10^3 t, was calculated to be lost as NO_3^-. This represented an increase of c. 37% in the amount of NO_3^- leached following the application of livestock manures, but only c. 4% of current total NO_3^- leaching in the UK. In this brief paper we are only reporting national averages. Clearly in some catchments, with dense concentrations of livestock or a large proportion of soils vulnerable to leaching, increase in NO_3^- losses will be greater and be cause for concern. Expressed as a % of the NH_3-N conserved by individual NH_3 abatement techniques, the increase in NO_3^- leaching ranged from +28 to –70%. The increases in NO_3^- leaching occurred following measures which conserved NH_3-N as TAN, which could subsequently be nitrified in soil and leached. The greatest losses of NO_3^- (20-30% of NH_3-N conserved) tended to occur following the application of measures to reduce NH_3 emissions from cattle slurries and manures, regardless of whether those measures were applied in buildings, to slurry stores or to slurry or manure application. This is because in the

UK cattle tend to be found in the west of the country where rainfall, and hence drainage over winter, are greatest. In contrast, similar measures, such as rapid incorporation of manures into soil, when applied to pig or poultry manure, led to losses of 10-20% of NH_3-N conserved. This reflects the greater concentrations of those livestock in the east where over-winter drainage is less. Two types of measure decreased NO_3^- leaching as well as NH_3 emission. Phase feeding of pigs, by reducing total N intake and hence N excretion, would be expected to decrease all forms of N loss. However, storing all FYM and poultry manure before spreading, instead of spreading immediately the manure is removed from buildings, might not be expected to reduce NO_3^- leaching. The reason for the subsequent decrease in NO_3^- leaching is that denitrification during storage emits c. 12-41% of TAN entering the store as N_2O or N_2 and hence less TAN remains when the manure is applied to land than would have been the case if the manure was spread when 'fresh'.

Nitrous oxide emissions: The increase in N_2O emissions was never more than 2% of the NH_3-N conserved. Phase feeding decreased N_2O emissions and storing manures decreased N_2O emissions following manure spreading. In total N_2O emissions were little changed by the adoption of NH_3 abatement techniques.

Conclusions

At the national scale, UK NH_3 emissions can be conserved without large increases in NO_3^- leaching. No method of abatement led to more than 30% of the NH_3-N conserved being lost as NO_3^-. Some NH_3 abatement methods also decreased emissions of both NO_3^- and N_2O. Thus it will be possible to formulate approaches to reducing emissions of NH_3 without axiomatically causing large increases in emissions of either NO_3^- or N_2O. However, in those catchments with the greatest potential for NO_3^- leaching, careful consideration will still need to be given to the implementation of an NH_3 abatement policy.

References

Chambers, B.J., E.I. Lord, F.A. Nicholson and K.A. Smith, 1999. Predicting nitrogen availability and losses following arable land application of manures: MANNER. Soil Use Man. 15: 137-143.

Webb, J. and T.H. Misselbrook, 2004. A mass-flow model of ammonia emissions from UK livestock production. Atm. Env. 38: 2163-2176.

Webb. J., M. Ryan, S.G. Anthony, A. Brewer, J. Laws, F.A. Aller and T.H. Misselbrook, 2006. Identifying the most cost-effective means of reducing ammonia emissions from UK agriculture using the NARSES model. Atm. Env. 40: 7222-7233.

Quantification of ammonia emission and nitrate leaching in EU-27 using MITERRA-EUROPE

G.L. Velthof[1], D. Oudendag[1], H.P. Witzke[2], W.A.H. Asman[3], Z. Klimont[3] and O. Oenema[1]
[1]Alterra, Wageningen UR, P.O. Box 47, 6700 AA Wageningen, The Netherlands
[2]Eurocare Bonn and Institute of Agricultural Policy,Market Research and Economic Sociology, Bonn Universit, Nussallee 2,53115 Bonn, Germany
[3]International Institute for Applied Systems Analysis (IIASA), Schlossplatz 1, 2361 Laxenburg, Austria

Introduction

The availability of nitrogen (N) fertilisers from the 1950s onwards has contributed to a boost in crop production and has also contributed to the increase in the number of farm animals and the production of N in animal manure. The difference between total N input and N withdrawal via harvested crop is the N surplus, which is an indicator for total N loss. The greater part of the N surplus is lost to the environment, either as ammonia (NH_3), nitrogen oxides (NO_x), nitrous oxide (N_2O) or dinitrogen (N_2) into the atmosphere, or as nitrate (NO_3) and other N compounds into groundwater and surface waters. These N emissions can lead to problems directly related to human health (e.g. pollution of ground water due to nitrate leaching), and ecosystem vulnerability (e.g. N deposition in natural ecosystems). A series of environmental policies and measures have been implemented at decreasing the emissions of NH_3 to the atmosphere (NEC Directive and IPPC Directive), the leaching of NO_3 to groundwater and surface waters (Nitrates Directive, Groundwater Directive and Water Framework Directive), and the emissions of the greenhouse gases CO_2, N_2O and CH_4 to the atmosphere (Kyoto Protocol). Pollution swapping is generally seen as a response to governmental policies that focus on one N loss form. The model MITERRA-EUROPE is a tool for integrated assessment of N emissions from agriculture at regional, country, and EU-27 levels. A study was carried out using MITERRA-EUROPE to quantitatively assess the effects of policies and measures on different N emissions and to identify risks of pollution swapping (Velthof *et al.*, 2007).

MITERRA-EUROPE

The starting point for MITERRA-EUROPE are the existing models CAPRI (http://www.agp.uni-bonn.de/agpo/rsrch/capri/capri_e.htm) and RAINS (http://www.iiasa.ac.at/rains) supplemented with data bases, soil data and expertise about emission processes. The data-base of MITERRA-EUROPE is on NUTS 2 level and includes data of fertiliser use, animal numbers, N excretion, land use, crop types and yields, soil type, topography, meteorological data, and emission factors for NH_3, N_2O, NO_X, and CH_4, and leaching factors for NO_3. Different measures to mitigate NH_3 emission and NO_3 leaching are included in the model, such as a decrease in protein content of feed, animal house adaptations, covered manure storage, air purification, efficient manure and fertiliser application techniques, balanced N fertilisation, bufferstrips near surface waters, cover crops, and limits to N application on sloping soils. The parametrisation and implementation of nitrate measures are presented in Velthof *et al.* (2007).

Results

Analysis of different scenarios on EU-27 level shows that:

- Denitrification to N_2 is the largest absolute source of N loss, followed by NH_3 emission, leaching, nitrous oxide emission and NO_X emission (Figure 1).
- The emission of NH_3 decreases on average (EU-27) by about 10% following national projections of agricultural development towards 2020 (Figure 1). The leaching of NO_3 decreased with 6 to 16 percent and N_2O emission with 1 to 4 percent. The decrease in the emissions was due to a decrease in the input of mineral fertiliser (4 to 8%) and of manure (1 to 2%) in these scenarios.
- Full implementation of the Nitrates Directive will decrease the emissions of both NH_3, NO_3, and N_2O on average with 8%, 36%, and 16%, respectively (Figure 1). This was because both the input of fertiliser (5-26%) and of manure (1 to 13%) decrease after full implementation of the Nitrates Directive.
- Risk on pollution swapping between NH_3 emission and NO_3 leaching is higher for NH_3 abatement techniques than for measures reducing nitrate leaching (not shown).
- There were large spatial differences in emissions in the EU, related to differences in input of fertilisers, animals number, soil types, and weather conditions (not shown).

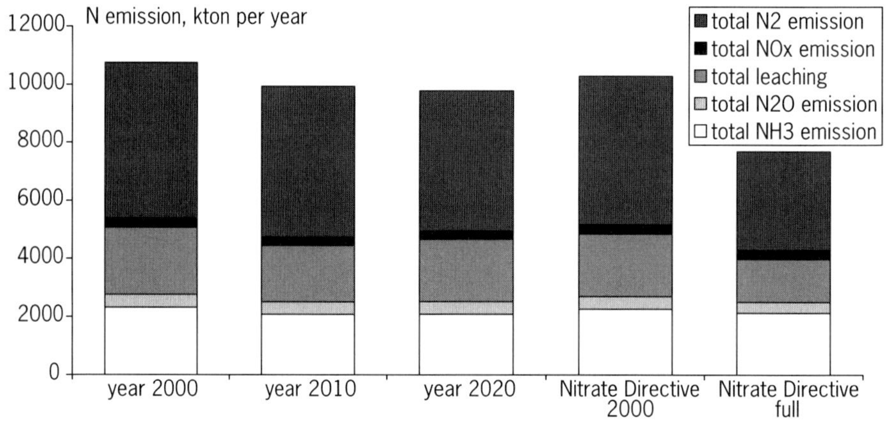

Figure 1. N losses (kton N per year) from animal housing, storage and soil in EU-27 for five scenarios. Year 2000 is the reference year and year 2010 and 2020 are the national agricultural activity projections for RAINS (Amann et al.*, 2006). Nitrates Directive 2000 is a scenario simulating implementation of NO_3 measures in Nitrate Vulnerable Zones (NVZ) in 2000 and Nitrates Directive full is a scenario simulating full implementation of these measures in NVZ (after Velthof* et al.*, 2007).*

References

Amann M., W. Asman, I. Bertok, J. Cofala, C. Heyes, Z. Klimont, M. Posch, W. Schöpp and F. Wagner, 2006. Emission control scenarios that meet the environmental objectives of the Thematic Strategy on Air Pollution. NEC Scenario Analysis report Nr. 2 IIASA, Laxenburg, 116 p.

Velthof, G.L., D. Oudendag and O. Oenema, 2007. Development and application of the integrated nitrogen model MITERRA-EUROPE. Contents of the report of task 1 of the service contract. 'Integrated measures in agriculture to reduce ammonia emissions'. Contract number 070501/2005/422822/MAR/C1. Alterra, Wageningen (in press).

Integrated assessment of atmospheric emissions, leaching and runoff of ammonia, greenhouse gases and nutrients at a landscape level

W. de Vries[1], J. Kros[1], G.L. Velthof[1], E. Gies[1], J.C. Voogd[1], A. Bleeker[2], J. Schröder[3] and M.P.W. Sonneveld[3]
[1]Alterra, Wageningen UR, P.O. box 47, 6700 AA Wageningen, The Netherlands
[2]Energy Research Centre of the Netherlands, P.O. Box 1, 1755 ZG Petten, The Netherlands
[3]Wageningen University, Department of Environmental Sciences, P.O. Box 47, 6700 AA Wageningen, The Netherlands

Introduction

Joined in an environmental cooperative, farmers in the Noordelijke Friese Wouden (NFW) in The Netherlands have made an agreement with the government about achieving environmental objectives at regional level in the next 5-10 years. One of the objectives is to develop some level of freedom regarding the measures, as long as the environmental objectives are attained. In the NFW, there is reluctance to the injection of animal manure, because of its negative impacts on soil fauna and soil structure. As an alternative, the presently forbidden use of above ground manure spreading under favourable weather conditions is suggested in combination with low protein feeding. Here, we present results of an integrated assessment of the environmental status for the year 2004 of the NFW area and of impacts of alternative management measures, using the model INITIATOR2 (De Vries *et al.*, 2005).

The model INITIATOR2

To gain insight in the environmental impacts of management measures on the environment, an integrated model system INITIATOR2 (Integrated Nutrient ImpacT Assessment Tool On a Regional scale) was applied to predict atmospheric emissions of ammonia and greenhouse gases (CO_2, CH_4 and N_2O) from housing, manure storage systems and soils, and to predict the accumulation, leaching and runoff of nitrogen (N) and phosphate (P) from soils. A database (GIAB) with spatially explicit data on animal numbers in many animal categories, agricultural practices and land management, such as manure application techniques, for each farm in the Netherlands was used (Anonymous, 2004). 2004 was used as a reference year for the calculations (see De Vries *et al.*, 2005). Alternative management measures, that are presently used in three sub areas of the NFW, include low protein feeding, lowering the N excretion, specifically of NH_4-N, and above ground spreading of animal manure. Here, ammonia concentrations are presently monitored to gain insight in the impact of alternative management measures (Bleeker *et al.*, 2007). In the model we assumed that the total N excretion is lowered by 18%, while the NH_4 fraction of N in the manure is lowered 13% by use of animal feed with 14% instead of 19% protein for cattle (Kebreab *et al.*, 2001). The standard average emission percentage for above ground spreading is set at 68% of the applied NH_4-N. When applied during favourable conditions (either during rain or by applying water after the spreading), we assume 35% emission (Sonneveld, pers. comm.). This is an average of values varying between approximately 20 and 68% that could be attained with this method.

Results

Model results are summarised in Table 1. The present NH_3 emissions exceed the landscape target of 1.8 kton NH_3-N (2.2 kton NH_3) for the year 2010 by approximately 10%. This target is derived from a back scaled NH_3 emission ceiling for the province of Friesland to the NFW area, based on the current NH_3 emissions in both areas. The ceiling for Friesland of 12 kton NH_3 is derived from a national emission ceiling (NEC) of 93 kton NH_3, while minimising the exceeding of critical loads during the disaggregation (Van Lent and Erisman, 2003). Low protein feeding reduces the NH_3 emission by 20% to a level near the landscape target. The combination of low protein feeding and above ground manure spreading leads to an increase in NH_3 emission using the reference value of 68% emission, but the lower emission percentage (35%) causes 10% less NH_3 emission than the present situation. However, more stringent reductions are required to reach the landscape target. E.g. the consequent application of injection attains this situation (Table 1). The various measures hardly affect the N deposition, which is largely influenced by emissions outside the NFW area. Furthermore, even though the NH_3 emission ceiling for the NFW is within reach, the critical N deposition of the nature target types occurring in the NFW is exceeded in 38-40% of the area for all situations. Only a complete emission reduction in the NFW area leads to exceed the critical N deposition by 6.1% (the target value is 10% for the year 2030). Low protein feeding and lowering the N input is also favourable to N_2O emission and N leaching. A decrease in NH_3 emissions is correlated with more N_2O emission (since injection and incorporation techniques lead to more anaerobic situations compared to above ground spreading) and more N leaching (pollution swapping).

Table 1. Predictions of the overall ammonia and nitrous oxide emission, the N deposition and the exceeding of critical N loads and critical nitrate leaching in the NFW in the present situation and after management changes.

Scenario	NH_3 (kton NH_3-N/yr)	N_2O (kton N_2O-N/yr)	N deposition (mol N/ha/yr)	Exceeding % CLN[1]	Exceeding % N leaching[2]
Present situation (2004)	2.2	0.46	1687	39.1	5.7
Low protein feeding (LPF)	1.8	0.41	1556	38.2	3.1
LPF + manure spreading 68%	2.7	0.32	1897	40.3	1.8
LPF + manure spreading 35%	2.0	0.35	1662	39.2	2.7
LPF + injection, 10-12%	1.6	0.51	1495	32.5	3.5
NH_3 emission NFW = 0	0	-	1040	6.1	-
Landscape objective	1.8	-	1500	10	0

[1]Area where the N deposition exceeds the critical load for nitrogen (CLN) of the nature in NFW.

[2]Area where the nitrate concentration in upper groundwater exceeds the target of 50 mg NO_3l^{-1}.

Conclusions

Model results do not show a significant difference in NH_3-emission for the combination of above ground manure spreading under favourable weather conditions in combination with low protein feeding (alternative) and the present N feeding and NH_3 application methods. This is in line with preliminary results of NH_3 measurements (Bleeker *et al.*, 2007). The model results imply, however, that measures which focus on NH_3 emission reductions to reach the emission target are insufficient to reach the target for exceeding critical load in the NFW area.

References

Anonymous, 2004. Statistics for agriculture and horticulture 2004 (in Dutch). Landbouw Economisch Instituut (LEI) en Centraal Bureau voor de Statistiek (CBS), Den Haag. 279 pp.

Bleeker, A., A. Hensen, E. Gies, W. de Vries, J. Kros and M.P.W. Sonneveld, 2007. Multi-purpose ammonia measurements on a landscape level. (this book).

De Vries, W., J. Kros and G. Velthof, 2005. Integrated evaluation of agricultural management on environmental quality with a decision support system. In: Z. Zhu, K. Minami and G. Xing (eds.) 3rd International Nitrogen Conference, 2004, Nanjing. China, Science Press: 859-870.

Kebreab, E., J. France, D.E. Beever and A.R. Castillo, 2001. Nitrogen pollution by dairy cows and its mitigation by dietary manipulation. Nutr. Cycl. Agroecosys. 60: 275-285.

Van Lent, A.J.H. and J.W. Erisman, 2003. De Stikstofmeetlat in de praktijk: resultaten van een verkennende studie. Eindrapportage IPO ML-06 Project, IPO, Den Haag.

Integral assessment of acidification of the animal sector as part of the total agricultural sector in the Netherlands

M.A. Thomassen and I.J.M. de Boer
Wageningen UR, Animal Production Systems Group, P.O. Box 338, 6700 AH Wageningen, the Netherlands

Introduction

Acidification is highlighted in legislation and reduction strategies by focusing on emission of ammonia (NH_3) on agricultural farms. As a result, NH_3 emission from agriculture reduced with 48% between 1990 and 2003, mainly due to emission reduction during manure application and a reduced herd size (Smeets *et al.*, 2004; MNP, 2006). Besides ammonia (NH_3), however, sulphur dioxide (SO_2) and nitrogen oxides (NO_x) contribute to acidification. Using on-farm emission data, Smeets *et al.* (2004) assessed that in 2000 agriculture contributed for around 90% to national ammonia emission (NH_3), for 3% to national nitrogen oxides emission (NO_x) and for 0.5% to national sulphur dioxide emission (SO_2). Emission of SO_2 and NO_x, however, mainly result from combustion of fossil fuels during, for example, transport of concentrates or the production of artificial fertiliser. To further reduce acidification from agriculture, insight is needed into the emission of NH_3, SO_2 and NO_x during the life cycle of the main agricultural products, especially animal production (Steinfeld *et al.*, 2006). The aim of this study, therefore, was to identify hotspots regarding the emission of NH_3, SO_2 and NO_x during the life cycle of the main animal products as part of the total agricultural sector.

Material and methods

Life Cycle Assessment (LCA) was used to assess acidification potential (in tonnes SO_2-equivalents) of the dairy cattle, pig and poultry sector, and the total agricultural sector based on national general data of the year 2003 (Guinée *et al.*, 2001). A cradle-to-farm gate LCA was performed, implying that all processes and transport until agricultural products leave the farm gate are included in the analyses. Processes included besides on farm processes, were transport and production of concentrates, artificial fertiliser, pesticides, rock wool and contract work. Emission of NH_3 during housing, grazing, and from manure storage facilities were based on national references related to nitrogen excretion of different animal categories (Van der Hoek and Van Schijndel, 2006). NH_3 volatilisation rates of the Europe Online RAINS model (2006) were used to assess NH_3 volatilisation during spreading of all types of manure in the total agricultural sector.

Results and discussion

The total acidification potential (AP) of the agricultural sector in the Netherlands equalled 643,765 tonnes SO_2-equivalents of which 25% occurred in foreign countries. The dairy cattle sector accounted for 23%, the pig sector for 15% and the poultry sector for 6% of the total AP. Other sectors, of which mostly arable production for human consumption and horticulture contributed for 56%. Figure 1 shows that on-farm emission of NH_3 remains the most important contributor to acidification, i.e. 50% for dairy cattle, 41% for pigs, 30% for poultry and 53% for total agricultural sector. On-farm emission of NH_3 included emission from manure during housing, grazing, and from manure storage facilities (i.e., 56% for dairy cattle, 99% for pigs and poultry and 30% for total agriculture), and emission during application of fertiliser (i.e. 44% for dairy cattle and 70% for total agriculture). On-farm SO_2 and NO_x emissions, mostly caused by direct energy use, contributed for 10% in the dairy cattle, 4% in the pig, 7% in the poultry and for 14% in the total agricultural sector. Figure 1 shows that off-farm acidifying emissions have a significant contribution to acidification, i.e. 40% for dairy cattle, 55% for pigs, 63% for poultry and 33% for the total agricultural sector. Off-farm acidifying emissions included mainly emission during production and transport of concentrates (i.e., 85% for dairy cattle, 99% for pigs, 100% for poultry and 82% for the total agricultural sector) and contract work (i.e. 11% for dairy cattle and 14% for the total agricultural sector).

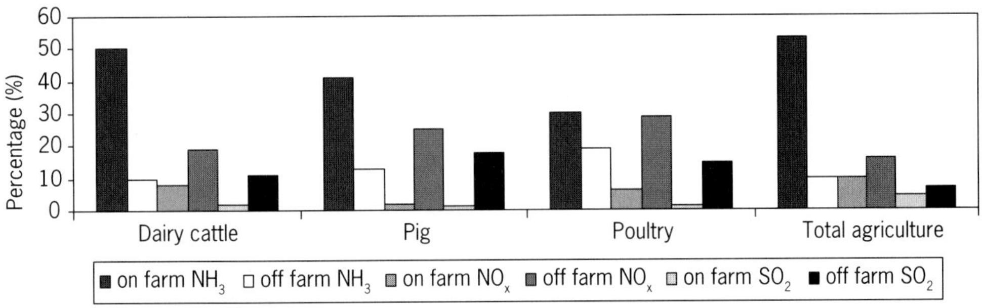

Figure 1. Contribution of the main acidifying elements to total acidification in different agricultural sectors.

Conclusions and recommendations

Results showed that the given policy attention to NH_3 emission reduction at farm level is justified due to the high contribution to acidification. Results also showed that NH_3, SO_2 and NO_x emissions of purchased inputs have a significant contribution to acidification. Besides the attention that has already been given to reducing NH_3 emission at farm level, we recommend researchers and policy-makers to focus on acidifying elements related to purchased inputs to reduce acidification of agricultural production.

References

Guinée, J.B., M. Gorrée, R. Heijungs, G. Huppes, R. Kleijn, A. De Koning, L. Van Oers, A. Wegener Sleeswijk, S. Suh, H.A. De Haes, H. De Bruijn, R. Van Duin, M.A.J. Huijbregts, R. Lindeijer, A.A.H. Roorda, B.L. Van der Ven and B.P. Weidema, 2001. Life Cycle Assessment; an operational guide to the ISO standards. Scientific background. Centrum voor Milieukunde- Universiteit Leiden, Leiden, The Netherlands, 314 pp.

Milieu- en Natuurplanbureau, 2006. Milieubalans 2006, MNP Report 500081001, Bilthoven, the Netherlands, 199 pp.

RAINS, 2006. Regional Air Pollution Information and Simulation model, Europe Online. International Institute for Applied Systems Analysis, Laxenburg, Austria www.iiasa.ac.at/rains/Rains-online.html?sb=8

Smeets, W.L.M., J.P. Beck, L.J. Brandes, R.M.M. Van den Brink, H.E. Elzenga, M.M. Van Eerdt, A. Hoen, P.H.M. Janssen, D.S. Nijdam, M.M.P. Van Oorschot, C.J. Peek and M.W. Van Schijndel, 2004. Actualisatie van de emissieraming van SO_2, NO_x, NH_3, NMVOS en fijn stof in 2010. RIVM Report 500037007,, Bilthoven, the Netherlands, 106pp.

Steinfeld, H., P. Gerber, T. Wassenaar, V. Castel, M. Rosales and C. De Haan, 2006. Livestock's long shadow, environmental issues and options. FAO, Rome, Italy.

Van der Hoek, K.W. and M.W. Van Schijndel, M.W. 2006. Methane and nitrous oxide emissions from animal manure management, 1990-2003. Background document for the Dutch national inventory report. RIVM report 680125002/2006, Bilthoven, The Netherlands, 50 pp.

Regional estimates of ammonia and nitrous oxide and their potential environmental impact: The case of dairy production in Czech Republic

M. Havlikova[1,2] *and C. Kroeze*[1]
[1]*Wageningen UR, Environmental Systems Analysis Group, The Netherlands*
[2]*Czech Hydrometeorological Institute, Na Sabatce 17, Prague 143 65, Czech Republic*

Introduction

Ammonia emissions (NH_3) may, by 2020, be a major contributor to acidification and terrestrial eutrophication in many European countries, including the Czech Republic (Amann *et al.*, 2005). Agriculture is the most important source of ammonia emissions, and many options to reduce emissions are available. However, several techniques to reduce ammonia emissions may, as a side-effect, increase emissions of nitrous oxide (N_2O), which is one greenhouse gases contributing to global warming (Brink, 2003). Integrated analyses are needed to identify effective strategies to reduce acidification, eutrophication and greenhouse gas emissions simultaneously. Detailed and consistent estimates of emissions are needed for such assessments. In an earlier study (Havlikova and Kroeze, 2006) we evaluated several methods with respect to quality, applicability to specific environmental and agricultural conditions and considerations of interrelations between sources, emissions, impacts and reduction strategies. In addition, we explored different ways to assess the potential environmental impact of agricultural emissions (Havlikova, unpublished data). In this paper we estimate emissions of NH_3 and N_2O from dairy production in the Czech Republic and assess their potential contribution to acidification, terrestrial eutrophication and global warming.

Emission estimates and impact assessment approaches

The Czech Republic is divided into fourteen administrative regions, specified into nine study areas based on (1) dairy farming intensity, (2) sensitivity of terrestrial ecosystems to acidification and eutrophication, (3) percentage of arable land in areas considered vulnerable to water pollution by nutrients, and (4) population density. For each study area the emissions of NH_3 and N_2O and their potential environmental impact are estimated by applying a simple process-based model. In Western European countries several methods have been applied. We adopt the German GAS_EM model developed by Dammgen *et al.* (2002). Emission reduction measures considered are those included in Code of Good Agriculture Practice (CGAP). We estimated that an applicability potential ranging between 4% and 84% for individual measures in the nine study areas, based on current policy plans of municipal governments in the Czech Republic. The potential environmental impacts are estimated by applying a set of site-dependent impact factors adopted from Life Cycle Assessment. These impact factors are combined with local environmental characteristics.

Results and discussion

Total emissions from dairy cattle are estimated at 13 kt NH_3 and 2 kt N_2O in year 2005. Preliminary results indicate that implementation of CGAP measures may reduce NH_3 emissions from dairy production by 10%. As a side effect, direct N_2O emissions from soils may increase by 20%. The potential environmental impact is based on characterisation factors for each of the environmental issues studied. The result indicates that the normalised potential impact of acidification is the highest in all study areas, whereas the contribution to global warming is the lowest. Application of CGAP resulted in a 6% reduction of the potential impact on acidification and terrestrial eutrophication in all study areas. Environmental problems may be considered more or less urgent by decision makers, as reflected by political reduction targets. Here we use a set of valuation factors based on the difference between environmental policy targets for the year 2020, and the current (2005) situation. In this set of factors acidification is assigned the highest value, i.e. it is considered as the most significant environmental problem. Using this set of valuation factors, we conclude that the most polluted area is characterised by high dairy production intensity, a high sensitivity to acidification and a large area of agriculture land vulnerable to excess nutrients, while the population density is low. This indicates that effective pollution reduction strategies focus not only on NH_3 and N_2O but also on nutrients such as nitrate and phosphate. The potential contribution of study areas to global warming increased linearly with the number of dairy cattle and application of CGAP, in line with Brink (2003). In future research, we will analyse the cost-effectiveness of interrelated emission strategies.

References

Amann, M., I. Bertok, R. Cabala, J. Cofala, C. Heyes, F. Gyarfas, Z. Klimont, W. Schöpp and F. Wagner, 2005. A final set of scenarios for the Clean Air for Europe (CAFE) programme. CAFE Scenario Analysis Report nr. 6. International Institute for Applied Systems Analysis, Laxenburg, Austria.

Brink, J.C., 2003. Modelling cost-effectiveness of interrelated emission reduction strategies, the case of agriculture in Europe. PhD thesis. Wageningen University, The Netherlands.

Dammgen U., M. Luttich, H. Dohler, B. Eurich-Menden and B. Osterburg, 2002. GAS-EM a procedure to calculate gaseous emissions from agriculture. Landbauforschung Volkenrode 52: 19-42.

Havlikova, M. and C. Kroeze, 2006. Evaluation of methods for quantifying emissions of air, water and soil pollutants. The Science of the Total Environment 372: 133-147.

A whole-farm strategy to reduce ammonia losses following slurry application

M.P.W. Sonneveld, J.J. Schröder, J.A. de Vos, G.J. Monteny, J. Mosquera, J.M.G. Hol, E.A. Lantinga, F.P.M. Verhoeven and J. Bouma
Wageningen University and Research Centre, Department of Environmental Sciences, P.O. Box 47, 6700 AA Wageningen, The Netherlands

Introduction

Literature suggests that the percentage of ammonium-N that volatilises to the atmosphere following different slurry application methods varies from 12% (injection) and 20-29% (trailing shoe) to 68% (surface spreading). Dutch regulations for ammonia emission now require farmers to inject slurry into the soil (shallow) or to apply it in small bands at the surface. For one particular dairy farm in the Netherlands (the 'Spruit' dairy farm) who has violated these regulations it was hypothesised that its alternative farm strategy could be equally effective for ammonia emission abatement as compared with the low-emission techniques required by law. The management strategy of this farm includes low-protein feeding, the use of bedding material, the application of slurry by surface spreading during rainy weather or adding water and sludge after slurry application. The overall objective of our research was to investigate for 2004 and 2005 how management at this farm is related to nitrogen (N) losses to the environment, including also ground- and surface water. However, in this paper we will limit ourselves to NH_3 emissions from slurry application alone.

Materials and methods

The Spruit dairy farm covers 37.1 ha of grassland on peat soil with a total of 79 milking cows. The farm furthermore holds almost 80 young animals and some bulls and sheep. More than 517,000 kg milk on average has been produced annually. The integrated horizontal flux method was used to measure the emission from one field (4 ha) following manure applications by the farmer. The total ammonia emission was determined over a period of 36 hours in different phases: a first phase during spreading, a second phase until the morning of the next day and a third phase till the end of the next day. Measurements of ammonia emission were effectively determined two times in 2004 and once in 2005. Two experiments were carried out with the micrometeorological mass balance method. In 2004, the slurry from the Spruit farm was applied on two grassland fields using broadcast spreading. In 2005, the slurry from the Spruit farm was applied using broadcast spreading on one field, whereas the slurry from the nearby Zegveld experimental farm was applied using band-application with a trailing shoe machine.

Results

Results of the ammonia measurements the flux-window experiments are presented in Table 1 and results of the ammonia measurements using the micrometeorological mass balance method are presented in Table 2.

Table 1. NH_3-emissions of the flux-window experiments.

Experiment	Emission kg NH_3-N ha^{-1}	Emission% applied NH_4-N	Emission% applied total-N
2 (2 phases)	3.7	18.5	6.5
3 (all phases)	3.5	17.8	6.9
4 (all phases)	10.9	68.4	27.3

Table 2. Ammonia emissions after slurry application measured with the micrometeorological mass balance method using different slurries and application methods.

Origin of slurry and application method	Year	Emission kg NH_3-N ha^{-1}	Emission % applied NH_4-N	Emission % applied Total-N
'Spruit'; surface spreading	2004	5.4	36.0	17.0
'Spruit'; surface spreading	2004	4.5	31.1	14.7
'Spruit'; surface spreading	2005	2.9	26.2	12.8
'Zegveld'; trailing shoe	2005	7.0	29.5	12.8

Discussion

The ammonia emission from the fourth experiment was higher than the previous experiments. Slurry application in this experiment was performed later than desired by the farmer because wind direction was not suitable for earlier measurements. Furthermore, anticipated rainfall did not take place which contrasts with the previous experiments where rainfall occurred (exp. 2) or water and sludge was applied (exp. 3). Experiments with the micrometeorological mass balance method showed that NH_3 emission was not significantly different between the two manure application techniques. Ammonia emission ranged from 26–36% of the applied NH_4-N (2.9–7.0 kg NH_3-N ha^{-1}).

Conclusions

The results for this specific farm suggest that farmers can reduce ammonia losses following slurry application via other measures than the ones embedded in current legislation.

Acknowledgments

The authors thank the Dutch Ministry of Agriculture and Wageningen University and Research centre for supporting the work. We also thank the Spruit family for their assistance.

The impact on ammonia emissions of strategies to reduce nitrate leaching losses following cattle slurry applications to grassland

J.R. Williams[1] E. Sagoo[1], B.J. Chambers[2], K.A. Smith[3], J.A. Laws[4]and D.R. Chadwick[4]
[1]ADAS Boxworth, Battlegate Road, Boxworth, Cambridge, CB3 8NN, United Kingdom
[2]ADAS Gleadthorpe, Meden Vale, Mansfield, Notts., NG20 9PF, United Kingdom
[3]ADAS Wolverhampton, Wergs Road, Woodthorne, Wolverhampton WV6 8TQ, United Kingdom
[4]IGER North Wyke, Okehampton, Devon, EX20 2SB, United Kingdom

Introduction

Around 90 million tonnes of farm manures supplying 450,000 tonnes of nitrogen (N) are applied to agricultural land in the UK each year (Williams *et al.*, 2001). Nitrate Vulnerable Zone legislation, which covers 55% of agricultural land in England, prevents autumn slurry and poultry manure applications on sandy and shallow soils in order to reduce nitrate leaching losses. This study was set up to investigate the effect that strategies to reduce nitrate leaching losses, (e.g. by moving slurry applications from autumn to spring and summer) on ammonia emissions from free-draining soils.

Methodology

Defra-funded experiments were set up on 2 commercial dairy farms in England (Table 1). Cattle slurry was applied at different timings between October and June/July in harvest season 2002/03 at the Cheshire farm and 2004/05 at the Somerset farm. At both farms, there were three replicates of each application timing and an untreated control arranged in a randomised block design. Slurry was applied to 24 x 24 m plots using commercial farm equipment. Ammonia emissions were measured for 7 days after each slurry application, using the micro-meteorological mass balance technique (Denmead, 1983). Nitrate leaching losses were measured following the autumn application timing and from an untreated control, using porous ceramic cups (10 per plot) installed at 60-90 cm depth. Nitrate-N concentrations in the porous cup water samples (collected after every 25 mm of drainage) were combined with estimates of over winter drainage volumes to calculate nitrate-N leaching losses (kg/ha). Crop dry matter yields and N uptake were measured at harvest (May for 1st cut silage and July for 2nd cut silage).

Results

At the Cheshire farm, ammonia emissions (Figure 1) were highest ($P<0.05$) following the slurry application in early June (before second cut) at 17% of the total N applied and lowest following the October timing at 4% of the total N applied. The higher ammonia losses following the early June timing were most probably due to a combination of higher soil temperatures (15 °C), 'dry' soil conditions and lack of grass cover (<5cm height), compared with the other application timings (soil temperature range 1-10 °C before first cut, and grass heights > 7.5 cm). At the Somerset farm, ammonia emissions were highest (P<0.05) when wet soil conditions in March (14% of total N applied) and dry soil conditions in June (26% of total N applied) inhibited the ability of the shallow injector to create sufficiently deep slots that will retain all of the applied slurry. Ammonia losses from the other timings, ranged between 5 and 9% of total N applied.

Table 1. Site details and cattle slurry application timings.

Harvest year	Site	Application method	Application timing
2003	Cheshire	Trailing hose (12 m boom)	Oct 2002, Feb, Mar and Apr 2003 (before 1st cut); Early June and late June 2003 (before 2nd cut)
2005	Somerset	Shallow injection (6m boom)	Oct 2004, Mar, early/late Apr 2005 (before 1st cut); June and July 2005 (before 2nd cut)

At both farms, nitrate leaching losses following the October application timings (208 mm of drainage at the Cheshire farm, and 170 mm of drainage at the Somerset farm) were less than 5% of total slurry N applied and were not different ($P>0.05$) to the untreated control. The low leaching losses were probably a reflection of the uptake of slurry N by the growing grass swards following application. There was no effect from the slurry applications ($P>0.05$) on either grass dry matter yields or N offtakes at first or second cut harvests.

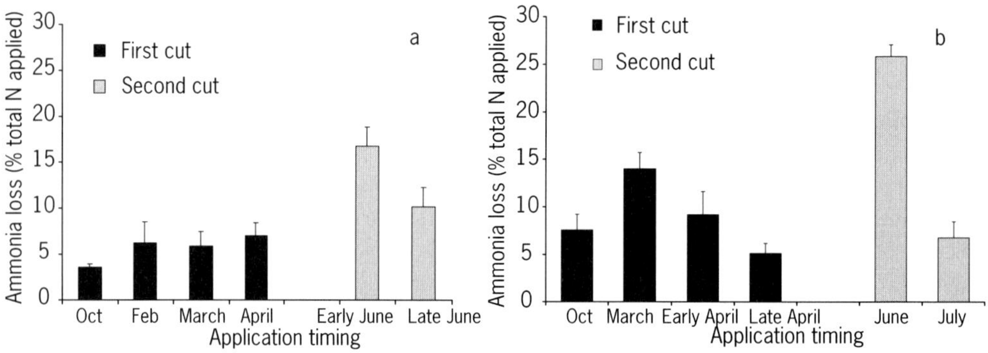

Figure 1. Ammonia emissions following different slurry application timings made with (a) a trailing shoe at the Cheshire farm in 2002/03 and (b) shallow injector at the Somerset farm in 2004/05.

Conclusions

These studies indicate that strategies to reduce nitrate leaching losses (e.g. moving slurry application timings from autumn to late spring/early summer) are likely to exacerbate ammonia emissions as a result of applications being made under warmer conditions and where slurry infiltration rates into soil are reduced (so called 'pollution-swapping').

References

Denmead, O.T. 1983. Micro-meteorological methods for measuring gaseous losses of nitrogen in the field. In: Gaseous Loss of Nitrogen from Plant -Soil Systems, Freney, J.R. and Simpson, J.R. (eds.), Martinus Nijhoff, The Hague, pp. 131-157.

Williams, J.R., B.J. Chambers, K.A. Smith and S. Ellis, 2001 Farm manure land application strategies to conserve nitrogen within farming systems. Proceedings of the SAC/SEPA Conference: 31 March- 2 April 1999, Agriculture and Waste Management for a Sustainable Future, pp. 167-179.

Agricultural scenarios for European air pollution policy

Z. Klimont, M. Amann, W. Asman, I. Bertok, J. Cofala, C. Heyes, M. Posch, W. Schöpp and F. Wagner
International Institute for Applied Systems Analysis (IIASA), Schlossplatz 1, 2361 Laxenburg, Austria

Background

In 2005, the Commission adopted a communication on a Thematic Strategy on air pollution (TSAP) (CEC, 2005). Among others, the Strategy also indicates the levels of ammonia emissions reductions and possible measures that may be required in order to meet the objectives. Currently, the work is ongoing to review the National Emission Ceiling Directive (NECD) and proposed targets will need to comply with the targets of TSAP (Figure 1). IIASA's RAINS/GAINS models (e.g. Amann *et al.*, 2004) have been used to support the discussion under these activities. This paper presents assumptions used for and results of the optimised scenario that meets Strategy targets, focusing on agriculture.

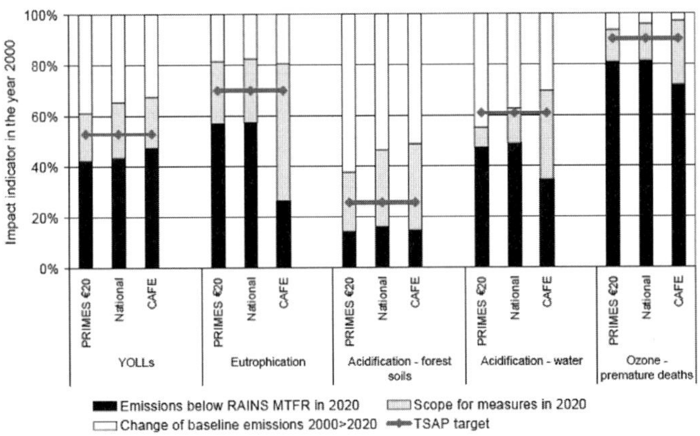

Figure 1. Changes in the impact indicators for 2020.

Principal assumptions

The NECD analysis makes use of national and European projections of livestock numbers and mineral fertiliser use. Although, on the EU level, the trends in the used activity scenarios are similar, i.e. decline in cattle and fertiliser use and slight increase in pig meat and poultry production, there are important differences for some countries that result in up to 10% variation in ammonia emissions between discussed scenarios. Similarly, significant differences in interpretation of penetration of abatement in the future, particularly of the IPPC directive, bring variation of up to 10% in estimated 2020 emissions for specific countries. On EU-25 level the difference between the interpretations has been estimated at about 150 kt NH_3 which is nearly 40% of the estimated change between 2000 and 2020 emissions in the current NEC baseline scenario (Klimont *et al.*, 2006). This has important implications on efforts and associated costs needed to reach the TSAP targets in specific countries as well as plans for revision of IPPC directive.

Results and discussion

Currently estimated development in air emissions (NEC baseline) will not allow achieving TSAP goals (Figure 1), although significant reductions of ecosystem and health related impacts are expected (indicated by light grey bars in Figure 1). According to the most recent calculations within the NECD review work (Amann *et al.*, 2006), achieving TSAP targets in the EU-25 would be associated with a reduction (compared to 2000 levels) of nearly 30% of NH_3 emissions as well as reductions of other air pollutants. The estimated burden varies from country to country from only few percent to well over 30% (Figure 2). Compared to the baseline scenario this represents additional reduction of 20% and could require significant investments in agricultural abatement technology.

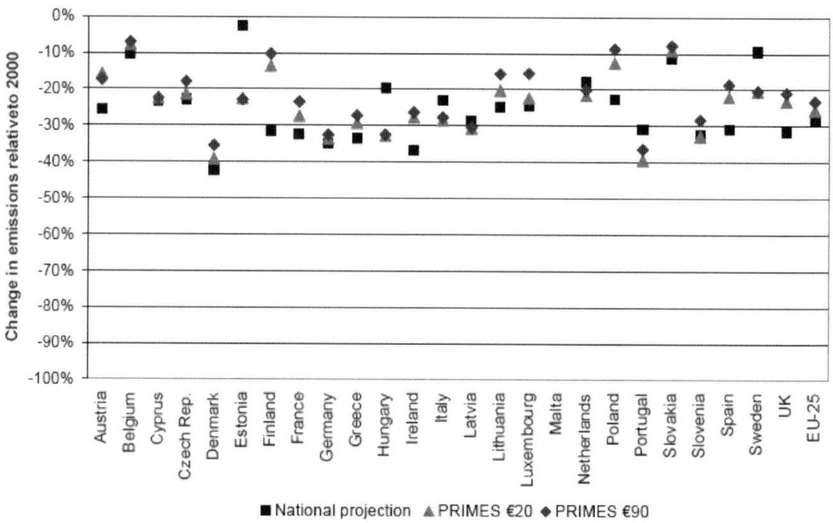

Figure 2. NH_3 emissions for the cost-optimised scenarios that meet in 2020 the TSAP environmental targets, for the three activity projections, relative to the emissions in 2000 (PRIMES scenarios differ in assumptions on carbon tax price but agricultural projections are the same for both of them).

References

Amann, M., W. Asman, I. Bertok, J. Cofala, C. Heyes, Z. Klimont, M. Posch, W. Schöpp and F. Wagner, 2006. Emission control scenarios that meet the environmental objectives of the Thematic Strategy on Air Pollution. NEC Scenario Analysis Report Nr.2. IIASA, Laxenburg, Austria.

Amann, M., J. Cofala, C. Heyes, Z. Klimont, R. Mechler, M. Posch and W. Schöpp, 2004. The RAINS model. Documentation of the model approach prepared for the RAINS review. International Institute for Applied Systems Analysis (IIASA), Laxenburg, Austria, www.iiasa.ac.at/rains/review/index.html

CEC, 2005. Communication from the Commission to the Council and the European Parliament on a Thematic Strategy on Air Pollution. SEC(2005) 1132. Commission of the European Communities, Brussels.

Klimont, Z., W. Asman, I. Bertok and F. Gyarfas, 2006. Measures in agriculture to reduce ammonia emisisons. Interim Report for the Commission Service Contract No 070501/2006/433072/FRA/C1. IIASA, Laxenburg, Austria.

Compliance with multiple EU directives when regulating farm intensification

N.J. Hutchings, I.S. Kristensen, M.S. Jørgensen, B.M. Petersen, F.P. Vinther and N. Detlefsen
Dept. of Agroecology, University of Aarhus, Research Centre Foulum, P.O. Box 50, 8830 Tjele, Denmark

Introduction

When farmers wish to intensify their operations, the authorities must consider compliance with a range of EU Directives. Regarding nitrogen (N), these include the National Emission Ceiling (NEC), Habitat, Nitrates and Water Framework Directives. On livestock farms, the N emission sources are losses of ammonia (NH_3), oxides of nitrogen (NO_x) and dinitrogen (N_2) from animal housing, manure storage, field-applied manure and fertiliser, crops and soil. N imported to the farm as fertiliser or animal feed passes through a variety of interlinked pathways, so emissions from one source can influence the emissions from other sources. This makes it difficult to assess the effect of increases in intensification or the application of abatement measures on N losses to the environment. The farm N surplus (import - export) includes these losses but the problem is whether it is feasible in practice to partition the surplus between losses. Here we describe a prototype model developed in Denmark.

The Model

The farm N surplus (N input – N output) can be more accurately estimated than the internal N flows, so the latter should be constrained within the former. The farm N inputs are imports of the purchased items; mineral fertiliser, animal feed, bedding, animal manure, livestock and seed, and the non-purchased items; N fixation and atmospheric N deposition. The farm N outputs are the crop and animal products sold, including any livestock manure. An N flow approach is then used in the calculation of internal N flows and emissions. The model can currently simulate pig, cattle and arable farms. The following inputs will be known for the farm after the proposed intensification; the number and type of livestock to be kept, the animal housing and manure storage facilities to be used, the area and soil type of the fields available to the farm, previous land use and proposed field management (cropping, fertilisation, manure application method), the proportion of the production of each crop to be sold, whether any straw produced is to be sold and whether a crop is to be grazed. For livestock, the production parameters will be known, e.g. the expected growth rate, annual milk production. For ruminants, an estimate of the proportion of feed from home-grown crops will be known. Standard values are available for the dry matter, energy and protein in crop production, depending on soil type and assuming the maximum N fertilisation permitted by national legislation. The model calculates the import of animal feed that is necessary to satisfy the livestock requirements from standard values or relationships. If crop production exceeds livestock requirements, the surplus will be sold. N excreted in faeces and urine is estimated, based on the feed ration and the N partitioned to animal products. The type of animal housing determines the type of manure produced and the addition of N in bedding. The emission of N as ammonia (NH_3) from animal housing, as NH_3, nitrous oxide (N_2O) and dinitrogen (N_2) from manure storage and as NH_3 following field application are then estimated using standard emission factors for each combination of manure type x application technique (Hutchings *et al.*, 2000). The maximum capacity of the crops to utilise N is calculated from the crop mixture and the maximum N fertilisation permitted for each crop. If insufficient manure N will be available, the deficit is filled using supplementary mineral N fertiliser. If the manure N available exceeds the permissible application, the surplus must be exported. The N input to the fields (mineral N + manure N – field NH_3 emissions) and the amount exported from the fields (from crop harvested) are now calculated. The difference is then partitioned between losses of N_2O, N_2, NO_3 and changes in the soil N. Simple models are used for soil N_2O and N_2 emissions via denitrification (Vinther and Hansen, 2004), NO_3 leaching (Simmelsgaard and Djurhuus, 1998) and the change in soil N (Petersen *et al.*, 2002). Since the sum of the total predicted N loss and change in soil N is inevitably either be greater or lesser than the farm N surplus, it is necessary to partition the error. A logical method might be to partition the error in proportion to the relative contribution of each loss to the total N surplus.

Discussion and conclusions

The tool has a number of advantages. Farmers can be reasonably expected to provide the data demanded. The model is simple so many scenarios for future management can be investigated quickly. Implementation via the internet (see demo version at www.farm-n.dk) means the model can be efficiently updated. However, it will not be easy to enable the model to simulate the effect of abatement measures. In particular, these abatement measures will alter the input of N to the field, so that the standard crop yields can no longer be used. Response functions are available but on farms where crops are used as animal feed (e.g. cattle farms), the protein content of the feed will also change. This means that the assumptions concerning the excretion of N will also change, so the model will have to iterate to achieve a closed N balance. The model attempts to achieve a balance between the desire to reflect the situation on individual farms and the desire to avoid demanding an excessive number of input data. Further development and testing will be required before it could be implemented and only after some time and experience will it become clear whether a satisfactory balance has been achieved. The principles underlying the model should be applicable to other countries but it must be remembered that the model is dependent on the availability of a range of standard data. This is the case in Denmark but may not be elsewhere.

References

Hutchings, N.J., S.G. Sommer, J.M. Andersen and W.A.H. Asman, 2001. A detailed ammonia emission inventory for Denmark. Atm. Env. 35: 1959-1968.

Petersen, B.M., J.E. Olesen and T. Heidmann, 2002. A flexible tool for simulation of soil carbon turnover. Ecol. Mod. 151: 1-14.

Simmelsgaard, S.E. and J. Djurhuus, 1998. An empirical model for estimating nitrate leaching as affected by crop type and the long-term N fertilizer rate. Soil Use and Manag. 14: 37-43.

Vinther, F.P. and S. Hansen, 2004. SimDen – A simple model to quantify N_2O emission denitrification. Danish Institute of Agricultural Sciences Report no 104, 47 pp. (in Danish).

Inventories and modelling - process models

Online monitoring and modelling of ammonia emission from pig units

B. Bjerg[1], H.B. Rom[2] and K. Sørensen[3]
[1]University of Copenhagen, The Faculty of Life Sciences, Department of Large Animal Science, Groennegaardsvej 2, 1870 Fredriksberg C, Denmark
[2]University of Aarhus, Research Centre Bygholm, Schüttesvej 17, 8700 Horsens, Denmark
[3]Danish Exergy Technology, Otterup Bakker 1, 9520 Skørping, Denmark

Introduction

It is known that ammonia emission from animal houses depends on a number of factors as room temperature, ventilation rate, air velocity above manure surfaces, manure removal frequency, manure temperature, floor design and feed composition. But, in general the animal producer does not know the actual magnitude of the ammonia emission from his production units, and consequently, his motivation and possibilities, to optimise these factors in order to reduce the ammonia emission, is limited. The objective of this work was to test a monitoring and modelling system that can give the producer online information on the actual ammonia emission from the production, and support his decisions on how pig units should be operated in order to limit the ammonia emission. The monitoring system implies continues measurement of ventilation rate and difference between ammonia concentration in inlet and exhaust. In addition the modelling system requires measurement of room and outdoor temperature and information on how the actual unit is used. Based on these measurements and information the monitoring and modelling system should be able to:

- Calculate the ammonia emission from one or a number batch of pigs in the herd.
- Compare ammonia emission from different batches of pigs in the herd.
- Reveal how ammonia emission depends of animal weight, ventilation rate, room temperature, outdoor temperature and changes in production conditions as feed composition.
- Notify for changes in ammonia emission.

Materials and methods

The tests were carried out from October to December 2005 in one of five equal sections in a commercial growing pig facility equipped with a central ventilations system. The ventilation control system included measuring vanes in the exhaust from each section, which were utilised to determine the ventilation rate from the studied section. In addition the tested monitoring system included ammonia and carbon dioxide measuring equipment from both Veng (based on Dräger Polytron electrochemical sensor, see www.vengsystem.com and www.draeger.com) and Innova (Photoacoustic Multi-gas Monitor 1314 in connection with Multipoint Sampler 1309, see www.innova.dk). Figure 1 illustrates the entire control, monitoring and modelling setup. The monitoring and modelling system were controlled by a PC with software entitled AME, which primarily:

- stored measurements;
- calculated hourly and daily averages;
- provided access to stated information on number and weights of animal;

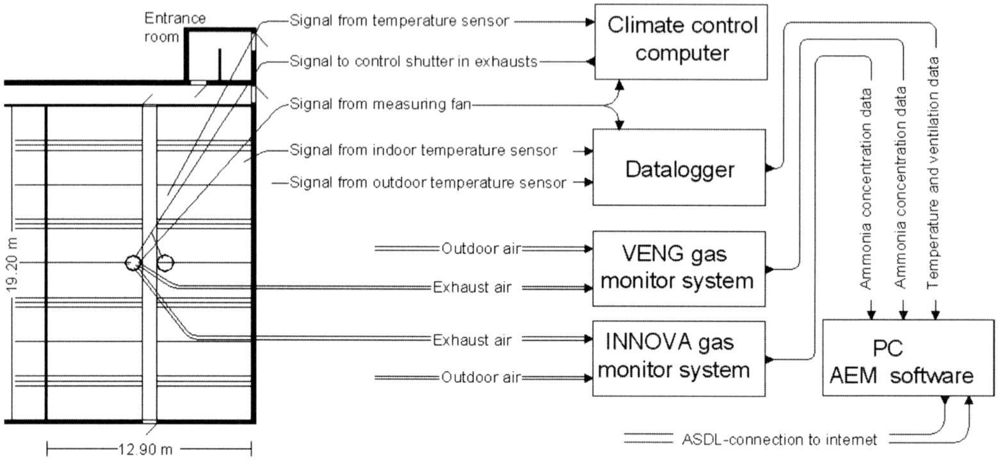

Figure 1. Control, monitoring and modelling setup.

- allowed utilisation of R-routines (www.r-project.org) to examine how the daily average ammonia emission depended on animal live weight, ventilations rate, room temperature, out door temperature and possible changes in how the section were operated (linear regression).

Results, discussion and conclusions

The test showed that it was possible to operate the ammonia monitoring and modelling system with two different gas monitors in a commercial pig production unit. Unfortunately the test period were marked by difference incidents that interrupted the measurements and created large gaps in the dataset, which the modelling system wasn't sufficiently prepared for. The collected data included 1616 hours where ammonia concentrations were measured with both Innova and Veng equipment. The R-square value for the correlation between the hourly values determined with the two type devices were 0.86. Among the 51 days where all 24 hours were represented the R-square value for the correlation between the daily average values determined with the two type devices were 0.93. These relative low correlations indicates that the accuracy of at least one of the used gas monitor was unsatisfactorily and, consequently, it is assessed that lack of reliable and inexpensive ammonia sensors is the largest obstacle for further development and spreading of the tested kind of ammonia monitoring and modelling systems.

Investigating ammonia reduction techniques using a dynamic model for swine barns

E.L. Cortus[1,2], S.P. Lemay[3] and E.M. Barber[4]
[1]Prairie Swine Centre Inc./ Department of Agricultural and Bioresource Engineering, University of Saskatchewan, 57 Campus Drive, S7N 5A9 Saskatoon, Canada
[2]Department of Agricultural and Biological Engineering, Purdue University, 225 South University Street, 47907 Lafayette, IN USA (current address)
[3]Research and Development Institute for the Agri-Environment, 120A chemin du Roy, G0A 1S0, Deschambault, QC, Canada
[4]College of Agriculture, 51 Campus Drive, S7N 5A8 Saskatoon, Canada

Introduction

Within swine barns, ammonia production is a function of the amount of ammonia within each urine puddle and the slurry, and the rate of convective mass transfer from each solution to the air. It is possible to measure ammonia emission from individual sources in smaller-scale studies or the cumulative emission from both sources in a room situation, but it is difficult to measure the contribution of each site to the total ammonia produced within a swine barn. This makes it difficult to quantify the contribution of source-specific factors on production rates. Considering all the factors involved in emission from urine puddles and slurry, the number of potential emission sites, and the interactions between all sites, mathematical models provide a time and cost-effective method to understand the various ammonia production methods, to develop reduction methods based on the understood principles or processes, and quantify the potential impact of ammonia reduction techniques or technologies. The objective of this paper is to demonstrate the application of a new dynamic ammonia production model for swine barns and the implications for identifying ammonia reduction techniques.

Dynamic ammonia production model for swine barns

Cortus (2006) developed and tested a dynamic, mechanistic, ammonia production model for grower-finisher swine barns, named the Ammonia Concentration and Emission Simulation (ACES) model. There are two main aspects that differentiate the ACES model from previous swine barn models. First, in the ACES model the room and slurry pit headspace are considered two separate control volumes linked by the air exchange rate through the slatted floor. When tested, this aspect showed that the majority of emissions (>95%) from a room could originate from the urine puddles on the solid and slatted floor surfaces under the right conditions. The second unique aspect of the ACES model is that in determining the emission rate from individual urine puddles and the slurry, the ammonia concentration of the surrounding air for each emission site is included in the convective mass transfer calculations. This aspect is important because urine puddles and slurry have the potential to absorb ammonia, as well as emit ammonia.

Model application

There are an endless number of factor combinations that can be tested with a model such as ACES, and some of these combinations can lead to effective ammonia reduction techniques. Three possible combinations were tested: (1) the pigs' diet was changed such that the urea concentration was reduced to 0.21 mol L^{-1} (from 0.30 mol L^{-1}), the total ammoniacal nitrogen concentration (TAN) of the slurry was reduced to 0.23 mol L^{-1} (from 0.30 mol L^{-1}) and the slurry pH was reduced to 6.3 (from 7.0); (2) the animals were trained to use only a small portion of the floor surface for dunging behaviour, reducing the slatted floor area to 10% of the total pen area (from 30%) with 50% of this floor area fouled, and the urine and faeces collected beneath the slatted floor were instantly removed (like a toilet system), theoretically making the slurry emission non-existent; or (3) a thin layer of oil was applied to the slurry surface which reduced the TAN content of the slurry to 0.05 mol L^{-1} (from 0.30 mol L^{-1}). The simulated emission rates resulting from these three combinations of factors were compared to simulated emission rates from a control set of conditions under a low ventilation rate (1.8 L s^{-1} pig^{-1}) typical of mechanically ventilated barns in cooler climates (Zhang, 1994). Table 1 illustrates that the two unique aspects of the ACES model discussed previously have important impacts on the simulated emission levels and evaluation of reduction techniques. First, because the majority of the total emission in both high and low ventilation conditions originated from the floor surface, reduction methods aimed at reducing the ammonia emission from urine puddles (Diet change and Pig toilet) were more effective than just reducing the slurry emission (Slurry cover). Second, even though changes to the slurry composition had small impacts on the total room emission (i.e. Slurry cover), a low ammonia concentration at the slurry surface resulted in the slurry absorbing ammonia from the air in the pit headspace under low ventilation (high concentration) conditions when the pit headspace concentration was high. These results highlight how the evaluation of ammonia reduction techniques requires an understanding of the ammonia production rates from the different sources. Also, an effective evaluation would combine the results of many simulations covering the expected range of all environmental variables.

Table 1. Simulated ammonia emission rates for a grow-finish swine barn and the percent reduction in emission levels using three possible ammonia reduction techniques.

	Emission, g NH_3-N d^{-1} AU^{-1}	Reduction in emission, %		
		Diet change	Pig toilet	Slurry cover
Floor	12.7	12	38	3
Slurry	0.5	146	100	152
Total	13.1	17	40	8

Conclusions

The impact of factors related to floor or slurry emission on the total emission rate from swine barns is dependent on the average simulated proportion of emission coming from each source, and increases in emission from one source can decrease the rate of emission from the other source. To be accurate, the assessment of reduction techniques should encompass all interactions between ammonia sources.

References

Cortus, E.L., 2006. A dynamic model of ammonia production within grow-finish swine barns. Ph.D. Thesis. Department of Agricultural and Bioresource Engineering, University of Saskatchewan, Saskatoon, Canada.

Zhang, Y. 1994. Swine Building Ventilation: A Guide for Confinement Swine Housing in Cold Climates. Saskatoon, SK: Prairie Swine Centre Inc.

Ammonia emission from sources in animal houses

H.-J. Müller and A. Kiwan
Leibniz-Institute for Agricultural Engineering e.V. (ATB), Department of Engineering for Livestock Management, Max-Eyth-Allee 100, 14469 Potsdam, Germany

Introduction

To limit the negative effect of the emitted ammonia from livestock buildings, emissions should be limited and minimum distances between livestock buildings and areas with sensitive plants and ecosystems must be established. Knowledge about ammonia emission behaviour in animal houses is needed to calculate the minimum distance and to find solutions for ventilation systems reduce the ammonia emissions. Therefore, basic research into emission streams of ammonia in special laboratories, in scale models and real livestock buildings is to be carried out by the Leibniz-Institute of Agricultural Engineering Potsdam-Bornim (ATB). A main aim is to investigate the influencing factors, such as air temperature, air humidity, air velocity and the pH-value of the sources, on the mass transport processes. The mass transfer coefficient is determined using special test facilities and scale models. One of the main influencing factors is the air velocity and the turbulence near the surface of the source.

Basics

The determining factor for the evaluation of emissions is the emission mass flow. This mass flow is strongly influenced by the air flow conditions inside the livestock building and especially by the conditions near the surfaces of the sources. The air flow conditions are mainly characterised from the air velocity, the flow direction and the turbulence of the flow. The basic equation for the mass transfer from the source into the air:

$$\dot{m}_\alpha = \beta \cdot \Delta c_\alpha \cdot A \quad (1)$$

$\dot{m}_\alpha$	kg/s	mass flow of substance α
β	m/s	mass transfer coefficient
Δc_α	kg/m^3	concentration difference of the substance α between surface of the source and the free atmosphere
A	m^2	the emitting area

To determine the mass transfer coefficient is a difficult task. Müller (1994) has described different methods to investigate the mass transfer coefficient β.

Materials and methods

Empirical models are used to determine the ammonia emission stream. The influence factors on mass transfer coefficient like temperature, pH-value and air velocity have been tested at ATB. This investigation takes place in the wind tunnel and in a stable model, furthermore ATB accomplishes measurements on ammonia emission streams in real animal stables.

Results and discussion

Many models have been reported in the literature that calculate the emission streams. Ni (1999) for example points at the importance of the convection mass transfer and the mass transfer coefficient, which ranges from 11.7×10^{-3} to 1.3×10^{-6} m/s. Figure 1 shows mass transfer coefficients determined by the ammonia absorption method. This method was in detail described by Hanel *et al.*(1990). The experiment took place in a ventilation model (Müller, 1994), where the emission source was simulated by using special filter paper, that was saturated in manganese (II) chloride solution. During the experiment and through the effect of air velocity a well known quantity of gaseous ammonia would be added to the supply air, therewith the used filter paper discolours and can be optically evaluated. So the different mass transfer coefficients at the interface of emission source can be determined. The investigation in real animal housing shows an increasing of emission flow by rising air flow through the building.

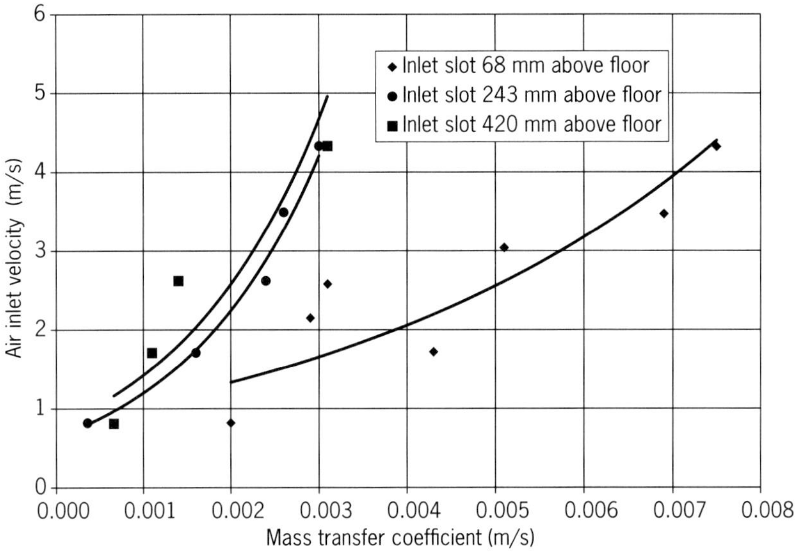

Figure 1. Connection between mass transfer coefficient and air inlet velocity model testing (different arrangement of air inlet) (Müller, 1994).

Conclusions

- The air flow near the surface of emission sources influence the emission mass flow.
- Reduction of emission demand low air velocity near the surface of sources.

References

Hanel, B., R. Osterburg and E. Richter, 1990. Untersuchung des Wärmeüberganges an komplizierten Geometrien mit der Ammoniakstoffübergangsmethode. Luft- und Kältetechnik, H. 4, S. 171-175.

Müller, S., 1994. Modelluntersuchungen zum Einfluss der Lüftung auf die Emission von Schadgasen aus Ställen. Grosser Beleg. TU Dresden. Institut für Strömungsmechanik.

Ni, J., 1999. Mechanics models of ammonia release from liquid manure: a review. J. Agric. Eng. Res. 72: 1-17.

How the use of a mechanistic model of ammonia volatilisation in the field may improve national ammonia volatilisation inventories

J. Rosnoblet[1], M. Theobald[2], S. Génermont[1], B. Gabrielle[1] and P. Cellier[1]
[1]INRA, EGC Unit, 78850 Thiverval-Grignon, France
[2]CEH, Penicuik, Midlothian, EH26 0QB, United Kingdom

Introduction

Most national inventories of ammonia emission from farm manure application to agricultural land use constant emission and abatement factors. However, ammonia volatilisation highly depends on the agri-environmental conditions, and is thus very variable. This presentation (based on the work of Theobald *et al.*, 2005) aims to show how the use of a mechanistic model of ammonia volatilisation, Volt'Air (Génermont and Cellier, 1997), may improve the UK ammonia emission inventory, NARSES.

Materials and methods

The analysis was carried out in three parts: (1) a 'sensitivity screening' of the model to assess which parameters and variables have the largest influence; (2) scenario analysis, to assess the variability of ammonia volatilisation for UK conditions and (3) a simulation for England and Wales to compare with the UK ammonia emission inventory. The sensitivity screening was done using typical ranges of UK input variables. The scenario analyses used meteorological time series averaged over 8-10 years. The driest, wettest, coldest and warmest individual years were also used to account for variability. The England and Wales simulation was done by applying the Volt'Air model to each 10 km grid square in the NARSES inventory for those countries (1,651 in total) and compared with the baseline inventory. NARSES calculates volatilisation by multiplying the mass of total ammoniacal nitrogen (TAN) applied by a fixed source-specific emission factor which is derived from field studies. We considered cattle slurry applied by broad-spreading (NARSES emission: 42% of TAN applied), for two application dates (January and July).

Results

Sensitivity screening showed that the emissions predicted by Volt'Air are most sensitive to soil pH, manure TAN and soil water content at the beginning of the simulation. Scenario analysis at the UK level showed that the model was markedly influenced by application date (CV 25.1-59.9%), soil type (24.2-27.1%), manure type (97.0-113.1%) and application/abatement technique (17.8-42.2%), and less influenced by the choice of meteorological station used for the input data (10.8-13.1%). In the England and Wales simulation, the proportion of TAN applied that is volatilised from each 10 km grid square differed greatly (spatially and temporally) between the two approaches (Figure 1). Volt'Air calculated a mean volatilisation of 20% of TAN applied in January (range: 0-60%) and 42% in July (0-84%), compared with the constant 42% of NARSES. The mapping of Volt'Air results shows a strong correlation between volatilisation rate and soil pH (Figure 2), which is not currently taken into account in NARSES. For January emissions, the predicted rates are generally lower than the fixed rate used in NARSES for soil pH less than 6.5

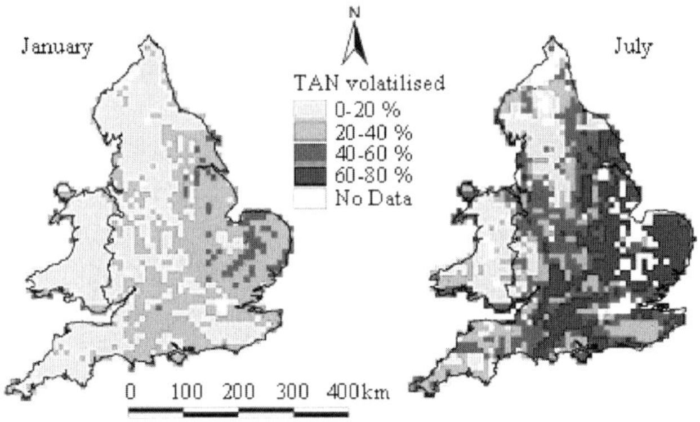

Figure 1. Grid-square % volatilisation from the Volt'Air model for January and July.

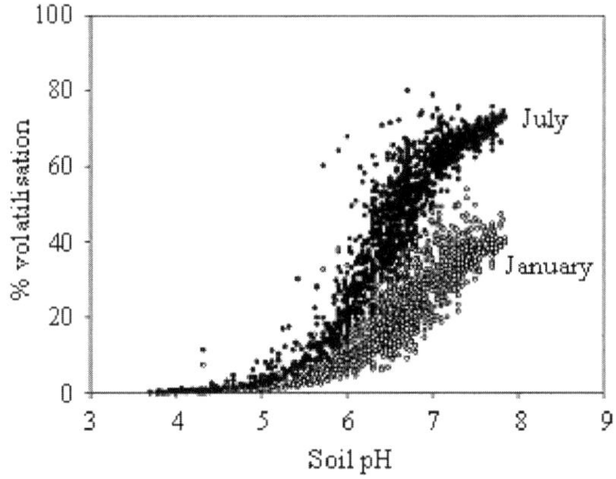

Figure 2. Dependence of the % volatilisation on soil pH (Volt'Air model, January and July).

and higher above this value. For July emissions nearly all of the predicted rates were lower than NARSES for all values of soil pH.

Conclusions

This analysis showed the potential of using a mechanistic model to account for the sources of variability in ammonia emission due to soil, climate and agricultural practices. However, care should be taken when using coarsely defined spatial and temporal data (e.g. mean monthly meteorological data and 10 x 10 km soil data).

References

Génermont, S. and P. Cellier, 1997. A mechanistic model for estimating ammonia volatilization from slurry applied to bare soil. Agric. and For. Meteorol. 88: 145-167.

Theobald, M.R., A.G. Williams, J. Rosnoblet, C. Campbell, T.R. Cumby, T.G.M. Demmers, B. Loubet, D.J. Parsons, B. Gabrielle, E. Nemitz, S. Génermont, M. Van Oijen, P. Cellier and M.A. Sutton, 2005. An assessment of how process modelling can be used to estimate agricultural ammonia emissions and the efficacy of abatement techniques. DEFRA, SID 5 Research Project, AM0130, 15 p. + appendices.

NH_3 emission from manure: physicists' view

Mikko Hautala
Department of Agrotechnology, P.O.Box 28 (Koetilantie 3), FI-00014 University of Helsinki, Finland

Introduction

The final aim in emission studies in agriculture is to find the mechanisms that affect the amount of emissions. This knowledge is needed in order to find the best available techniques in farm level to decrease the emissions. Presently inventories have mainly focused on determining the total amounts of emissions from various sources. This is the first step in finding ways to decrease the emissions. However, there are various reasons, why experimental data and therefore inventories may be very inaccurate. Fresh manure with urine contains about ten folds more ammonia than dryer old manure. NH_3 content depends on the temperature and pH of the manure. Furthermore emission depends on air velocity v above the manure. The mass transfer coefficient which determines the flux through the surface boundary layer is proportional to $v^{4/5}$ in normal turbulent conditions. The challenge is to measure v correctly. Finally, the emission factors for NH_3 emission are often given per cow or livestock unit even though emission does not depend on these but solely on the area of the manure. Emissions are possible only if ammonia molecules are transported from manure to air. Physics tells that there are two and only two possibilities: convection and diffusion. If one somehow could prevent the presence of these transport processes, we would have no emissions. This is theory. Trying to minimise these processes is practice. Trying to find practical ways to minimise emissions should be based on physics and understanding the basic physical processes is to be encouraged. Preventing or minimising the reactions that generate ammonia is chemistry and microbiology.

Theory

The NH_3 emission from manure can be separated into different processes. NH_3 molecules are first created in the manure in various chemical and microbiological processes. Then the molecules diffuse to the surface of the manure. From the surface they further diffuse in air through the laminar boundary layer and finally by turbulent convective motion into all parts of the building. Turbulent motion is assumed to be fast enough to yield space independent concentrations. This is the so called ideal mixing model. NH_3 emission rate is theoretically modelled using information from literature. First, the surface concentration of NH_3 is calculated using equations adopted from Zhao and Chen (2003): the amount of NH_3 dissociation in the manure, the fraction of NH_3-N concentration in the total ammoniacal nitrogen C_{TAN} (kg/m^3) and the ratio of the NH_3 concentration at the manure side of the interface between the manure and air and the NH_3 concentration at the air side of the interface between the liquid manure and air. This comes from chemistry and microbiology. Then physics, i.e. Fick's diffusion law and boundary layer theory are used for mass flux calculation from the manure surface. The final approximative equation (deviates less than 50 % from exact calculations) is:

$$emission\,flux\,(g/m^2/h) = 0.02 \cdot 10^{T(°C)/20+pH-8} \cdot C_{TAN}\,(kg/m^3)\,/\,\delta\,(mm) \quad (1)$$

where δ (in mm) is the thickness of the laminar boundary air layer and presents the physics part of the equation in addition to the fact that emission rate is strictly proportional to emitting area, when diffusion plays a role. δ varies from 2 to 20 mm depending on wind. 20 mm calm, 2 mm strong wind. δ is pretty well known for flat surface with uniform wind (Incropera and de Witt,1990). Neither of these assumptions is strictly valid in practical cases. Equation (1) is valid for any cover. For a porous cover δ is the thickness of the cover. For liquid or solid cover flux should be divided by at least 10^4 since diffusion coefficient is so much smaller in liquid than in air. If the manure is put into soil δ is the relevant depth of soil above the NH_3 level. This is what physics tells. Equation 1 predicts that if manure temperature changes by 20 degrees or pH changes by one unit, the emission rate changes by one magnitude. These predictions come from the microbiological and chemical equations (Zhao and Chen, 2003). Probably the dependences are not too badly wrong. However, the magnitude may be less accurate (e.g. Ni, 1999).

Discussion

We have tested the simple Equation (1) by emission measurements in numerous free stalls in Finland and Estonia both in summer and winter (Hautala and Teye, unpublished). The ammonia emissions were measured using three partly independent methods, i.e. using a properly modelled and calibrated closed chamber technique and using methods based on carbon dioxide and methane balances. All three experimental methods and theory (Equation 1) were mostly within an order of factor two agreement when the NH_3 emission rates varied by three orders of magnitude, from 0.001 g $m^{-2}h^{-1}$ to 1 g $m^{-2}h^{-1}$. The time dependence of ammonia concentration in the closed chamber in chamber measurements further indicates that the ammonia molecules originated from the very surface (less than 1 mm). The proper modelling of the emission process and the proper size of the chamber were essential in acquiring this information. The moderate agreement between theory and the three experimental methods gives some support that the theory would have the essential phenomena included at least in case of cow manure. However, it is to be noted that the variation in emission mainly came from the variation of temperature. The small emission rates are from winter and the large ones from summer. Equation (1) should be a useful aid when discussing the practical ways of ammonia abatement. Also closed chamber measurements with good enough resolution of time dependence of gas concentrations are very useful in understanding the physical processes relevant in ammonia emissions. These will be discussed in detail separately (Hautala and Teye, unpublished).

References

Incropera, F.P. and D.P. De Witt, 1990. Fundamentals of heat and mass transfer. Third edition, Wiley and Sons, New York, pp 439.

Ni, J., 1999. Mechanistic models of ammonia release from liquid manure: a review. J. Agr. Eng. Res. 72: 1-17.

Zhao, B. and S. Chen, 2003. Ammonia volatilization from dairy manure under anaerobic and aerated conditions at different temperatures. ASAE Annual International Meeting. ASAE Paper Number 034148.

Inventories and modelling - national inventories

Spanish methodology for the improvement of ammonia emissions inventories from livestock

M. Cambra-López[1], F. Estellés[1], S. Calvet[1], A.G. Torres[1], N. Alonso[2] and A. Cristóbal[2]
[1]Universidad Politécnica de Valencia. Animal Science and Technology Institute, Camino de Vera, s/n 46022, Valencia, Spain
[2]Ministerio de Medio Ambiente. Air Quality Unit, Plaza de San Juan de la Cruz, s/n 28071, Madrid, Spain

Introduction

Intensive livestock operations are the main source of atmospheric ammonia, which is emitted from animal houses, manure storage systems and when manure is spread on land (NRC, 2003). European regulations concerning ammonia imply that all Member States must control and reduce ammonia emissions from all sources, comply with international thresholds, and report on their emissions in the mandatory emission inventories. In Spain, livestock farming accounted for up to 90% of the total ammonia emissions in 2004 (EPER, 2006). These emissions are quantified using default values for activity data and emission factors taken from EMEP-CORINAIR (2001), which are based on the average European farming situation. However, countries are always encouraged to use their own national data to obtain accurate national emission estimates. In this context, the Spanish Ministry of the Environment, in collaboration with the Universidad Politécnica of Valencia, with the intervention of the Ministry of Agriculture, Fisheries and Food, developed a working plan to improve background information on the key factors influencing ammonia emissions from livestock systems, to obtain realistic emission estimates and improve the quality of the National Emission Inventory.

Material and methods

Considering the aim of this project, three different tasks were carried out. In a first step, technical data on representative animal housing systems and manure management systems (MMS) was collected. A questionnaire was designed for this purpose and key experts in the field were surveyed in order to identify the most common systems for each rearing system. Secondly, a methodology for on-site measurements of ammonia emissions from commercial livestock facilities was developed. A mass balance method was applied to measure ammonia emissions in the inside of animal houses using a photoacoustic gas monitor. Measurements devices were also designed to register gas concentrations at other farm locations (inside and outside). Measurements were carried out for representative livestock systems identified from the results from the questionnaire. Finally, in parallel, an update of the National Emission Inventory was addressed. All activity data from livestock and emission factors in use at present in the inventory were reviewed and certain aspects were modified in order to obtain a new methodological proposal adapted to the Spanish reality.

Results and discussion

The results from the questionnaire showed predominance of certain MMS in Spanish farms (Table 1). Other interesting conclusions were derived from the questionnaire, like the fact that specific MMS must be clearly defined to be able to match each one with its equivalent in other

European countries, and that these types of systems must be considered as open categories, to be able to represent the sequential nature of them. A descriptive guide on characterisation of Spanish farms and manure management systems is being written as a result. Regarding the on-farm ammonia measurements, although the project is still on an early stage, preliminary results have been obtained in poultry and in rabbit farms, showing consistency with the results obtained in similar situations in other Mediterranean countries. Moreover, a technical report named 'Methodology for the estimation of atmospheric emissions from the agrarian sector for the national emission inventory' (UPV, 2006) was elaborated, as a product of the deep revision of all data involved in the estimation of emissions. This new methodology has been incorporated in the last National Emission Inventory for the time series 1990-2004.

Table1. Major manure management systems (MMS) for different animal species.

Animal species		Identified MMS
Cattle	• Dairy cattle	Management in liquid or solid state + composting + land application
	• Other cattle	Solid manure + composting + land application
Swine		Slurry channel or pit + aerobic lagoons or tanks + land application
Sheep and goats		Management in solid state
Poultry	• Broiler	Broiler litter + solid storage + composting + land application
	• Laying hens	Layer manure + manure belts with or without air drying + solid storage + composting + land application
Rabbits		Management in solid state + composting + land application

Conclusions

Overall, important improvements in the National Emission Inventory for ammonia from livestock systems have been achieved, especially the methodological guide, the revision of activity data and emission factor allocation. The trials carried out so far will certainly allow, in the medium-term, the use of emission factors derived from experimental work under Spanish conditions and in the end, the proposal of realistic and adapted abatement measures in order to reduce ammonia emissions from livestock in Spain.

Acknowledgements

This work has been funded by the Spanish Ministries of the Environment and Education and Science (project AGL 2005-07297), and by the Consellería de Empresa, Universidad y Ciencia, co-funded by FEDER European funds.

References

EMEP-CORINAIR, 2001. Atmospheric Emission Inventory Guidebook, Agriculture. Third Edition, European Environment Agency, Copenhagen.

EPER, 2006. EPER-España. http://www.eper-es.es. Accessed 28th December, 2006.

NRC, 2003. Air emissions from animal feeding operations: current knowledge, future needs. The National Academy Press, Washington, D.C., 286 pp.

UPV, 2006. Methodology for the estimation of atmospheric emissions from the agrarian sector for the national emission inventory. In press.

Animal husbandry in Germany: national evaluation for animal housing systems - assessment of environmental impacts

B. Eurich-Menden[1], W. Achilles[1], H. Döhler[1], F. Eichler[2], S. Fritzsche[1], E. Grimm[1], W. Hartmann[1], A. Hielscher[1] and H. Lutz[1]
[1]Association for Technology and Structures in Agriculture (KTBL e.V.), Bartningstr. 49, 64289 Darmstadt, Germany
[2]Federal Environment Agency, Wörlitzer Platz 1, 06844 Dessau, Germany

Introduction

The goal of the 'national evaluation for animal housing systems' was the development and application of a method which allows the evaluation of the impacts of animal housing systems on the environment and animal welfare as two equally important criteria. In contrast to the 'Best Available Techniques' (BAT) for intensive livestock rearing focusing exclusively on installations for the intensive rearing of poultry or pigs, the new project takes into account smaller installations and, in addition, housing systems for cattle and horses.

Methods

Based on the technical description, production-specific performance levels (emission and consumption levels, etc.) were determined for 139 housing systems for cattle, pigs, poultry and horses. The environmental impacts of the different housing systems were assessed taking into account the emission of ammonia, odour, dust, nitrous oxide and methane, as well as nitrogen and phosphorous inputs into the soil, technical energy requirements in the housing and process water demand. Based on expert's knowledge and literature research, quantitative and qualitative evaluation schemata were developed separately for each indicator and each animal species to be assessed. The actual evaluation of the housing systems was carried out based on the quantitative and qualitative evaluation schemata using a five-step range of evaluations from 'very low' to 'very high'. The selected environmental indicators were evaluated individually. This assessment was carried out based on quantitative data, if available. If no data were available, a qualitative evaluation schema was used as a basis of the evaluation. Afterwards, the individual evaluations of the environmental indicators were summarised in an overall rating of the individual housing system. The systems were grouped into the following three categories: A – particularly advantageous, B – satisfactory, and C – sufficient for existing facilities; for new facilities and alterations, other housing systems are recommended. Since ammonia and odour are used as indicators in the permit procedure for housing facilities, they were chosen as leading indicators for the overall rating. Less importance was given to energy requirements in the housing and process water demand. In most cases the overall rating is the product of a qualitative conclusion of the experts, because no well-founded data basis was available for many of the indicators to be evaluated.

Quantitative evaluation schemata

Table 1 gives an example for a quantitative evaluation schema.

Table 1. Part of the quantitative evaluation schema for NH_3 using laying hens as an example.

Indicator	Unit	Grading				
		Very low	Low	Medium	High	Very high
NH_3 from housing	kg TP^{-1} a^{-1}	< 0.05	0.05-0.1	0.1-0.15	0.15-0.2	> 0.2

Qualitative evaluation schemata

Due to the gaps in the data, quantitative evaluation schemata were supplemented with qualitative evaluation schema (Table 2). Based on the different factors which have an influence on the emission potential of the selected indicator within a housing system, numerous schemata were developed which were adapted to the animal species and the direction of production. In these schemata, the most important influencing factors were listed, whose different variations with regard to possible emission potential were classified.

Table 2. Part of the qualitative evaluation schema forNH$_3$ using laying hens as an example.

Influencing factor	Emission potenial NH_3		
	Low	→	High
		measures and variations	
Dung remains in the stable for	≤ 1 week		≥ 1 week
N/P-reduced feeding	Available		Not available
Wintergarden (outside scratching area)	Available/no indoor scratching area	Not available/only indoor scratching area	Available/with additional indoor scratching area

Results and conclusions

Table 3 gives an example for an overall rating of a selected laying hen housing system. Due to the very high ammonia and the high odour emission potential in this housing system, the overall rating leads to category C. Of the 139 housing systems evaluated, 5.8% were evaluated as particularly advantageous (category A) with regard to their environmental impact, 87.8% were considered satisfactory (category B), and 6.4% were regarded as sufficient (category C).

Table 3. Example overview of the emission potential for different substances and the energy and water demand and the resulting category of a selected laying hen housing system.

Housing system laying hens	Emission (air)					Emission (soil) N and P	Demand		Category
	NH_3	Odour	Dust	N_2O	CH_4		Energy	Water	
Deep litter housing for laying hens	Very high	High	High	Medium	Medium	Not available	Medium	Low	C

References

KTBL, 2006. Nationaler Bewertungsrahmen Tierhaltungsverfahren. KTBL Schrift 446, Darmstadt, www.ktbl.de, 778 pp.

Agriculture emission inventory in Italy: synergies among conventions and directives

R.D. Cóndor and R. De Lauretis
Agency for the Protection of the Environment and Technical Services (APAT), Via V. Brancati 48, Rome 00144, Italy

Introduction

In the last years, the agriculture emission inventory has been updated and improved. As a result, different aspects have been considered and probably the most important one is keeping consistent methodologies for the preparation of inventories under the Convention on Long-Range Transboundary Air Pollution and the United Nations Framework Convention on Climate Change. The following sections provide an overview of ammonia emissions and projection scenarios, and describe activities which contribute with improving the emission inventory in Italy.

Ammonia emission inventory and projection scenarios

In 2004, 93.9% of the Italian ammonia emissions (98.0% in 1990) originated from the agriculture sector, 171,712 t were *agricultural soils* emissions coming from soils with fertilisers (50%) and losses from spreading (44%), and 225,264 t were *manure management* emissions coming from cattle (61%), swine (16%) and poultry (15%) livestock categories. According to Art. 3.2 of the Kyoto Protocol, we have prepared a consistent projection scenario of greenhouse gas emissions for the Italian Demonstrable Progress document. In this context also ammonia emission projections were performed for the agriculture sector. We could forecast, with a business as usual scenario, that agricultural ammonia emissions are expected to increase, by 5.0% in 2010 (410,739 t) and 3.4% in 2020 (404,589 t), both with respect to 2005. Most of the activity data used from 1990 till 2005, as the number of animals, use of fertilisers and cultivated surface/production, reflects the effects of the Common Agricultural Policy. Furthermore, in the framework of the National Emission Ceiling Directive and in cooperation with research institutions, a scenario with measures has been developed. Under this scenario ammonia emissions are expected to be 392,432 t in 2010 and 383,821 t in 2020.

Improving the ammonia emission inventory

The centralised institutional arrangement of the Italian emission inventory allows a complete control and coordination with national institutions. In addition, the emission inventory has been complemented with the National Inventory System for quality control and quality assurance (QA/QC), as required by the UNFCCC. Instead, at national level, the emission inventory is part of the Italian National Statistics System, which makes available a network of national statistics. In particular, for the ammonia emission inventory, new country-specific parameters, such as average weight, nitrogen excretion rates and the liquid and solid manure production have been updated with outcomes from the MeditAIRaneo Agriculture project. In this context, recommendations from the Integrated Prevention Pollution and Control Directive for swine and poultry categories have been used for the estimation of ammonia emission factors (APAT/CRPA, 2006a). Other specific activities will be described below.

National agriculture statistics

We have assesed and identified agricultural statistics useful for improving emission estimations, specially those coming from the Farm Structure Survey (FSS). Consequently, important steps have been achieved in cooperation with the National Institute of Statistics (Agriculture Service). National information related to the type of housing and the duration of the grazing period, are provided by the FSS 2005. Instead, queries related to manure and slurry storage and application, will be considered for future surveys.

Research activities

Two different conventions signed between APAT and the Ministry of Environment will contribute to improve emission estimates. The '*Climate Convention*', will provide data related to particulate matter measurements from the agriculture sector, for swine and poultry (APAT/ Università di Milano, 2006). On the other hand, the '*NEC Convention*' has as objective to assess the potential reduction of ammonia emissions from livestock and fertiliser emissions. For this purpose, the distribution and potential of swine and poultry farm systems, subject to the IPPC Directive, have been evaluated (APAT/CRPA, 2006b). Furthermore, the impact of the use of slow release fertilisers, organic agriculture and the use of inhibitors on ammonia emissions have been assessed (APAT/ENEA, 2006).

Conclusions

In summary, national research activities and agricultural statistics have improved the ammonia emission inventory. In the future, national agricultural surveys and other research results will contribute with future improvements. In this framework, we consider very important to implement synergies among the different international conventions and european directives, when preparing national emission inventories.

References

APAT/CRPA, 2006a. Progetto MeditAIRaneo: Agricoltura. Final Report, Reggio Emilia, Italy.

APAT/CRPA, 2006b. Predisposizione di scenari di emissione finalizzati alla progettazione di interventi per la riduzione delle emissioni nazionali di ammoniaca ed alla valutazione di misure e di progetti per la tutela della qualità dell'aria a livello regionale. Final report. Reggio Emilia, Italy

APAT/ENEA, 2006. Valutazione della possibilità di sostituzione dell'urea con altri fertilizzanti azotati. Final report. Rome, Italy.

APAT/Università di Milano, 2006. Valutazione dei fattori di emissione di particolato e dei gas serra ed ammoniaca, in relazione alle techniche di abattimento di inquinanti atmosferici. Report N°3 period 07/07/2006-07/11/2006. Milano, Italy.

Assessment of manure management systems in Austria and improvement of the emission inventory

B. Amon, M. Fröhlich and T. Amon
University of Natural Resources and Applied Life Sciences, Department of Sustainable Agricultural Systems, Division of Agricultural Engineering (DAE), Peter Jordan-Strasse 82, 1190 Vienna, Austria

Introduction

Under the UNECE Convention on Long-range Transboundary Air Pollution (CLRTAP), the Gothenburg Protocol has been set up to abate acidification, eutrophication, and groundlevel ozone. Austria has a national emission ceiling of 66 Gg NH_3. The Protocol requires best available techniques to be used to keep emissions down. Emission inventories must provide transparent, consistent, comparable, complete, and accurate data on sources and sinks of national emissions and must evaluate the progress towards meeting the reduction commitments. Emission inventories must not only estimate national emissions as accurately as possible, but should as well be able to show the annual trend in emissions, and the effect of mitigation measures. Agricultural emissions depend to a large extent on the animal housing, and on the manure management system (MMS). Data on MMS are a mandatory pre-requisite for accurate emission estimates. Mitigation measures can only be identified, if representative data on the MMS distribution are available. The project aims at the following: Detailed overview on Austrian animal husbandry, improvement of the Austrian emission inventory, modelling of typical farms and estimation of their emissions, development of emission scenarios, and proposal of feasible mitigation measures.

Survey and questionnaire

The questionnaire assesses relevant parameters in all stages of animal husbandry systems: housing and exercise yards, grazing, waste and washing water, manure storage, manure application, animal feeding, and mineral fertiliser application. The Austrian questionnaire is based on the questionnaire that was used in the Swiss DYNAMO project (Menzi *et al.* 2003). The questionnaire was sent to a representative sample of 5,000 Austrian farms. The sample design and the subsequent drawing of the sample were done in close cooperation with Statistics Austria. For the sample design, the Statistics Austria proposed the following criteria:

- NUTS (Nomenclature des unités territoriales statistiques) 1 region: (1) Eastern Austria (Burgenland, Lower Austria, Vienna), (2) Southern Austria (Carinthia, Styria), (3) Western Austria (Upper Austria, Salzburg, Tirol, Vorarlberg).
- Weighing factor 'hv': weighted sum from arable area and the number of livestock units scaled with 1.21.

Farms with animal husbandry play a greater role in the emission inventory than farms without animal husbandry and should be more often represented in the survey sample. Thus, the Statistics Austria weighed arable land with the factor 0.2 and livestock numbers with the factor 0.8. The survey aimed to achieve a rate of questionnaire return of 40–50%. This made a range of accompanying measures necessary. Special attention was given to an early and comprehensive information of Austrian farmers. Project background and details were published in a range of farmers´ journals and on a project web site. Project preparation and questionnaire development were done in close cooperation with the Austrian Chamber of Agriculture and with the Regional Chambers of Agriculture. Members of the Regional Chambers of Agriculture directly contacted farmers and informed them about the project. In the case of no return of the questionnaire, farmers were contacted again and asked to fill in the questionnaire. 2,060 questionnaires were returned to DAE which corresponds to a rate of return of 41%.

DYNAMO – Dynamic Ammonia Emission Inventory

The returned questionnaires were fed into a data template by the Statistics Austria. On the basis of this template, a data base was created that contains the questionnaire information. Anonymity of the farms that supplied data is guaranteed. The data base was checked for representativeness and plausibility prior to the emission calculations. Emissions are calculated with the help of the computer based program 'DYNAMO' (Dynamic Ammonia Emission Inventory) (Menzi *et al.*, 2003). 'DYNAMO' estimates ammonia losses from the whole manure management continuum. Emission estimates are based on the amount of N in the sections housing, grazing, exercise yard, manure storage and manure application. 'DYNAMO' differentiates animal categories, manure management systems and a range of management parameters. Ammonia losses from mineral fertiliser application are estimated, as well. 'DYNAMO' emission estimation procedures are based on the N flow model as it has been described in the CORINAIR Guidelines (EMEP/CORINAIR 2005). Ammonia emissions mainly depend on the amount of total ammoniacal nitrogen (TAN) rather than on the amount of total N. Ammonia emissions are estimated as a percentage of TAN present in each stage of the manure management continuum. Only for fresh excreta in the animal house or during grazing, 'DYNAMO' estimates ammonia losses on the basis of the total nitrogen content. In a first step, model farms are calculated from the survey results. Then, emissions will be estimated for the Austrian provinces and for the whole of Austria. Abatement potentials will be shown and measures for a reduction of ammonia and greenhouse gas emissions will be derived from the results. The project lasts until the end of February 2007 and is financially supported by the Austrian Federal Ministry for Agriculture and Forestry, Environment and Water Management.

References

EMEP/CORINAIR, 2005. Atmospheric Emission Inventory Guidebook. Third edition. http://reports.eea.eu.int/EMEPCORINAIR/en

Menzi, H., L. Ruettimann and B. Reidy, 2003. DYNAMO: A new calculation model for dynamic emission inventories for ammonia. Proc. Internat. Symposium 'Gaseous and odour emissions from animal production facilities', Horsens, Denmark, June 1-4 2003.

The 'Swiss Ammonia GAP'? Comparison of emissions and air quality trends

B. Reidy[1], R. Ballaman[2], H. Menzi[1] and B. Achermann[2]
[1]Swiss College of Agriculture (SHL), Laenggasse 85, 3052 Zollikofen, Switzerland
[2]Federal Office for the Environment, 3003 Berne, Switzerland

Introduction

Agricultural activity is well recognised as the major contributor of ammonia (NH_3) to the atmosphere. Leading to eutrophication and acidification of natural ecosystems and to the formation of secondary aerosols, NH_3 is part of those major air pollutants in the 1999 Gothenburg Protocol for which Parties to the protocol have to achieve emission ceiling values. In the context of international efforts to reduce the impacts of atmospheric pollutants it is therefore important to demonstrate that achieving emission ceilings is also resulting in corresponding air quality improvements, and to analyse obvious discrepancies.

Comparison of calculated NH_3 emissions with measured air quality trends

In Switzerland, agricultural ammonia emissions have been shown to account for over 90% of the total ammonia emissions, with livestock production as the major contributor to the emissions. Based on detailed inventory calculations, total annual NH_3 emissions underwent a substantial reduction of 17% within the period from 1990 to 2000. Whereas the NH_3 emissions from non-agricultural sources remained largely constant, the calculated emission reduction from agricultural sources could mainly be explained by reduced livestock numbers (contributing approximately two thirds of the reduction), improved livestock and manure management (e.g. increased utilisation of low emission manure spreading techniques, low protein feed for pigs, increased grazing) and reduced mineral fertiliser use and sewage sludge applications (Reidy, unpublished data). Since 2000 agricultural NH_3 emissions remained largely unchanged. In order to evaluate the estimated decrease of the NH_3 emissions, the inventory data was compared with NH_3 and NH_4^+ air concentrations and with NH_4^+ depositions monitored at national (network NABEL) and cantonal sites in Switzerland. Interestingly, the observed emission pattern from the inventory data could be neither confirmed by the development of the NH_3 and NH_4^+ air concentrations nor by the NH_4^+ deposition measurements. Despite considerable annual variations the available measurements for the sum of NH_4^+ and NH_3 concentrations of two representative stations remained largely unchanged since 1993 (Figure 1a). Moreover, since 2000 NH_3 concentrations measured at several locations in Switzerland with passive sampling technique tend to increase despite unchanged NH_3 emissions (Figure 1b). This pattern can hardly be explained with reduced dry deposition rates of NH_3 due to lower SO_2 emissions. The most significant SO_2 emission reductions occurred before 1995 and since 2000 the SO_2 concentrations remained almost stable (Figure 1c). The findings of the NH_3 and NH_4^+ air concentrations are supported by NH_4^+ wet deposition measurements from several locations in Switzerland. Although NH_4^+ concentrations in precipitations slightly decreased during the last 15 years and considerable annual and site specific fluctuations were observed, the deposition remained largely constant over the whole measuring period (Figure 1d).

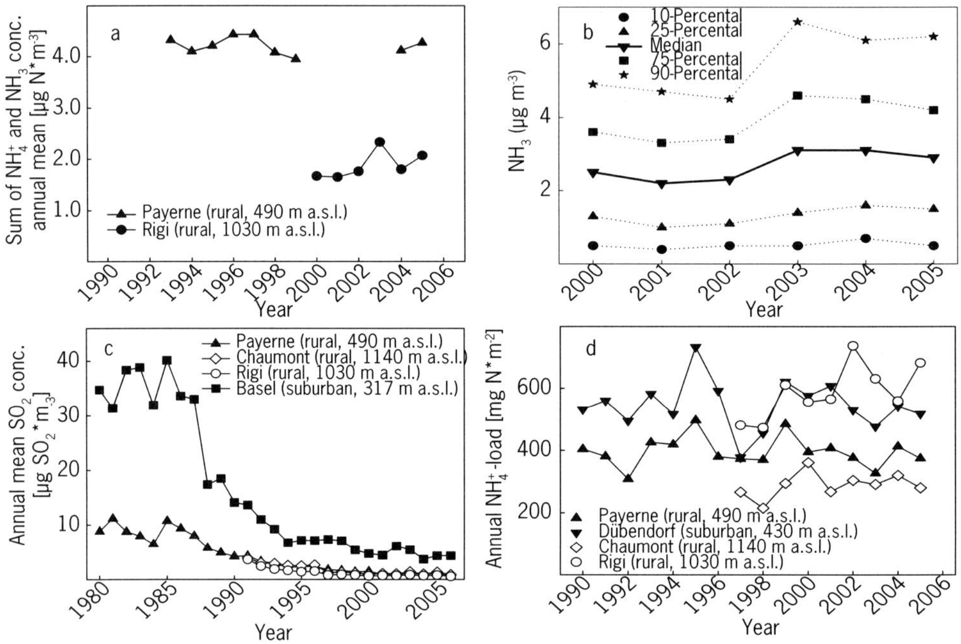

Figure 1. (a) Sum of NH_4^+ and NH_3 concentrations of two representative measuring stations in Switzerland. (b) NH_3 concentrations of 16 different measuring stations in Switzerland. Data is shown as percentiles. (c) Annual NH_4^+ wet depositions at different locations in Switzerland. (d) Development of SO_2 concentrations since 1980 at different locations in Switzerland.

Conclusions

In contrast to the calculated emission reduction, no significant trends towards lower air concentration levels or deposition loads of reduced N compounds are apparent since the beginning of the monitoring activities in Switzerland. Additional studies are therefore required to analyse this discrepancy. From an emission point of view, the emission factors for agricultural and non-agricultural sources should be critically reviewed. Particularly emissions from the recently and widely introduced animal friendly husbandry systems may be uncertain because the emission factors are only based on estimates. Moreover, the efficiency of the improved livestock and manure management may have been overestimated. With respect to the atmospheric measurements it is well known that atmospheric processes (meteorology and chemistry) can influence the partitioning between gaseous NH_3 and aerosol NH_4^+. Since Figure 1a shows the sum of the gas and the aerosol phases, changes in single compounds could be hidden. However, the results of an equilibrium modelling study indicate that the concentrations of gaseous NH_3 remained at the same level since 1993 (Spirig and Neftel, 2005). Additional studies including other sites could therefore be helpful to explain in more detail the behaviour of N compounds in the atmosphere.

References

Spirig, C. and A. Neftel, 2006. Influence of trace Gases from Agriculture on Particulate matter concentrations. Internal Report. Bundesamt für Umwelt (BAFU) 2006.

An ammonia emission inventory model for Irish agriculture

B.P. Hyde
Environmental Protection Agency Regional Inspectorate, McCumisky House, Richview, Clonskeagh Road, Dublin 14, Ireland

Introduction

Agriculture accounts for approximately 99% of ammonia (NH_3) emissions in Ireland. As a signatory to both The Gothenburg Protocol and the National Emissions Ceilings Directive (NECD), Ireland is required to produce and update an annual inventory of NH_3 emissions to the atmosphere. An accurate and transparent inventory is required to identify the major sources and hence develop effective mitigation polices that may be required to meet agreed targets. The work presented here is an update of that published by Hyde *et al.* (2003) who developed an inventory model based on that then used for UK agriculture (Misselbrook *et al.*, 2000). The recent completion of a national farm facilities survey (Hyde, unpublished data) and the publication of revised emission factors required an updating of the Irish inventory. The results of the revision are presented here.

Emission inventory calculation

The construction of the Irish inventory and the derivation of revised emission factors have been previously described (Hyde *et al.*, 2003; Misselbrook *et al.*, 2000, 2004). However, Hyde *et al.* (2003) noted the absence of detailed data on manure management practices and that Ireland had fallen behind many of its European counterparts in terms of NH_3 emission research. The recent completion of a national farm facilities survey (Hyde, unpublished data) and the publication of revised emission factors presented an opportunity to update the Irish inventory. Where possible emission factors were linked with country specific information in relation to farming practice. Ammonia emission inventory totals were calculated using the revised inventory for the period 1990-2010 based on national animal census and statistical data and national projections of animal numbers and fertiliser use (Binfield *et al.*, 2006). Cattle were sub-divided into eight sub-categories and two housing types based on animal census data, information on manure management practices and fertiliser nitrogen application rates for grazing. Seven pig sub-categories with seven pig-feeding practices in each sub-category were identified. Poultry were sub-divided into seven sub-categories based on detailed registries of poultry numbers. Sheep were sub-divided into 10 sub-categories, which reflects the upland and lowland rearing of sheep practiced in Ireland. In addition emissions from silage, hay, maize and arable crops were included in the inventory calculations.

Results

The emission inventory presented is divided into six major categories (Table 1), each of which was separately modelled. In total 82 different animal/housing categories, three forage crops and six tillage crops are modelled. Emission totals for each of the six major source categories have been calculated for the period 1990-2010. Emission totals for 1990, 2004 and 2010 (assuming business as usual) are presented in Table 1. On average cattle account for approximately 80% of total emissions with pigs, poultry and sheep accounting for approximately 6%, 2% and 5% of total emissions, respectively. Conserved grassland and maize and arable crops account for 6.5% and 0.5%, respectively, of emission totals.

Table 1. Ammonia emission totals (kt NH_3) for the six major categories for 1990, 2004 and 2010.

Source category	1990	2004	2010
Cattle	87	86	81
Pigs	5	7	7
Sheep	6	5	4
Poultry	2	3	2
Conserved grassland & maize	6	6	6
Tillage crops	1	1	1
Total	107	108	100

Conclusions

The revised inventory model has been adopted nationally and is currently in use for reporting purposes to the UNECE. The key focus remains meeting the challenges of the reductions required for the period 2010 to 2020. National research is focussing on evaluating low emission-spreading techniques with particular emphasis on the trailing shoe. The potential of out wintering pads to reduce emissions from cattle housing is also being investigated. In addition, The Irish Department of Agriculture and Food is preferentially grant aiding low emission spreading technologies. Based on the results presented the target of 116 kt NH_3 set for 2010 under the Gothenburg Protocol and National Emissions Ceilings Directive may be met. This is largely driven by projected reductions in cattle numbers as a result of Common Agricultural Policy reform. This is in direct comparison to the large increase in cattle numbers that occurred in the 1990's reaching peak levels in 1999. Further reductions will be possible dependent on the level of infiltration of low emission landspreading techniques at farm level. However, future emission targets are likely to require further substantial reductions in emission levels. The results presented herein differ to those calculated previously as they are based on the latest available data. Inventory calculations will be updated as the results of Irish NH_3 emission research become available.

References

Binfield, J., T. Donnellan, K. Hanrahan and P. Westhoff, 2006. World Trade Reform: and the WTO Doha Development Round: Analysis of the Impact on EU and Irish Agriculture. Rural Economy Research Centre, Teagasc, Athenry, Galway, Ireland. 79 pp.

Hyde, B.P., O.T. Carton, P. O'Toole and T.H. Misselbrook, 2003. A new inventory of ammonia emissions from Irish agriculture. Atmospheric Environment 37: 55-62.

Misselbrook, T.H., T.J. Van der Weerden, B.F. Pain, S.C. Jarvis, B.J. Chambers, K.A. Smith, V.R. Philips and T.G.M. Demmers, 2000. Ammonia emission factors for UK agriculture. Atmospheric Environment 34: 871-880.

Misselbrook, T.H., D.R. Chadwick, B.J. Chambers, K.A. Smith, J. Webb, T.G.M. Demmers and R.W. Sneath, 2004. Inventory of Ammonia Emissions from UK Agriculture, 2003. Inventory Submission Report, DEFRA Contract AM0127. 35pp.

The model for agricultural mineral flows for policy support

L.J. Mokveld
Wageningen UR, Agricultural Economics Research Institute (LEI), The Hague, The Netherlands

Introduction

This paper presents the representation of the Model for Agricultural Mineral Flows for Policy Support (MAMBO). In the past 15 years the Manure and Ammonia Model (MAM) has been used to monitor and predict the level and allocation of ammonia emission from farming in the Netherlands. MAMBO is the successor of MAM. The objective of MAMBO is the calculation of NH_3 emission and other gaseous emissions from animal manure production, processing and application.

The scientific base of MAM/MAMBO

By the development of MAMBO, a generic formulation was chosen to facilitate the use of data with a deviating structure (i.e. animal categories, crops, manure categories, housing types). Furthermore, adjustments to incorporate the policy concerning manure and emissions in MAMBO were made. MAMBO can be used to calculate the total of nutrient flows and ammonia emissions. To establish this, data on five key processes regarding animal manure are gathered and processed in this model:

1. Manure production on farm.
2. Manure application room on farm (potential maximum application of manure within statutory and farm level constraints).
3. Manure excess at farm level (production minus application room).
4. Manure distribution between farms (transport).
5. Soil loads with minerals.

The calculations take place on three spatial levels (Figure 1). The first three processes, manure production, application room and manure excess, are calculated at farm level. The manure distribution is calculated at the level of 31 predefined manure regions (area level) and soil loads are calculated at municipality level (applied).

Manure production

The primary calculations of manure production and application room take place at animal level, whereas information that's not available at that level is disintegrated. The ammonia emission (Figure 1) takes place in stables (1), the pasture (2), the storage (4) and with transformation (3) at firm level.

Application room and excess

In order to determine whether a farm has a manure surplus or room for off-farm manure, the manure produced on the farm is balanced against the application room of the farm. In a case of manure surplus, the economic consequences of the surplus are minimised by finding the most appropriate type of manure for each particular farm. The application room is based on policy as well as good agricultural practices.

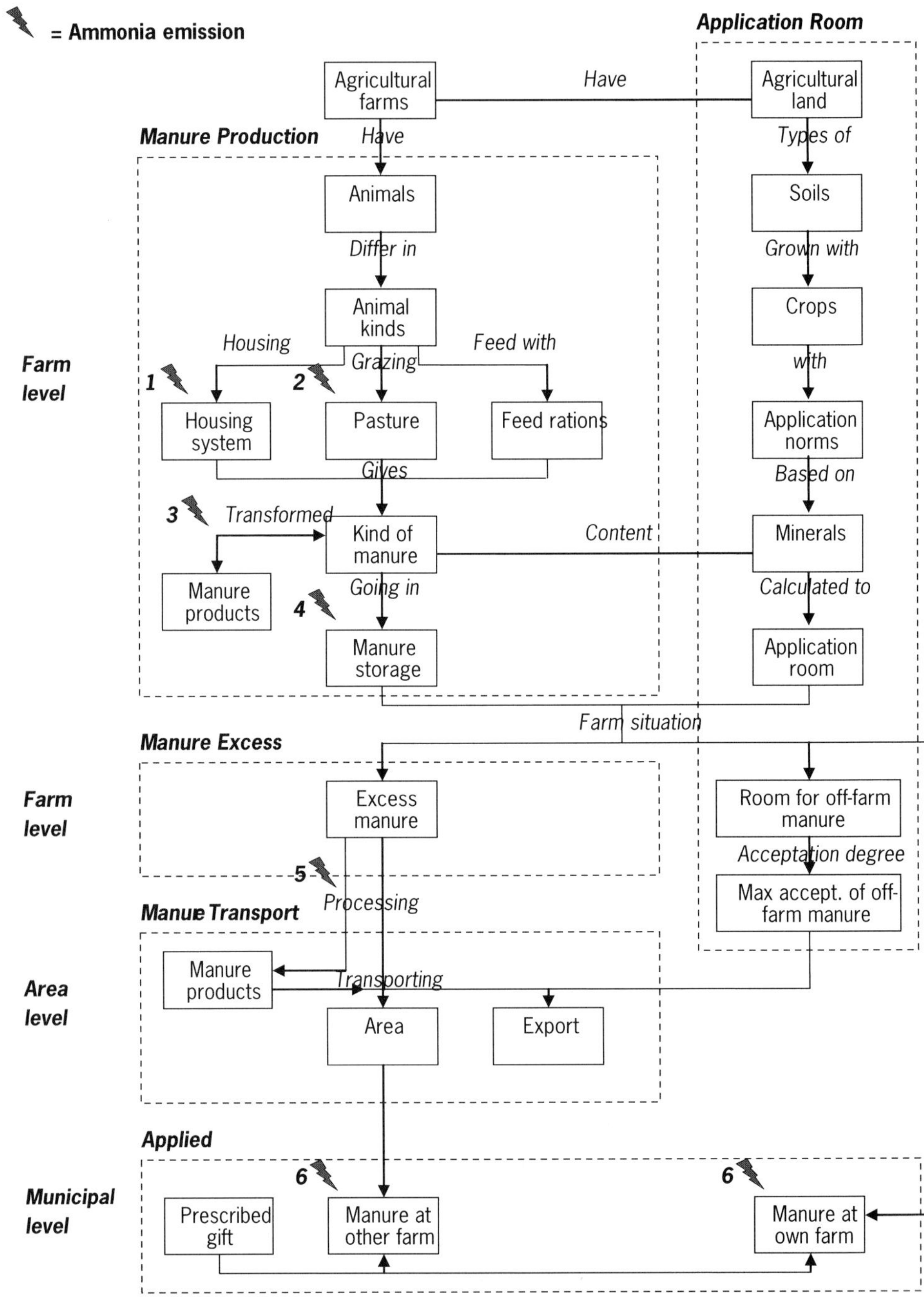

Figure 1. Structure of the manure and ammonia emission model (Groenwold et al.*, 2002).*

Manure transport

MAMBO includes three options for manure that cannot be applied at farm level: it can be transported to other farms within the same region, it can be transported to other regions or it can be exported. This can all be done either processed or unprocessed. Given the necessity for a farm to transport manure, the main drive for transport of any type of manure is minimising manure transfer costs. Receiving farms aim to maximise crop returns. Only when manure is processed (5), ammonia emission takes place (Figure 1)

Soil load

In MAMBO, the total mineral load of the soil depends on three factors: the application of on-farm manure, the application of off-farm manure and the application of mineral fertiliser. The Dutch farm accountancy data network provides data and statistics available about the use of mineral fertilisers at a regional level. These are divided at municipal level with a distributive code. The distributive code holds data on the time of manure application, the effectiveness of the nutrients and the amount of nutrients in the applied manure. For this purpose, the manure transfers on municipality level are calculated from the results of manure transfers on regional level by disaggregating these to municipality level. The ammonia emission takes place with the application of manure on farm land.

Results

The results of the calculation of ammonia emissions in The Netherlands is presented in the paper of Luesink (2007).

References

Groenwold, J.G., D. Oudendag, H. Luesink, G. Cotteleer and H. Vrolijk, 2002. Mest- en Ammoniakmodel. LEI, Den Haag, Rapport 8.02.03.

Luesink, H.H., 2007. Calculation of the ammonia emission in The Netherlands. Paper for the International conference on Ammonia in Agriculture, Wageningen, March 2007.

Calculation of the ammonia emissions in The Netherlands

H.H. Leusink
Wageningen UR, Agricultural Economics Research Institute (LEI), The Hague, The Netherlands

Introduction

In the second national environmental outlook the national ammonia emissions were stipulated to be 251 million kg in 1980, 285 million kg in 1986 and 234 million kg in 1989. In that same report a forecast was made for the ammonia emissions in The Netherlands. National ammonia emissions of 152, 114 and 104 million kg were predicted for the years 1994, 2000 and 2010 respectively. This triggered the yearly calculation of the ammonia emissions in The Netherlands. Calculation and publishing of the ammonia emission is the task of two institutes: the Agriculture Economics Research Institute (LEI) and the Netherlands Environmental Assessment Agency (MNP). LEI is concerned with the gathering of activity data for the calculations and the calculation of the ammonia emissions from agriculture. NMP has a coordination role, calculates emission factors as well and publishes the results and the background information.

Process and organisation

Calculation of the ammonia emissions of a certain year takes place in the two following years. At first, a preliminary amount is calculated, and a year later when all activity data are available the final amount is calculated. The time schedule (Table 1) of the process and organisation of the yearly calculation is based on the time of the debate (second half of May) in parliament. The Working group Emission Monitoring (WEM) coordinates, controls planning and advises on proposed monitoring programs for a great number of emissions from all sectors.
Not only improving the way that national ammonia emissions are calculated is a time consuming process. Improving the collection of the necessary or new data is a process of three to four years. In this process, researchers from LEI and MNP suggest proposals to the different groups who decide if improvements or new data should be adopted.

Activity data

Besides the organisational efforts for the ammonia emission calculation, the actual data gathering process (Table 2) involves many more parties to ensure timely availability and quality of the data at hand. The data sources presented in Table 2 depend on the level of origin at which the data is

Table 1. Present process and organisation of the yearly calculation of the Dutch ammonia emission.

Month	Organisation	Kind of work
June/August	MNP/LEI/Others	Proposals for improvement
September	WEM/MNP	Advice and decisions
October/may	MNP	Coordination
October/December	LEI/MNP	Collecting data
January	LEI	Calculation with MAM/MAMBO
February/May	MNP/LEI	Report of results

collected and the used model (MAM or MAMBO). MAMBO eliminates the influence of origin in such a way that inputs can vary from an individual animal level to national level.

Table 2. Data source, gathering method, frequency of collecting and prior indices to calculate ammonia emissions in The Netherlands.

Emmission type – Parameters	Lvl[1]	Freq.[2]	Source[3]	Index	Remarks
Manure production					
– Number of animals	F	Y	LBT	Animal kind	
– Excretion	N	Y	WUM	Ration+mineral	Dairy: 4 rations
Emission from housing					
– Housing system	R/F	4Y	LBT	Animal kind	Dairy: Lvl. = F
– Grazing system and housing period	N	Y	BIN	Animal kind	WUM equivalent
– Emission factors	N	Occ.	MNP	Housing system	
Emission from grazing					
– Grazing system and period	N	Y	BIN	Animal kind	WUM equivalent
– Emission factor	N	Occ.	MNP	No index	
Emission from storage					
– Amount	N	Occ.	LBT	Manure type	Dated
– Storage period	N	Occ.	Research	Manure destination	Dated
– System	N	Occ.	LBT	Manure type	Dated
– Emission factors	N	Occ.	MNP	Storage system	
Application of manure and fertiliser					
– Crop area	F	Y	LBT	Crop kind	
– Limits	R	Y	LNV, BIN, Research	Crop/soil kind, mineral	Application gifts
– Distribution between farms	P	Y	LNV	Manure kind, mineral	Only manure
– Application systems	R	5Y	LBT	Manure type	Only manure
– Total available N	N	Occ.	Research	Manure kind	
– Emission factors	N	Occ.	MNP	Manure type, application system	
– Acceptation degree	R	Y	BIN	Crop kind	Application gifts
– Export	N	Y	LNV	Manure type	

[1]Levels are: Farm (F), National (N), Regional (R) and Provinces (P)

[2]Frequencies are: Yearly (Y), 4-Yearly (4Y), 5-Yearly (5Y), Occasionally (Occ.)

[3]Sources are: LBT = annual agriculture census; WUM = working group uniform mineral- and manure excretions; BIN = Dutch farm accountancy data network; Research = different *kind of research reports; LNV = Ministry of Agriculture, Nature and Food quality; MNP = Netherlands Environment Assessment Agency.*

Comparison of models used for national ammonia emission inventories within the EAGER workgroup

H. Menzi[1], U. Dämmgen[2], H. Döhler[3], B. Eurich-Menden[3], F.K. van Evert[4], N.J. Hutchings[5], H.H. Luesink[6], T.H. Misselbrook[7], G.-J. Monteny[8], J. Webb[9] and B. Reidy[1]
[1]*Swiss College of Agriculture, Laenggasse 85, CH-3052 Zollikofen, Switzerland*
[2]*Federal Agric. Res. Cent., Inst. of Agroecology, Bundesallee 50, 38116 Braunschweig, Germany*
[3]*Association for Technology and Structures in Agriculture (KTBL), Bartningstrasse 49, 64289 Darmstadt, Germany*
[4]*Plant Research International, P.O. Box 16, 6700 AA Wageningen, The Netherlands*
[5]*Danish Institute of Agricultural Sciences, Research Centre Foulum, Tjele, Denmark*
[6]*LEI, P.O. Box 29703, 2502 LS The Hague, The Netherlands*
[7]*Inst. of Grassland and Env. Research, North Wyke, Okehampton, Devon EX20 2SB, United Kingdom*
[8]*Wageningen UR, Agrotechnology. and Food Innovations B.V., P.O. Box 17, 6700 AA Wageningen, The Netherlands*
[9]*ADAS Research, Woodthorne, Wergs Road, Wolverhampton WV6 8TQ, United Kingdom*

Introduction

The Gothenburg Protocol of the UN Convention on Long-range Transboundary Air Pollution (UNECE 1999) requires the reporting of national annual emissions of ammonia (NH_3). Accurate inventories of agricultural NH_3 emissions are required to calculate the total national emissions, since they commonly account for more than 80 % of the total emissions (EMEP 2005). To allow a co-ordinated implementation of the Protocol, different national inventories should be comparable; at present they are not. A core group of emission inventory experts has therefore inaugurated the EAGER network (European Agricultural Gaseous Emissions Inventory Researchers Network), with the aims of achieving a detailed overview of the present best available inventory techniques, compiling and harmonising the available knowledge on emission factors (EF) and initiating a new generation of emission inventories that satisfies protocol requirements. As a first step in summarising the available knowledge, the objective of the work reported in this paper was to determine the degree to which results obtained with different NH_3 emission models currently used for inventory calculations agree, and to evaluate any larger disagreements.

Material and methods

The following models used in the framework of the national NH_3 emission inventory calculations and manure policy analyses in different countries of Europe were included in the comparison: (1) DYNAMO, Switzerland (Reidy and Menzi, 2006), (2) DanAm, Denmark (Hutchings *et al.*, 2001), (3) GAS-EM, Germany (Daemmgen *et al.*, 2002), (4) NARSES, United Kingdom (Webb and Misselbrook, 2004), (5) MAM, Netherlands (Luesink *et al.*, 2004) and (6) FARMMIN, Netherlands (van Evert *et al.*, 2003). The models all use a mass-conservative N-flow approach starting with a specific amount of nitrogen (N) excreted by a defined livestock category and simulate the total ammoniacal nitrogen (TAN) flow over the different stages of emissions (grazing, housing, manure storage and application). NH_3 emissions are generally calculated with EF, where the EF is the percentage of the respective TAN pool emitted. Emissions were compared for a dairy cattle and a pig scenario, with different levels of model standardisations: (1) to test the congruency of the underlying N flow, the national specific N excretions rates, TAN contents and EFs were replaced by a set of standardised values (FF scenario). (2) Only the N excretion and TAN contents were standardised whereas the national EFs were used (FN scenario). (3) National values were used for N excretion, TAN contents and EFs (NN scenario).

Results and discussion

The FF scenario (standardised N excretion and EF) resulted in very similar estimates of the NH_3 emission for the respective emission stages as well as for the total emissions. This indicates that the underlying N flows of the different models are highly comparable. The small differences observed could largely be explained with slight modifications of the N flow, either related to an altered partitioning of the N deposited during grazing and housing or to the extent other N transformations are taken into account (e.g. mineralisation or denitrification processes) in the different models. Differences between the models were more pronounced when the emissions were calculated with national emission factors and/or national N excretion rates (FN and NN scenario). The variation in the calculated emissions was primarily the result of the distinct national emission factors and N excretion rates. Both parameters reflect the specific livestock and manure management systems of the different countries and can be explained by: (1) Differences in N excretion that result from differences in feeding practice (e.g. protein concentration in the diet) or from different production intensities (e.g. milk yield per cow, growth rate per pig). (2) Variations in the types of animal housing, storage and technology used for field applications. (3) Variations in animal management (e.g. duration of the grazing period). (4) Climatic factors. The variations in the calculated emissions of the FN and NN scenarios are therefore real, fully valid and are the reason why emission inventories are most appropriately constructed at a scale that reflects the heterogeneity in agriculture and climate.

Conclusions

The models compared are generally well congruent and existing differences in the results obtained can mostly be explained by existing differences of natural conditions and farm management in the countries from which the models originate. The congruency exercise has led to a greater harmonisation of the structure and function of the models tested, because the scientific debate necessary to understand the variation in results from the different models generated awareness and consensus concerning the importance of some processes (e.g. mineralisation).

References

Dämmgen, U., M. Lüttich, H. Döhler, B. Eurich-Menden, and B. Osterburg, 2002. GAS-EM – a procedure to calculate gaseous emissions from agriculture. Landbauforschung Völkenrode 52: 19-42.

EMEP. 2005. Database of the national submissions to the UNECE LRTAP Convention maintained at EMEP. http://webdab.emep.int/ (February 14, 2006).

Hutchings, N.J., S.G. Sommer, J.M. Andersen and W.A.H. Asman, 2001. A detailed ammonia emission inventory for Denmark. Atm. Env. 35: 1959-1968.

Luesink, H.H., C.H.G. Daatselaar, G.J. Doornewaard and H. Prins, 2004. Sociaal-economische effecten en nationaal mestoverschot bij varianten van gebruiksnormen. LEI, Den Haag, Rapport 3.04.08.

Reidy, B. and H. Menzi, 2006. DYNAMO: An ammonia emission calculation model and its application for the Swiss ammonia emission inventory. Proceedings of the Workshop on Agricultural Air Quality: June 5-8, Potomac, MD, USA.
UNECE, 1999. Protocol to the 1979 Convention on Long-Range Transboundary Air Pollution to abate acidification, eutrophication and ground-level ozone. UNECE, Geneva.
Van Evert, F., H. Van der Meer, H. Ten Berge, B. Rutgers, T. Schut and J. Ketelaars, 2003. FARMMIN: Modeling crop-livestock nutrient flows. Agronomy Abstracts 2003, ASA/CSSA/SSSA, Madison, WI.
Webb, J. and T.H. Misselbrook, 2004. A mass-flow model of ammonia emissions from UK livestock production. Atm. Env. 38: 2163-2176.

Nitrogen in The Netherlands over the past five centuries

H. Noordijk
Netherlands Environmental Assessment Agency (MNP). P.O. Box 303, 3720 AH Bilthoven, The Netherlands

The history of nitrogen emissions

Eutrophication through nitrogen compounds is a major environmental problem in the Netherlands. Especially the yearly deposition of NH_x is high, with an average of 1200 mol/ha; however, the deposition of NO_y at 680 mol/ha is also considerable. This is due to the high population density, and traffic and industry, combined with an extremely intensive cattle breeding sector. The forces driving this intensive land use are determined by geography, history and culture. The dynamics behind Dutch ammonia emissions were already present in the Dutch 'Golden Age' (1580-1680). Until about 1560, farming focused primarily on the needs of farmers and their families. But then a revolt against the Spanish rule resulted in a severe disruption of the prosperous Flanders. People, capital and knowledge shifted to the northern Dutch provinces, especially Holland, where an age of unprecedented prosperity followed. The main driving force was trade, creating a unique proto-capitalistic culture of private and common enterprise. Profitable markets were explored and were soon dominated by the Dutch naval and economic power structures. This development turned Holland into the main European trade centre. This caused low price levels in the young Dutch Republic, which increased the prosperity of the growing city population. Farmers responded with a reduction in the cultivation of corn due to cheap imports from the Baltic region, and a considerable rise in stockbreeding. Farmers became agricultural entrepreneurs, specialised and efficient, and geared to the demands of city populations and international markets. On the coast, a diary belt emerged, with specialised young cattle breeding areas appearing inland. The quantity of Dutch cattle multiplied, cheap cattle fodder was imported and more profitable meat and dairy products were exported. Manure production increased rapidly and a lively manure trade followed, with the resultant growth in ammonia emissions. The wage, prosperity and energy demand of the growing Dutch population was at a worldwide unprecedented and unrivalled level. However, extreme prosperity never lasts forever. Wars with envious major European powers made an end to the Golden Age. Prosperity declined until far in the 19th century; city populations decreased and farmers shifted from stockbreeding to arable farming. Just after 1880 cattle breeding increased again due to cheap American imports of food and fodder, fertiliser and a growing and increasingly prosperous city population. And once more, a century of unrivalled growth in Dutch cattle breeding followed that was primarily based on the same driving forces as in the 16th century.

Calculating historical emissions

Emissions and depositions over 1981-2004 have already become available (De Ruiter *et al.*, 2006). For this study, the emissions have been scaled back into the past. For ammonia, the main scaling parameters are the amount of livestock and the nitrogen excretion per animal, which is the highest in Europe. Livestock numbers for 1900-1980 are easily available, but there are fewer data for the 19th century. Changes in the nitrogen excretion per animal are expected to follow the recorded milk production per cow. This results in a 60% excretion decrease per animal for 1800-1980. The calculated ammonia emissions for the entire livestock decrease by 95% over the same period. Before 1800, population numbers and a few data on livestock determined the scaling procedure. Annual anthropogenic emissions are estimated to cause a deposition of 25 mol/ha for 1500, together with 40 mol/ha from natural sources (Figure 1). Emissions and depositions of NO_x are scaled over 1900-1980 using statistics for several industrial sectors, number of cars, the number of kilometres driven and the population growth. Electricity emissions follow the same pattern as industrial emissions. An efficiency correction is applied which halves the emission per unit output in 1980, compared to 1900. For cars it halves when compared to 1936. Foreign emissions before 1950 are expected to follow the Dutch pattern. Before 1900, the population growth is used as the main parameter. Annual anthropogenic deposition for 1500 is about 20 mol/ha, with an estimated natural deposition of 4 mol/ha.

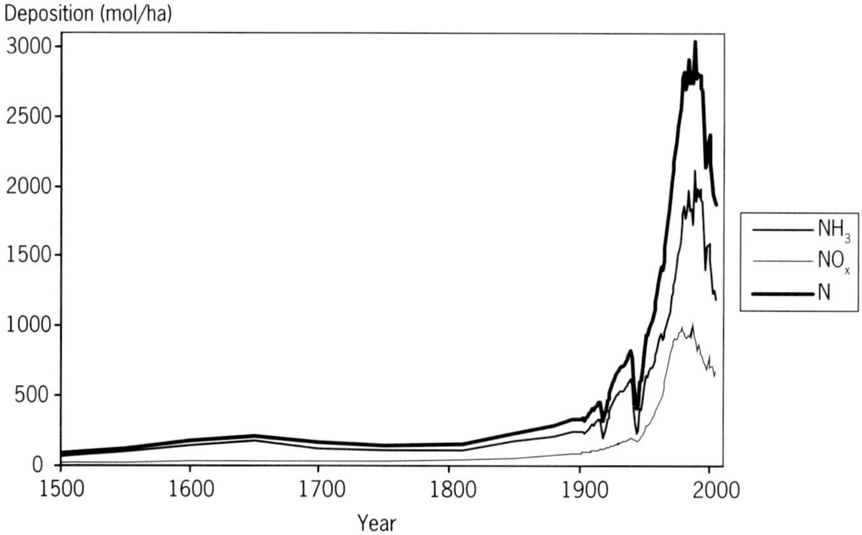

Figure 1. The annual average deposition of nitrogen compounds for The Netherlands.

References

De Ruiter, J.F., W.A.J. Van Pul, H.A. Van Jaarsveld, E. Buijsman, 2006. Zuur- en stikstofdepositie in de periode 1981-2002. MNP report 500037005/2006, Bilthoven, The Netherlands (in Dutch).

Inventories and modelling - emission factors

PM emissions in and from force-ventilated turkey and dairy cattle houses as factor of health and the environment

T. Hinz[1], S. Linke[1], J. Karlowski[2], R. Myczko[2], T. Kuczynski[2] and J. Berk[1]
[1]Federal Agricultural Research Centre, Bundesallee 50, 38116 Braunschweig, Germany
[2]Institute for Buildings Mechanisation and Electrification of Agriculture, ul. Biskupinska 67, 60-463 Poznan, Poland

Introduction

There is an increasing need to control emissions of particulate matter (PM). Agriculture is a substantial source of PM emissions. PM also reduced air quality within livestock buildings with implications for the health and welfare of farmers and animals. Effects of particles on individuals and their dispersion in the air depend on different parameters but strongly on their size and mass. Different target–oriented definitions are used (ISO, 1996; US EPA, 2001). In two bilateral German- Polish research projects, measurements were carried out in force-ventilated buildings housing turkeys and dairy cows. Concentration of total dust (TSP) and the PM10-fractions inside the houses and the air flows through the exhaust ducts were measured and emission factors calculated. These values may be useful to estimate the size of emissions and the influencing parameters.

Materials and methods

Investigations were done on a dairy farm in the Konin region of Poland and in an experimental turkey buildings of the Institute of Animal Welfare and Animal Husbandry in Celle, Germany.64 cows are kept in a building of 46.25x12 m equipped with four temperature controlled axial fans with a nominal maximum flow of 5950 m^3/h each. Fresh air entered the cow house through windows with variable opening. The degree of declination of the windows was adjusted manually upon evaluation of atmospheric conditions. There was a passage in the corridor used for forage distribution. On both sides of it are forage tables and two rows of chained cows' boxes with straw bedding. Straw manure is removed by mechanical grabbers of an electrically-powered machine once a day. Average milk yield is 9200 kg milk per year per cow. The experimental turkey set-up covers 12 single compartments with 4.7 birds/ m^2 each (Hinz *et al.*, 1999). Two compartments form one unit with respect to the ventilation system. While in a previous study environmental enrichments for the birds were in the focus, in this study different types of litter were investigated: wood shavings, straw and cellulose. Independent from the following method of mass detection samples inside the buildings were collected with a sampling velocity of 1.25 m/s through a circular slot, but isokinetic in the emission flow. Conventional gravimetric filter procedure served as reference of total dust. To determine the complete size distribution a high-volume sampler with a pre-separator was used to separate large particles. The coarse fraction was analysed with light diffraction method. A scattering light monitor was installed in the flow behind the cyclone separator to get knowledge about the passing through fine particles. For control and calibration purpose an absolute filter collected the fine particle fraction.

Results and discussion

Table 1 gives the concentration of total dust (TSP) inside the stables, in the exhaust and the resulting emissions of total dust and PM10. In the cow stable the concentration was generally low with values <1 mg/m^3. Depending on the climate conditions PM concentration in November was lower than in June. In both cases the concentration inside the stable at 1.5 m above ground was 3 times higher than the concentration in the exhaust flow. Particle size distribution differs essentially between the air inside the stable and in the exhaust. Particle emission created nearly PM10 only (100% TSP), but on a very low level. These figures differ from those from turkeys. Concentration inside the stable reached 7.5 mg/m^3 with the greatest values for cellulose but similar values for straw and wood shavings. In contrast to the dairies the greater values were measured in winter. Again concentration in the exhaust flow was less than inside the stable but not so clear as found for the cows. One reason may be the size of emitted particles which included a high quantity of large particles up to some hundred μm. The share of PM10 is about 25% of total dust. Highest PM emissions occur using cellulose.

Table 1. Concentration and emissions in and from the stables.

Type of animal / season / litter	C_{inside} [mg/m³]	$C_{exhaust}$ [mg/m³]	[g/animal/h]	[g/animal/h]
cow / summer / straw	0.550	0.188	0.033	n.a.
cow / winter / straw	0.198	0.064	0.009	0.008
turkey / summer / cellulose	3.814	n.a.	n.a.	n.a.
turkey / summer / wood shaves	1.981	1.172	0.096	0.024
turkey / summer / straw	1.265	1.189	0.095	0.024
turkey / winter / cellulose	7.492	4.271	0.120	0.3
turkey / winter / wood shaves	2.543	1.681	0.044	0.011
turkey / winter / straw	2.609	1.790	0.047	0.012

Conclusion

On a dairy cattle farm in Konin, Poland and in an experimental turkey house in Celle, Germany, PM emissions were investigated. In both cases same instrumentation was used to measure total dust and PM10. An excellent management strategy by the farmer resulted in a very clean dairy house with very low concentration and emissions, which may be not the normal case. The situation in the turkey house is comparable with other studies and confirmed previous investigations. Depending on the ventilation system, the species of animal and their activity the ratio PM10/TSP varies from 25% (turkeys) to nearly 100% (cows) in the emission flow.

References

Hinz, T., J. Berk, J. Hartung and S. Linke, 1999.Environment enrichments and dust emissions in turkey production. In: Congress proceedings, dust control in animal production facilities.

ISO, 1996. ISO 7708 Publication date: 1996-01 Air quality- Particle size fraction definitions for health-related sampling.

US EPA, 2001. Code of Federal Regulations; PM10, 2001a PM2.5, 2001cl.

Ammonia emissions in and from force-ventilated turkey and dairy cattle houses, validation of emission factors by direct measurements

T. Hinz[1], *S. Linke*[1], *J. Karlowski*[2], *R. Myczko*[2], *T. Kuczynski*[2] *and J. Berk*[1]
[1]*Federal Agricultural Research Centre, Bundesallee 50, 38116 Braunschweig, Germany*
[2]*Institute for Buildings Mechanisation and Electrification of Agriculture, ul. Biskupinska 67, 60-463 Poznan, Poland*

Introduction

Agriculture, and especially animal production, is the main source of ammonia emissions. Therefore limits on ammonia emissions have been agreed in the Convention of Long Range Transboundary Air Pollution Gothenburg protocol. National emission ceilings are given and inventories required. Two of the major sources are dairy cattle and minor poultry production. Annual emissions will be calculated from animal places and specific emission factors. These emission factors will be measured or determined by models. Models are based on data but verification by measurements is essential. In two bilateral German-Polish agricultural research projects, measurements were made turkey and dairy cow production units. For force-ventilated stables air flow and ammonia concentration in the exhaust were measured and emission rates estimated. These values cannot be used to calculate annual emissions but may be used to assess models and uncertainties.

Materials and methods

Investigations were done on a dairy farm in the Konin region of Poland and in an experimental turkey unit of the Institute of Animal Welfare and Animal Husbandry in Celle, Germany. 64 cows were kept in a building of 46.25 x 12 m equipped with four temperature-controlled axial fans with a nominal maximum flow of 5950 m^3/h each. Fresh air entered the cow house through declined window openings. Degree of declination of the windows was adjusted manually upon evaluation of atmospheric conditions. Using a flow direction sensor it was checked that air flowed into the stable through the windows. So its sure that the total flow leaves the building by the exhaust chimneys. There is a passage corridor in the middle used for forage distribution. On both sides of it are forage tables and two rows of chained cow stalls with straw bedding. Straw manure is removed by mechanical grabbers of an electrically powered machine once a day. Average milk yield is 9200 kg milk per year per cow. The experimental turkey set-up covers 12 single compartments with 4.7 birds/m^2 each (Hinz and Berk 2002). Two compartments form one unit with respect to the ventilation system. Depending on the temperature air flow varied between 1160 m^3/h and 8850 m^3/h. While in the previous study environmental enrichments for the birds were the focus, in the present study different types of litter were investigated: cellulose, wood shavings and straw. In both stables air flow was monitored using the computer-aided control systems. Ammonia concentration was measured using opto acoustic multi gas monitors. In the turkey barns the samples were taken from the exhaust duct of the compartment concerned. In the cattle house one instrument was placed directly in the opening of a chimney in the middle of the stable as reference. To find out possible local inhomogeneity of ammonia concentration, a second monitor of the same type was used. Both instruments were calibrated against each other.

Results and discussion

Table 1 gives averaged values of measured ammonia concentration in the exhaust and the resulting emissions from the dairy cows and the turkeys on measuring days in summer and winter. The concentration in the cattle house was very low and similar in winter and summer. Greater emissions were measured in summer due to the greater air flow. But nevertheless these measured values were less than those published. Dämmgen (personal communication, 2006) used an emission factor of 0.62 g/animal/h for dairies with 9000 kg annual milk yield kept in houses on straw. Concentration in the exhaust flow from the turkey building were greater than from the dairies, but due to lesser air flow, the emission factors for the turkey buildings were less, with one exception. The values correspond with those given in the literature, e.g. Müller *et al.* (2006).

Table 1. Averaged concentration in the exhaust flow and emission factors.

Type of animal / season / litter	$C_{exhaust}$ [mg/m³]	[g/animal/h]
cow / summer / straw	1.259	0.221
cow / winter / straw	1.275	0.183
turkey / summer / cellulose	1.544	0.128
turkey / summer / wood shavings	2.764	0.252
turkey / summer / straw	1.716	0.151
turkey / winter / cellulose	3.736	0.109
turkey / winter / wood shavings	5.383	0.153
turkey / winter / straw	3.170	0.094

Conclusion

On a farm in Konin (Poland) the opportunity was given to measure ammonia emissions from a dairy house with force-ventilation. The measurements were carried out in June and November 2006 on two days each. Emission factors per animal were much smaller than usually used. An excellent management strategy by the farmer resulted in a very clean stable and the emissions not be typical. Ammonia emissions from the turkey house were comparable with values given in the literature. The measurements are part of a long-term study and the results in table 1 are averages of fattening periods. Emissions increased with the age of the birds. Wood shavings caused the greatest emissions. In a next step these results from the experimental turkey house will be measured on a commercial farm.

Refernces

Hinz, T. and J. Berk, 2002. Airborne contaminants in tom-turkey production under enriched husbandry conditions. Ann. Anim. Sci. Suppl. 1: 39-41.

Müller, H.-J., R. Brunsch and W. Berg, 2006. Ammoniakemissionsmassenströme in und um Tierhaltungsanlagen. KTBL- Schrift 449.

Aerial pollutants emissions from confined animal buildings

L.D. Jacobson[1], A.J. Heber[2], S.J. Hoff[3], Y. Zhang[4], D.B. Beasley[5] and J.A. Koziel[6]
[1]University of Minnesota, BAE Dept., St. Paul, MN 55108, USA
[2]Purdue University, ABE Dept, West Lafayette, IN 47907, USA
[3]Iowa State University, ABE Dept, Ames, IA 50011, USA
[4]University of Illinois, ABE Dept, Urbana, IL 61801, USA
[5]North Carolina State University, BAE Dept, Raleigh, NC 27695, USA
[6]Ag. Research & Extension Ctr, Texas A&M University, Amarillo, TX 79106, USA

Introduction

U.S. livestock and poultry producers and state and federal regulators have become more concerned over gases and particulate matter (PM) that are generated and emitted from their animal operations. To address the need for gas, odour, and PM emission from actual animal production buildings, funding was secured by a six-state research team for a USDA funded project entitled: Aerial Pollutants Emissions from Confined Animal Buildings (APECAB). The objective of APECAB was to quantify long-term (yearly) air pollutant emissions from confined animal buildings and establish methodologies for real time measurement of these emissions and build a valid database of air emissions for US livestock and poultry buildings.

Materials and methods

The APECAB project measured baseline emission rates of odour, ammonia (NH_3), hydrogen sulphide (H_2S), carbon dioxide (CO_2) and particulate matter (including total suspended particulate (TSP) and PM_{10}) from six types of animal confinement buildings (2 barns for each type) located in different states in the United States. The following types of animal housing systems were studied: lactating sow (farrowing) with pull-plug gutters (Illinois), dry sow (gestation) with pull-plug gutters (Minnesota), pig fattening (finishing) with deep pits (Iowa), pig fattening with pull-plug gutters (Texas), high-rise cage laying hen houses (Indiana), and broiler houses on floor litter (North Carolina). Air sampling from two adjacent identical buildings for each type was conducted from the winter 2003 through spring 2004.

Results and conclusions

A small portion of the results from this study are shown in Figure 1. Average NH_3 concentrations in the pig fattening barns were twice as high in deep pit barns (20 ppm) compared with the pull plug (9 ppm) facilities. NH_3 emissions from the deep pit fattening barns were also higher (50 to 60 g NH_3/d-AU) than from the pull plug buildings (35 to 40 g NH_3/d-AU). The pull plug fattening barns produced higher PM_{10} concentrations (450 to 500 vs. 150 to 170 μg PM_{10}/m^3) and emissions (3.0 vs. 0.75 g PM_{10}/d-AU) than the deep pit barns. The cage layer barns in this study had both the highest annual average concentrations and emissions of NH_3 and PM_{10}. NH_3 concentrations and emissions were found to be 36 ppm and 425 g/d-AU respectively, along with PM_{10} levels of 800 μg/m^3 and 19.2 g/d-AU emitted. Other results from the study found the deep pit pig fattening barns with the highest odour concentration and emissions, 1490±640 OU/m^3 and 90±83 OU/s-AU respectively. The pull plug fattening barns had a lower average odour concentration of 680±200 OU/m^3, however the emission rate for odour (73±79 OU/s-AU) was only slightly lower than the deep pit site. The sow lactation rooms had the lowest odour concentration and emission rates with averages of 620±570 OU/m^3 and 43±29 OU/s-AU, respectively. The dry sow barns had an average odour concentration of 970±630 OU/m^3 and an emission rate of 45±38 OU/s-AU.

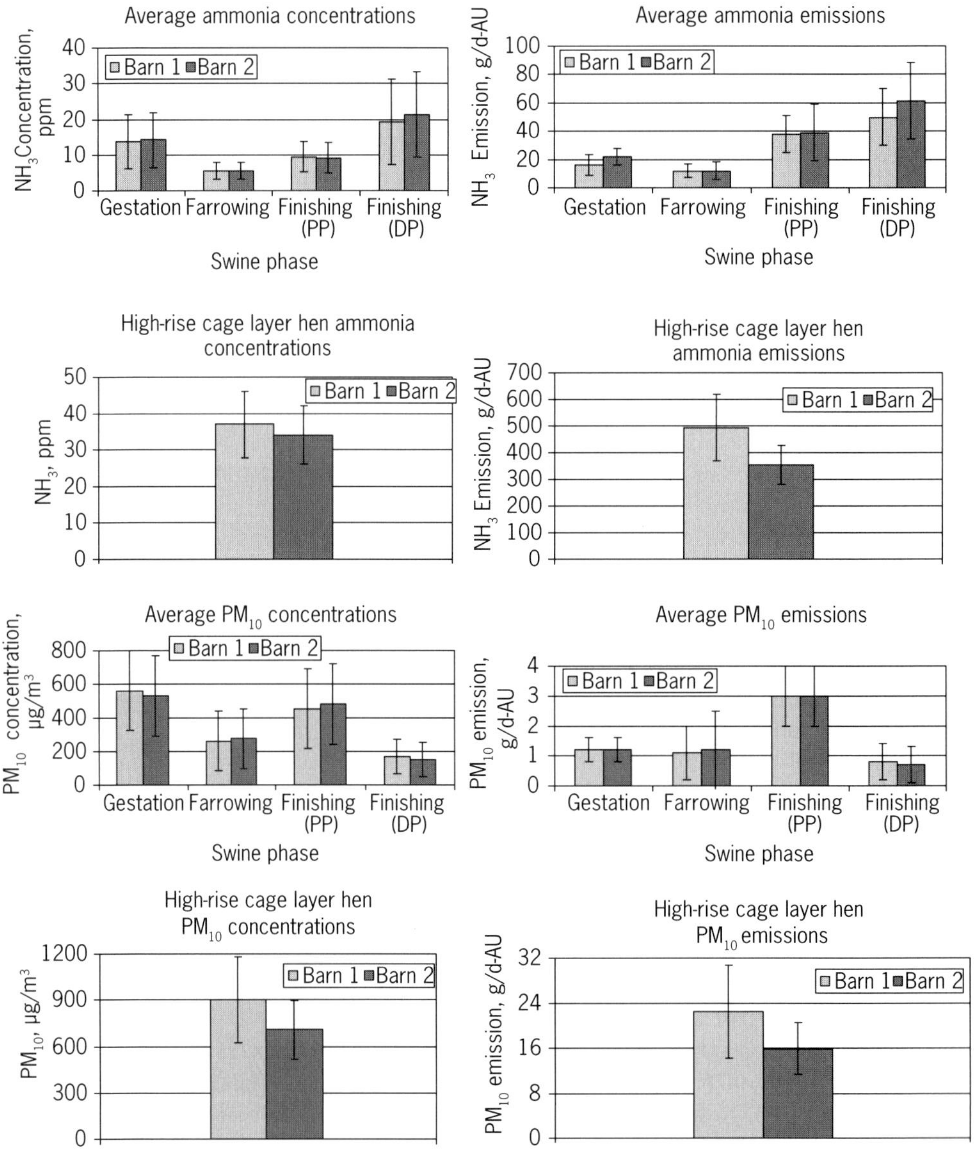

Figure 1. Average Annual Concentrations and Emissions of NH_3 and PM_{10} for Pig and Layer Buildings. Error bars represent one standard deviation of the data. (PP) refers to a pull-plug system; (DP) refers to deep-pit system.

Multi-purpose ammonia measurements on a landscape level

A. Bleeker[1], A. Hensen[1], E. Gies[2], W. de Vries[2], J. Kros[2] and M.P.W. Sonneveld[3]
[1]Energy Research Centre of the Netherlands, P.O. Box 1, 1755 ZG Petten, The Netherlands
[2]Alterra, Wageningen University and Research, P.O. Box 47, 6700 AA Wageningen, The Netherlands
[3]Wageningen University and Research, P.O. Box 47, 6700 AA Wageningen, The Netherlands

Introduction

The rural region of the Northern Friesian Woodlands (NFW-region) in The Netherlands is known for its cultural landscape value which combines dairy farming with historic landscape elements such as hedgerows. At both farm and regional level, a struggle exists to find a balance between economy (dairy farming), ecology (biodiversity) and social heritage. Over a period of 3 years, intensive measurements of ammonia (NH_3) will be conducted in this region. Since different research programmes are running simultaneously in the area, different objectives with respect to these measurements have been defined. Firstly, measurements are performed to investigate the impacts of different management approaches. Secondly, the extent to which environmental targets set for the region can be met will be evaluated. Finally, measurements at a landscape level are made, in order to bridge the gap between local-scale high resolution (both in space and time) measurements and large-scale (National, European) model calculations. The third objective is related to a European scale Integrated Project, called NitroEurope, that investigates the link between nitrogen budgets and greenhouse gas emissions. The measured NH_3 concentrations will thus be used to validate emission models, that will be applied from landscape to European scale (e.g. De Vries *et al.*, 2007). In this paper we present monitored concentrations of NH_3 for the first year and preliminary conclusions based on these measurements. Possible improvements for the coming years are also evaluated.

Study area and measuring network

Figure 1 shows the overall study area of the NFW-region (in grey), as well as the location of three sub areas. The research in the sub areas focuses on the impact of new and conventional methods of animal feeding and manure application, distinguishing between soil types (clay and sand). The new methods include the, presently forbidden, use of above-ground manure spreading under favourable weather conditions in combination with low protein feeding. (see also De Vries *et al.*, 2007). The question is whether this combination leads to similar or even better results than the present application techniques, such as shallow injection, which in itself is an efficient measure to reduce NH_3 emissions but has negative impacts on soil fauna and soil structure. For more information on the conventional manure application methods see De Vries *et al.* (2007). NH_3 passive sampler measurements are used on a monthly basis at 10 locations in each sub area to determine the spatial distribution of the NH_3 concentrations. More intensive measurements are performed by means of the new high precision NH_3 measurement system called Ammonia (in an extended way used by Wichink Kruit *et al.*, 2007) in the two contrasting sandy areas, for determining the temporal distribution. Furthermore, passive samplers are used at 30 locations in the whole NFW-region as a first step in evaluating environmental targets within the region.

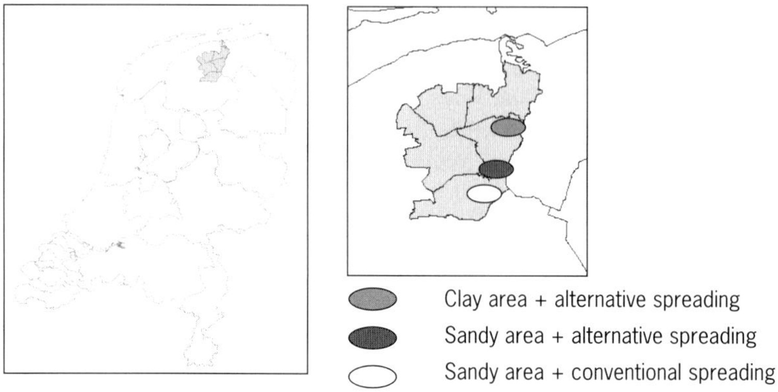

Figure 1. Location of NFW-region and monitoring sub areas within the NFW-region.

First year of measurement results

Figure 2 shows monthly averaged NH_3 concentrations for different areas within the NFW-region. The areas on sandy soil show a small (not significant) difference for the new and conventional feeding/spreading methodologies. The clay area shows greater concentrations than those in the sandy areas. At the moment, the reason for this difference is not clear and has to be investigated in more detail at a later stage of the measurement programme. For the entire NFW region, the average concentrations are clearly less than those in the sub areas (especially in the spreading period), reflecting the spatial distribution of concentrations within the NFW region.

Preliminary conclusions based on the measurements

The first results for the sandy areas do not show a significant difference for the combination above ground manure spreading under favourable weather conditions in combination with low

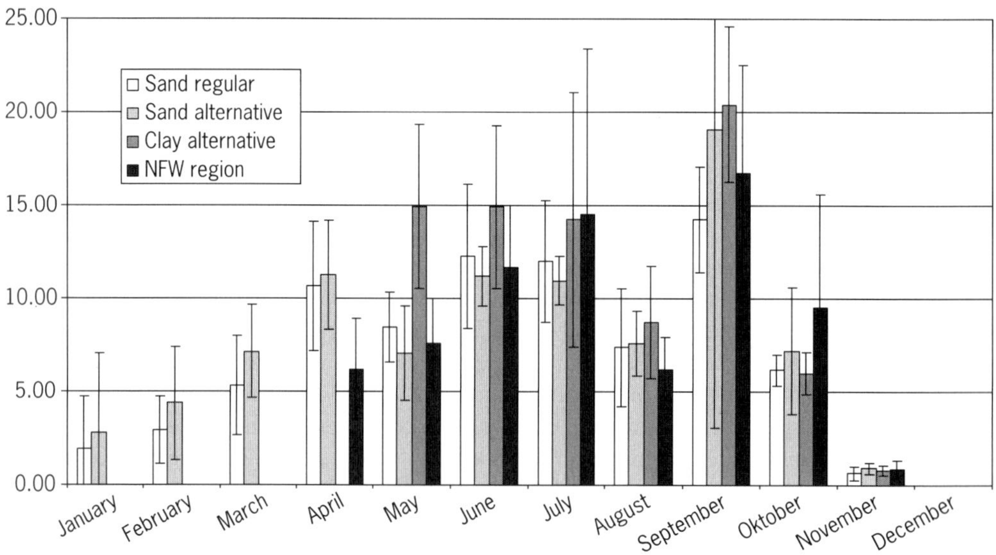

Figure 2. Monthly average NH_3 concentrations in 2006 for the areas within the NFW-region.

protein feeding (alternative) and the present NH_3 application methods without low protein feeding. Using information from intensive monitoring stations, together with detail information from individual farmers on e.g. farm/land management is needed to get a better understanding of observed concentrations within the region. Model calculations will be done to facilitate this process. More measurements on the spatial distribution of the NH_3 concentrations within the entire NFW-region are needed to more accurately determine if environmental targets within the region can be met.

References

De Vries, W., J. Kros, G. Velthof, E. Gies, J. C. Voogd, A. Bleeker and M.P.W. Sonneveld, 2007. Integrated assessment of atmospheric emissions, leaching and runoff of ammonia, greenhouse gases and nutrients at a landscape level. Proceedings International Ammonia Conference (this issue).

Wichink Kruit, R.J., W.A.J. Van Pul, R.P. Otjes, P. Hofschreuder, A.F.G. Jacobs and A.A.M. Holtslag, 2007. Ammonia fluxes and derived canopy compensation points over non-fertilized agricultural grassland in The Netherlands using the new gradient ammonia - high accuracy - monitor (GRAHAM). Atm. Env. 41: 1275-1287.

Ammonia, trimethylamine, and other volatile organic compound emissions from selected animal buildings

N.M. Ngwabie[1,2], G.W. Schade[1,3], T.G. Custer[1,4], S. Linke[5] and T. Hinz[5]
[1]University of Bremen, NW1, Otto-Hahn-Allee 1, D-28359 Bremen, Germany
[2]Swedish University of Agricultural Sciences, Alnarp, Box 86, 230 53 Alnarp, Sweden
[3]Texas A&M University, College Station, 1104 Eller O&M Building, TX 77843-3150, USA
[4]Max Planck Institute for Chem., Johann-Joachim-Becher Weg 27, 55128 Mainz, Germany
[5]Federal Agricultural Research Centre, Bundesallee 50, 38116 Braunschweig, Germany

Introduction

Livestock management is a known source of methane, ammonia, nitrous oxide, and dust. Volatile organic compounds (VOCs) are also emitted, but we do not have enough information on the type, factors affecting their release and the effects on the environment. To address this, we conducted a study on aerial pollutants from livestock. This report identifies some major VOCs emitted in sheep and pig buildings in Germany, following our previous report from a dairy cowshed (Ngwabie *et al.*, 2007).

Materials and methods

VOCs were measured with a commercial PTR-MS (Lindinger *et al.*, 1998), while the concentrations of methane, nitrous oxide, ammonia and carbon dioxide were monitored in 2-minute intervals with a PAS instrument (Brüel and Kjaer model 1302). Measurements were carried out inside a sheep barn and from a pigsty flue in Mariensee near Neustadt-Hannover, Germany, in January and February 2005 respectively. The pigsty had both weaner and finishing pigs kept in a slated floor system. Liquid manure collected in a pit beneath the floor was channelled to a storage tank outside the pigsty. The barn for the sheep contained 120 sheep on a slatted floor with the manure removed underneath the floor twice a day. Air was pulled either from inside (*'barn'*) or outside (*'out'*) of the animal sheds through filters and via a three-way valve to the PTR-MS and photo-acoustic monitoring devices. Animal shed air was sampled for 30 minutes at the beginning of each hour while outside or reference air was sampled for the last 30 minutes of each hour.

Results and discussion

VOC identification using the PTR-MS was based on information in the literature (Schade and Crutzen, 1995; Hobbs *et al.*, 2004; McGinn *et al.*, 2003), isotopic signatures, and fragmentation ions. Table 1 shows results for selected trace gases identified and the ions used for the calculation of their mixing ratios (μ). Trace gas emissions showed periodic spikes with emission surges coinciding with manure removal and animal feeding. Mass emission ratios, E_V (Equation 1), for VOCs to ammonia (or methane) were calculated where significant correlations where observed (R^2>0.5). These were then multiplied with the methane (sheep: 0.023 Tg yr^{-1}, swine: 0.553 Tg yr^{-1} for 2002) or ammonia (sheep: 0.002 Tg yr^{-1}, swine: 0.123 Tg yr^{-1} for 2002, Dämmgen, 2002) mass release rates to estimate nationwide VOC fluxes as presented in Table 2. Trimethylamine (TMA) was the only N-containing VOC identified. Mixing ratios concurred with earlier findings (Schade and Crutzen, 1995), identifying TMA as a significant odour carrier.

$$E_v = \frac{\mu(VOC)_{barn} - \mu(VOC)_{Out}}{\mu(NH_3)_{barn} - \mu(NH_3)_{Out}} * \frac{g/mol\ N\ in\ VOC}{17\ g/mol\ NH_3} \quad [g\ N/g\ NH_3] \tag{1}$$

Table 1. Some trace gases identified in the livestock with associated mass to charge ratios, mixing ratios and observed thresholds.

VOC and other gases	Cowshed	Sheep barn	Pigsty
Data from multi-gas monitor	µmol mol^{-1} (ppm) Median (Range)		
Ammonia	4.12 (0.6-10.30)	6.39 (0.92-19.6)	4.17 (1.16-7)
Nitrous oxide	0.51 (0.38-0.64)	0.47 (0.4-0.60)	0.4 (0.34-0.5)
Methane	77.3 (30-208)	71.1 (35.2-336)	4.7 (3.01-9)
Data from PTR-MS	nmol mol^{-1} (ppb) Median (Range)		
Trimethylamine (TMA) (m/z 60)	16.55 (2.8-69.9)	14.2 (2.6-35.4)	7.7 (2.5-18)
Dimethyl sulphide (m/z 63)	6.89 (0.8-30.8)	1.6 (0.5-5.3)	2.9 (0.9-5.9)

Table 2. Selected VOC emissions from livestock in Germany in Gg C yr^{-1}.

VOC	Dairy cows	Sheep	Pig	Total
Trimethylamine (TMA)	0-3	0-0.02	0.3-0.9	0.3-4 [Gg N yr^{-1}]
Methanol	0.5-2.5	0-0.4	3-9	3.5-11.9
Ethanol	4-31	1.5-9	<0.1	5.5-40
Dimethyl sulphide	0-0.25	<0.01	0.3-1.3	0.3-1.6 [Gg S yr^{-1}]

Conclusions

Trace gases emissions, including previously unidentified VOCs, have been quantified from livestock buildings, with major daily variations due to activities in the sheds. More research into VOC production mechanisms and reduction strategies during these periods is needed.

Acknowledgements

This work was partially funded by the Bundesministerium für Verbraucherschutz, Ernährung und Landwirtschaft, and the German Research Society.

References

Dämmgen, U., 2004. Nationaler Inventarbericht 2004 - Berichterstattung unter der Klimarahmenkonvention der Vereinten Nationen - Teilbericht für die Quellgruppe Landwirtschaft, Landbausforschung Völkenrode Sonderheft 260.

Hobbs, P.J., J. Webb, T.T. Mottram, B. Grant and T.M. Misselbrook, 2004. Emissions of volatile organic compounds from UK livestock agriculture. J. Sci. Food Agric. 84: 1414-1420.

Lindinger, W., A. Hansel and A. Jordan, 1998. Online monitoring of volatile organic compounds at the ppt levels by means of Proton-Transfer-Reaction Mass Spectrometry (PTR-MS), Medical applications, food control and environmental research. Int. J. Mass Spect. & Ion Proc. 173: 191-241.

McGinn, S., H. Janzen and T. Coates, 2003. Atmospheric Ammonia, Volatile Fatty Acids and other Odorants near Beef Feedlots. J. Env. Qual. 32: 1173-1182.

Schade, G.W. and P.J. Crutzen, 1995. Emissions of Aliphatic Amines from Animal husbandry and their Reactions: Potential source of N_2O and HCN. J. Atm. Chem. 23: 319-346.

Cost of ammonia abatement techniques in Spain

C. Pineiro[1], P. Illescas[2], G. Montalvo[2], M. Herrero[3], R. Giraldez[4] and M. Bigeriego[5]
[1]PigCHAMP Pro Europa, S.A., Ctra. San Rafael, 42, 40006, Segovia, Spain
[2]Tragsega, S.A. Conde de Peñalver, 84, 8 planta, 28006, Madrid, Spain
[3]FEASPOR, los Coches, 5, 4C, 40002, Segovia, Spain
[4]Consultant
[5]Ministry of Agriculture, Fisheries and Food, Alfonso XII, 62, 28071, Madrid, Spain

Introduction

The Council Directive 96/61/EC of 24 September 1996 concerning Integrated Pollution Prevention and Control defined the Best Available Techniques (BAT) as: 'The most effective and advanced stage in the development of activities and their methods of operation which indicate the practical suitability of particular techniques for providing in principle the basis for emission limit values designed to prevent and, where that is not practicable, generally to reduce emissions and the impact on the environment as a whole:

- Techniques shall include both the technology used and the way in which the installation is designed, built, maintained, operated and decommissioned.
- Available techniques shall mean those developed on a scale which allows implementation in the relevant industrial sector, under *economically* and technically viable conditions, *taking into consideration the costs* and advantages, whether or not the techniques are used or produced inside the Member State in question, as long as they are reasonably accessible to the operator.
- Best shall mean most effective in achieving a high general level of protection of the environment as a whole.'

Therefore, it is necessary to have a common methodology to calculate cost of abatement techniques to decide if a technique is BAT or not.

Objective

The objective of this study was to present a calculation on cost of every Best Available Techniques under Spanish conditions. The information provided will allow defining the most cost-effective methods for reducing ammonia emission from Spanish farms.

Material and methods

The calculation was carried out according to the methodology set out in BREF, using current costs of equipment, labour and feed, taking into account the economic life of the investment, deducting grants, including changes in performance and annual interest rate of 5%. The costs were calculated for feeding techniques, animal housing, slurry storage and spreading techniques. Units used for assessing costs were:

- € per place per year for feed and housing techniques.
- € per m^3 or tonnes per year for manure or slurry storage and manure or slurry spreading categories.

All these costs have been expressed also as € per kg pig produced, because in the pig sector it is more easily understood than € per place per year or € per m^3 per year, and it is easier to calculate the cost for all of the production process. The basis for this calculation was 20 marketed pigs of 100 kg per sow per year. Further adjustments can easily be undertaken to reflect local conditions.

Results and discussion

Extra costs calculated for abatement techniques are listed in Table 1. The standard concepts used and the transparency of the proposed methodology allows its implementation in other EU-27 MS just using the appropriate figures for local conditions.

Table 1. Extra cost for abatement techniques, regarding reference system.

		(€/place per year)	(€/t pig produced per year)
Feeding techniques	Phased feeding of growing pigs	1.52	5.2
	Low protein diet supplemented with amino acids	0.39 – 2.61	1.3 – 8.8
Housing: gestating sows	Reduced manure pit	5.69 – 6.83	2.1 – 3.0
	Littered system	47.61 – 80.45	17.9 – 30.2
	Frequent manure removal	0	0
Lactating sows	Manure pan underneath	17.52 – 37.18	2.2 – 4.6
Weaners	Manure channel with sloped	0.23 – 2.67	0.3 – 4.6
	Partially slatted floor	0 – 2.25	0 – 3.9
	Frequent manure removal	0	0
Growers-finishers	Manure channel with sloped	0.73 – 7.74	2.5 – 26.3
	Partially slatted floor	0 – 4.33	0 – 14.7
	Frequent manure removal	0	0
		(€/m^3 per year)	(€/t pig produced per year)
Spreading (pig slurry)	Trailing shoe	0.92 – 1.41	11.5 – 17.6
	Band spreader	0.79 – 1.21	9.9 – 15.1
	Incorporation	0.23 – 0.61	2.9 – 7.6

Acknowledgements

This work is part of a project funded by the Spanish Ministry of Agriculture, Fisheries and Food. We also thank Martin Ryan (DEFRA) for his help in cost calculation.

References

European Commission, 2003. Integrated Pollution Prevention and Control. Reference Document on Best Available Techniques for Intensive Rearing of Poultry and Pigs (BREF).

Ammonia-based air permits for Idaho dairies

R. E. Sheffield[1] *and B. Louks*[2]
[1]*University of Idaho Extension, Box 1827, 83301, Twin Falls, Idaho, USA*
[2]*Idaho Department of Environmental Quality, Boise, Idaho, USA*

Introduction

The *Rules for the Control of Ammonia from Dairy Farms* are intended to provide a Permit by Rule to those dairies that may emit more than 100 tons of ammonia per year from their facility. The DEQ uses the annual 100 ton emission level as a threshold for requiring Permits to Construct for industrial, municipal and agricultural sources in Idaho. The threshold values for this Rule were derived from manure excretion tables prepared by the American Society of Agricultural and Biological Engineers and the USDA – Natural Resources Conservation Service (NRCS) and were agreed upon by IDA and ICL. In addition to manure excretion, the IDA and ICL took into account their best professional judgment to account for differences in ammonia losses from various dairy housing types, manure storage systems, and land application practices. The threshold values represent differentiations in the type of dairy housing system as well as the type of liquid manure application system that is employed on the farm. Differences in threshold levels for drylot and freestall scrape dairies account for how each farm land-applies their manure and the relative amount of ammonia loss that typically occurs on the facility, as well as during the application of manure and wastewater. Greater thresholds are allowed for dairies which export all of their manure and wastewater off the farm (Drylot = 5,063 head; Freestall Scrape = 2,781 head). Lower thresholds are allowed for dairies which apply their wastewater with center pivot drop nozzles or drag-hose (ground level) application systems. The lowest thresholds are used for openlot and freestall scrape dairies which apply wastewater with overhead pivot sprinklers or other conventional sprinkler systems. No differentiation is made between the type of farms and how solid manure is applied and/or incorporated. Regardless of the farm's land application practices, freestall flush facilities greater than the threshold value of 1,638 head (2,293 AU) are subject to this Rule. During the development of this Rule, various manure treatment systems and handling practices were evaluated for their effectiveness in reducing ammonia emissions (Table 1). Many scientific studies, extension bulletins, NRCS handbooks, and EPA guidance documents were referred to when preparing the professional judgment toward relative effectiveness in reducing ammonia and the allocation of points. An arbitrary point system with a maximum of 20 points was assigned to each practice. A practice receiving 20 points would equate to a system or practice that would result in a major reduction, approximately 70 percent, in ammonia emissions for that specific process. Each practice was rated on a year-round basis, and as if all of the manure practically available for the practice was handled by that practice. Variations due to normal seasonal use of each practice was taken into account in the points awarded to each BMP. Variations due to seasonal practices (such as corral harrowing or direct land application of liquid manure) and expected weather conditions have been factored into these ratings. Points awarded to land application practices assume that the practice is utilised on all manure that is applied. Points are allowed to be pro-rated to reflect the actual waste treatment or handling that is occurring on each farm. Inspections will be made to ensure the BMP allocations totalling 27 points are employed on the farm. Inspections will be

Table 1. Ammonia control practices for Idaho dairies.

System	Component	Ammonia Control Effectiveness [1]			Compliance
		Open Lot	Freestall Scrape	Freestall Flush	Method [3]
Waste Storage and Treatment Systems	Synthetic Lagoon Cover	15	20	20	1
	GeoteXtile Covers	10	13	13	1
	Solids Separation	3	3	3	3, 4
	Composting	4	4	4	1
	Separate Slurry and Liquid Manure Basins	6	10	-	1
	In House Separation	0	12	0	1
	Direct Utilization of Collected Slurry	6	10	-	1, 3, 4
	Direct Utilization of Parlor Wastewater	10	10	10	1
	Direct Utilization of Flush Water	8	0	13	3, 4
	Anaerobic Digester	-	-	-	-
	Anaerobic Lagoon	-	-	-	-
	Aerated Lagoon	10	12	15	2
	Sequencing-Batch Reactor	15	20	20	2
	Lagoon Nitrification/Denitrification Systems	15	20	20	2
	Fixed-Media Aeration Systems	15	20	20	2
General Practices	Vegetative or Wooded Buffers (established)	7	7	7	1
	Vegetative or Wooded Buffers (establishing)	2	2	2	1
	Alternatives to Copper Sulfate	-	-	-	-
Freestall Barns	Scrape Built Up Manure	-	3	3	1
	Frequent Manure Removal	UD	UD	UD	-
	Tunnel Ventilation	-	-	-	-
	Tunnel Ventilation w/Biofilters	-	10	10	1
	Tunnel Ventilation w/Washing Wall	-	10	10	3, 4
Open Lots and Corrals	Rapid Manure Removal	4	2	2	1, 2
	Corral Harrowing	4	2	2	1
	Surface Ammendments	10	5	5	2
	In-Corral Composting / Stockpiling	4	2	2	1
	Summertime Deep Bedding	10	5	5	1
Animal Nutrition	Manage Dietary Protein	2	2	2	2
Composting Practices	Alum Incorporation	12	8	6	2
	Carbon:Nitrogen Ratio (C:N) Ratio Manipulation	10	7.5	5	2
	Composting with Windrows	-	-	-	-
	Composting Static Pile	6	4.5	3	1
	Forced Aeration Composting	10	7.5	5	1
	Forced Aeration Composting with Biofilter	12	8	6	1
Land Application [2]	Soil Injection - Slurry	10	15	7.5	2
	Incorporation of Manure within 24 hrs	10	10	10	2
	Incorporation of Manure within 48 hrs	5	5	5	2
	Nitrification of Lagoon Effluent	10	10	15	3, 4
	Low Energy/Pressure Application Systems	7	7	10	1
	Freshwater Dilution	5	8	8	1, 2
	Pivot Drag Hoses	8	8	10	1
	Subsurface Drip Irrigation	10	10	12	1

Notes:

1. The ammonia emission reduction effectiveness of each practice is rated numerically based on practical year-round implementation. Variations due to seasonal practices and expected weather conditions have been factored into these ratings. Not implementing a BMP when it is not practicable to do so, does not reduce the point value assigned to the BMP, nor does it constitute failure to perform the BMP. UD indicates that the practice is still under development.

2. Land application practices assume practice is conducted on all manure; points will be pro-rated to reflect actual waste treatment; points can be obtained on exported material with sufficient documentation.

3. Method used by inspector to determine compliance:
 1=Observation by Inspector
 2=On-Site Recordkeeping Required
 3,4=Deviation Reporting Required. Equipment upsets and/or breakdowns shall be recorded in a deviation log and if repaired in a reasonable timeframe does not constitute non-compliance with this rule.

made by personnel from the Idaho Department of Agriculture. Follow-up inspections will be conducted as part of ISDA's regular dairy inspection program.

2006 Inspections

Thirty-six dairies which registered for the Rule were inspected during the summer of 2006, and were awarded an average of 32.9±6.1 points of BMP utilisation (Table 2). Ninety-five percent (95%) of the dairies were found to be in compliance with the Rule, demonstrating at least 27 points of BMP utilisation. All of the open lot dairies were found to be in compliance, with an average BMP score of 34±6.2. One freestall scrape and one freestall flush facility were found not to meet the 27 point threshold during the initial inspection. Freestall scrape dairies had an average BMP utilisation score of 31.6±6.3 while flush dairies scored an average 32.9±5.2 points. The five most common BMPs to be installed on all of the dairies were found to be solid separation, corral harrowing, low pressure irrigation, composting and rapid manure removal from outdoor lots. Manure injection and incorporation was found not to be widely practiced. Only one freestall scrape dairy reported to inject its manure, and only four freestall scrape dairies (11% of all dairies) incorporated manure within 24 hours of application. Twenty-six percent (26%) of all dairies reported to incorporate manure within 24 hours of application. Freshwater dilution of stored manure and wastewater was found to be practiced on 61% of all dairies, with the most common dilution rates ranging from 5:1 to 9:1 of fresh to wastewater. Management of the level of dietary protein fed to cattle was found to be implemented on 61% (23 dairies) of the permitted dairies. Seventy-one percent (71%) of all freestall flush dairies reported managing protein levels, compared to 59% of open lots and 57% of freestall scrape dairies.

Table 2. Compliance of Idaho dairies to ammonia control rule during 2006.

	All Dairies	Open Lot	Freestall Scrape	Freestall Flush
Number of Dairies	38	17	14	7
Number of Dairies in Compliance	36 (95%)	17 (100%)	13 (93%)	6 (86%)
Average Points[1]	32.9 ± 6.1	34.0 ± 6.2	31.6 ± 6.3	32.9 ± 5.2

[1] ± Standard Error.

EU Project 'BAT SUPPORT': 'Best Available Techniques for European Intensive Livestock Farming - Support for the Implementation of the IPPC-Directive'

H. Eckel, H. Döhler, B. Eurich-Menden and E. Grimm
Association for Technology and Structures in Agriculture e.V., Bartningstrasse 49, 64283 Darmstadt, Germany

Objectives of the project

The overall objective of this project, starting in February 2007, is to develop an integrated and consistent methodology for the classification of livestock housing systems including storage, treatment and spreading of the manures produced, in terms of 'Best Available Techniques.' This includes the secondary objectives:

- the development and application of systems to evaluate the environmental and economic performance of different technologies in use in intensive livestock farming including cost/ benefit considerations regarding emission reduction;
- the assessment of animal health and welfare implications that are specific for the intensive livestock farming sector;
- the development of a documentation system for BAT techniques to support the implementation of new and effective techniques, very much depending on the possibilities of use of documentation across borders;
- to prepare a suitable 'Glossary of Terms on Livestock Manure Management' to support the harmonised description of techniques in Europe;
- to disseminate the results widely including all relevant stakeholders in the Livestock Farming Sector.

Background

The relevance and purpose of this project relates to the implementation of the European IPPC Directive (Council Directive 96/61/EC on Integrated Pollution Prevention and Control of 24 September 1996). The purpose of the Directive is to achieve integrated systems to prevent polluting emissions to air, land and water, including measures concerning waste, in order to achieve a high level of protection of the environment taken as a whole. One crucial sector is livestock production due to its very large impact on the wider environment, such as emissions to the air (ammonia, greenhouse gases, dust and bioaerosols), discharges to soils and surfaces water (nitrogen, phosphorus, heavy metals, organic compounds) as well as noise and odour nuisance. The successful application of a BAT (best available technology) policy requires a high degree of support for both the regulator(s) and industry. Comprehensive advice and guidance notes are essential for effective implementation of the integrated pollution control regime in the Member States. For that purpose, the European IPPC Bureau (EIPPCB) in Seville is charged by the European Commission with the organisation of a European information exchange in order to establish special Reference Documents (known as BREFs) for each of the categories of industrial activities listed in the Directive. The first BREF for Intensive Livestock Farming (ILF) was published in 2003 (European Commission 2003: 'Reference Document on Best Available Techniques for Intensive Rearing of Poultry and Pigs'). The BREF documents assist the regulatory authorities by describing reference techniques and reference levels for each sector. BREFs are uniformly defined on a Europe-wide basis and are crucially intended for regular updating. The procedure of BAT classification as described in the BREF requires a harmonised approach on European level. In the development of the first BREF for the intensive livestock farming sector in 2003, this harmonisation was lacking and many European countries used their own system and their own criteria for the evaluation of the best technologies. This lead to differing results and to the call for harmonised guidelines for the formulation of classification methods of BAT for intensive livestock farming. The availability of field data relating to livestock and livestock waste management systems in current use varies across Europe. There is an urgent need for harmonisation of both the required data acquisition and of the procedure to identify and define systems as 'Best Available Technique'. For some of the New Member States, IPPC is still being implanted on an informal basis due to a lack of available data.

Approach

The project consists of three phases: (1) the establishment and update of data on production systems and techniques in intensive livestock farming in use in Europe, (2) the development of a documentation and assessment system for the classification of those techniques identified as 'Best Available Technique' and (3) a coordination process with experts from the livestock farming sector, included in the project as 'Advisory Group' in order to achieve a widespread awareness and acceptance of the outputs and findings from the project.

Target groups

The developed system will be made widely available in form of guidelines to all stakeholders working with BAT definition and it will greatly assist in their work by providing a common platform and a harmonised and consistent approach for description and assessment of the livestock systems. The main target group for the results of BAT-SUPPORT project will be the experts in the Member States involved in preparing the national contributions for the updating of the BREF for Intensive Rearing of Poultry and Pigs.

Acknowledgements

The project partners are the Association for Technology and Structures in Agriculture, KTBL, Germany; Research Centre for Animal Production, CRPA, Italy; ID-Lelystad, Animal Sciences Group, ID-Lelystad, Netherlands; Danish Agricultural Advisory Service, National Centre, DAAS, Denmark; PigCHAMP Pro Europa S.A., PigCHAMP, Spain; Institute of agricultural engineering, IBMER, Poland; Institute of Agricultural and Environmental Engineering Research, CEMAGREF, France; Research Institute of Agriculture Engineering, VUZT, Czech Republic; and Swiss College of Agriculture, SHL, Switzerland.

Measurement methods - sensors and technology

Measuring atmospheric ammonia concentrations using differential optical absorption spectroscopy

D.E. Lolkema, J.B. Bergwerff, A.J.C. Berkhout, M. Haaima, G.R. van der Hoff, C.J.M. Potma and D.P.J. Swart
RIVM, Laboratory for Environmental Monitoring, P.O. Box 1, 3720 BA Bilthoven, The Netherlands

Introduction

Most traditional techniques to measure atmospheric ammonia concentrations are sampling techniques. They therefore involve direct contact with the highly adhesive ammonia. Except for passive samplers, this results in memory effects which may lead to erroneous measurements. This problem can be avoided using optical remote sensing techniques. Differential optical absorption spectroscopy (DOAS) is such a technique. The National Institute for Public Health and the Environment (RIVM) has developed a DOAS system to measure atmospheric ammonia concentrations: the RIVM Ammonia Measurement System by UV-Spectroscopy (RAMSUS). Other advantages of the system are the low maintenance demands and the high up-time. This abstract describes the system, presents results of a comparison with a more traditional denuder system and gives an outlook to a system to measure ammonia deposition.

Measurement principle

The DOAS technique is based on the wavelength dependent absorption of light over a specified light path. The absorption depends on the concentrations of the different trace gases present in this light path. This is expressed by Lambert-Beer's law:

$$I_\lambda = I_{0,\lambda} * e^{-\alpha(i,\lambda)*C(i)*L} \quad (1)$$

with I_λ the received light intensity at wavelength λ, $I_{0,\lambda}$ the emitted light intensity at wavelength λ, $\alpha(i,\lambda)$ the absorption coefficient for trace gas i at wavelength λ, $C(i)$ the concentration of trace gas i and L the length of the light path. Ammonia has strong absorption lines with narrow features in the wavelength range 200-230 nm. Other trace gases with narrow absorption features in this wavelength range are nitrous oxide and sulphur dioxide. The absorption features of those three trace gases are least squares fitted to the ratio of emitted and received light in the wavelength range 200-230 nm using the standard DOAS technique. This results in the determination of the concentration of those trace gases.

System set-up

The system consists of a combined sender/receiver unit. The sender is a 150W Xenon lamp and the receiver a telescope. A corner cube retroreflector directs the lamplight directly back to the telescope. The retroreflector is placed at approximately 50 meters from the sender/receiver unit. This results in an effective light path of approximately 100 meters. A spectrograph separates the light from the telescope by wavelength. A charge coupled device (CCD) detector measures the light intensities at the different wavelengths. Data is averaged for five minutes to increase the

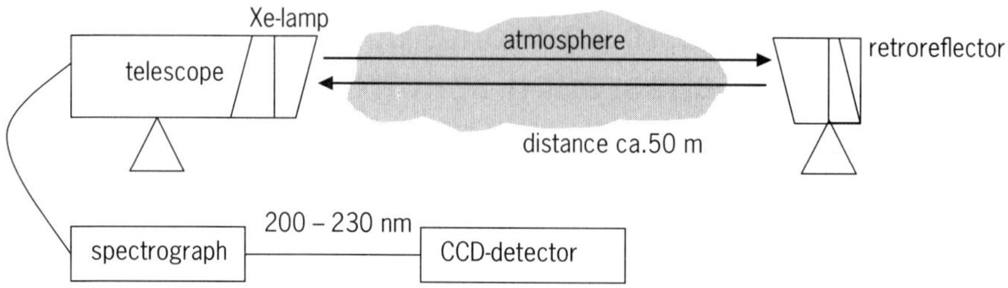

Figure 1. System set-up of the RIVM Ammonia Measurement System by UV-Spectroscopy (RAMSUS).

signal-to-noise ratio. All system control and data storage is managed by one computer. Figure 1 shows a schematic set-up of the system.

Field experiments

During a one-year period, RAMSUS measured at the weather station at the Haarweg in Wageningen. Comparisons were made with the measurements of a GRadient Ammonia - High Accuracy - Monitor (GRAHAM), located at the same site. The GRAHAM measured at three altitudes: 4.35, 1.95 and 1.00 meter. RAMSUS measured at an altitude of 3.00 meter. Figure 2 gives an example of the measurements.

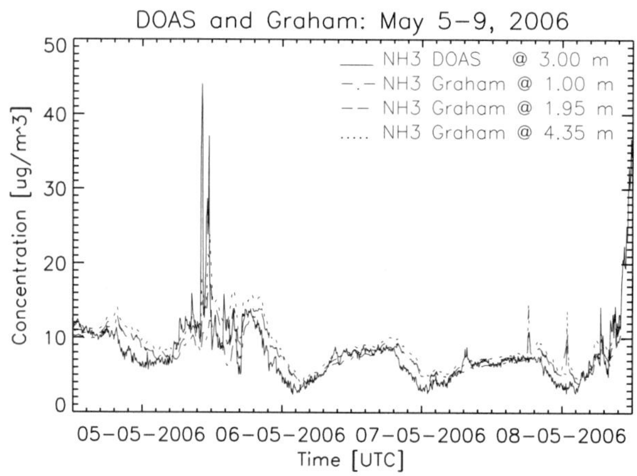

Figure 2. Typical period of ammonia concentrations measured by the RIVM Ammonia Measurement System by UV-Spectroscopy (RAMSUS) and by the GRAHAM, a more traditional denuder system.

Conclusions and outlook

It has been shown that RAMSUS has (1) a fast response (seconds to minutes) and no memory effects; (2) a precision better than 0.3 $\mu g/m^3$ for a 5-minute average; and (3) low maintenance demands and a high up-time (better than 85%). In 2007 we will develop an operational ammonia deposition system. To this end, three of these systems will be placed above each other, each system measuring at a different altitude.

Semiconductor sensor in continual measuring of different ammonia concentration

P. Zabloudilová
Czech University of Life Sciences, Technical Faculty, Dept. of Physics, Kamýcka 872, 165 21 Prague 6, Czech Republic

Introduction

Determination of ammonia emissions occurrence and proposals for appropriate measures for elimination depends largely on reliability of the methods of determination of ammonia concentrations. Nowadays, emphasis is placed especially on methods of continual measurement of ammonia emissions because the process of liberation changes in time and thus single-shot determinations of concentrations can be affected by error. A potential method for continual monitoring of ammonia emissions is a semiconductor sensor with solid phase and surface (adsorption) detection.

Material and methods

Experimental measuring of ammonia emissions was conducted in a laboratory. Wooden waste (shavings and saw-dust) was used as a sorption material. Individual measured samples (600 g) had combinations of two starting moistures of the sorption material (10±1% and 30±1%) and three initial ammonia concentrations (prepared by adding 1.5, 3 and 6 g of 24% ammonia liquor/sample). Emissions from individual material samples were measured in a plexiglass container (air-temperature 20±1 °C, relative humidity was 56±1%, constant air flow through the fermenter). Ammonia concentration was measured continually with five electronic sensors fitted with five semiconductor sensors SP-53 (FIS Inc.) and simultaneously with the 1312 Photoacoustic Multi-gas Monitor (PMM) with the Multipoint Sampler 1309 (INNOVA Air Tech Instr.). All five sensors were positioned in parallel in the container outlet, in its upper part in which a PMM filter was positioned as well. The values of the sensor and the PMM were recorded in 2-minute intervals. The PMM also measured water vapour. In total, 122 samples were measured, each sample for 12 hours at minimum, with at least 10 repetitions for a given combination of moisture and ammonia.

Semiconductor sensor SP-53: The sensor SP-53 is made by a planar technology. The sensor is based on pressed tin dioxide (SnO_2) put on Al_2O_3 substrate (a carrier of the active layer). On the other side of the substrate a flat heating device is impressed which maintains a constant temperature necessary for reaction of ammonia with SnO_2. If the ambient air is without ammonia adulterant, the active layer shows a constant resistance value during heating to a stable temperature. Ammonia accumulated on the active material surface of the sensor cause a change of the material structure, specific resistance between electrodes decreases which manifests as a change in voltage during supplying from the source of constant voltage. The sensor provides a continual signal which can sampled with regard to the technical parameters of the sensor (response time, correction). The sensor construction enables, in the environment without ammonia molecules, return of its signal to the initial value of clean air. The sensors used for measuring were calibrated repeatedly with the PMM.

Photoacoustic Multi-gas Monitor: The principle of operation of the PMM is based on photoacoustic infrared spectroscopy. A sample of analysed air is sucked with a tube through filters to a measuring chamber in which it is hermetic sealed using valves. Pulsating rays of appropriate wavelength are emitted from an infra-red source to the measuring chamber through an optical filter. The gas molecules absorb the infrared, and its temperature increases and decreases. Temperature waves are recorded as pressure changes which are detected in a form of acoustic signal directly proportional to the concentration of the gas being measured. The PMM can measure concentrations of five gases from one sample as well as water content.

Results

The measurement results were statistic-processed and average values of ammonia concentrations were processed graphically in dependence on time by reason of a good clarity. The courses of ammonia emissions of the samples of wooden saw-dust are similar. Comparing the measured values the values measured by the PMM were considered objective because this principle of measuring of gas concentrations shows precision and reliability which is by a log higher than that of the sensors used.

The figures show differences between the concentration values measured by the sensors and the PMM. The differences are changing in time and for both moistures of the sorption material and higher ammonia concentrations they are lowest approx. between the 2nd and 7th hour after the start of measuring. Using the initial ammonia concentration of 1.5 g/sample the desorption course has a similar trend but differences in measured values of ammonia concentrations are lowest at the beginning of measuring and over time they only increase. Gradual, even if uneven increase in differences between measured values may be caused by a long-term stability of the sensors. Generally, we can say that differences between the values measured by the sensors and the Multi-gas Monitor are higher if the initial moisture of the material is rather high and ammonia concentration rather low. While measuring low concentrations of ammonia higher differences even between the values measured by individual sensors were found. The differences were partially demonstrably different ($P<0.01$) only if moisture of the material was 30%.

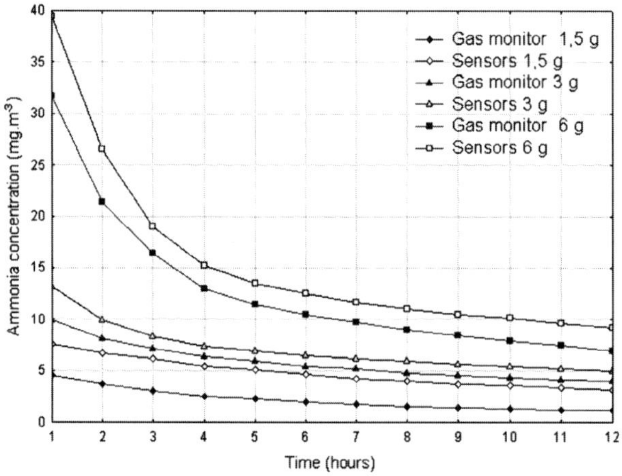

Figure 1. The curves of average ammonia concentrations - initial sorptive material humidity 10 ± 1% (wood shavings).

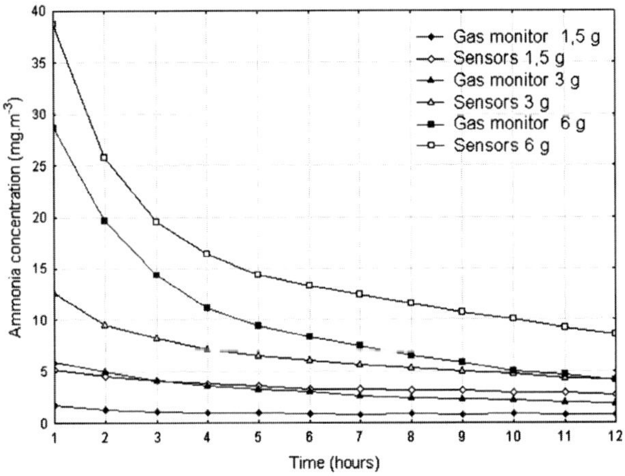

Figure 2. The curves of average ammonia concentrations - initial sorptive material humidity 30 ± 1% (wood shavings).

Resolution of the sensors may also have effect on the differences between individual sensors and values measured by the sensors and the PMM because the resolution of semiconductor sensors with an active layer of SnO_2 (without structural or chemical modification of this layer) against ammonia is relatively lower than that against other reducing gases. Differences may be also influenced by possible calibration inaccuracy, or non-homogeneity of the sorption material.

Conclusion

Differences between ammonia concentrations measured by the sensors SP-53 and the PMM show non-linear change in time and depend on, among others, the initial moisture of the sorption material and the initial ammonia concentration in this material. Precision of the sensors is especially low in measuring of low concentrations of ammonia in combination with rather high moisture of the sorption material. Long-term stability of the sensors SP-53 is worse.

Acknowledgements

This work was supported by grant of Czech University of Life Sciences Prague, CIGA, No 31120/1313/313131 and by grant of Ministry of Agriculture Czech Republic, NAZV, No QF 3140.

References

Kawashima, S. and S. Yonemura, 2001. Measuring ammonia concentration over a grassland near livestock facilities using a semiconductor ammonia sensor. Atm. Env. 35: 3831–3839.

SnO_2 solid state gas sensors and changes of some their properties for measurement of NH_3 concentration

J. Pecen
Czech University of Agriculture Prague, Technical Faculty, Department of Physics, Kamýcká 129, 165 21 Prague, Suchdol, Czech Republic

Introduction

Ammonia (NH_3) is an important substance in the nitrogen cycle. Agricultural emission NH_3 has become one of the most important problems causing the worldwide air pollution (Ni and Heber, 2001). To understand and control of NH_3 emission production depends on sampling and measurement techniques that includes devices, instruments and procedures. Suitable sensors with excellent properties are the first and very important part in detection and quantification of NH_3. Solid state gas sensors are very often used for this purpose. Their basic function principle performs a change of their conductivity depending mainly on the interaction of measured gas and sensitive material (the layer on the sensing element of the sensor) such as In_2O_3, Ga_2O_3, TiO_2, WO_3, SnO_2, ZnO, CeO_2, CuBr, NiO, ZrO_2, $In_xO_yN_z$, etc (Bendahan *et al.*, 2002). Nowadays the *ZnO and* SnO_2 are preferred in gas sensors` design and thin film metal oxide devices are the most promising. Conventional polycrystalline thin film sensors have the lack of long-term stability related to the porous nature of layers and large amount of grain boundaries. The electric resistance of n-type semiconductor under the high oxygen partial pressure is high because electron transfer from the semiconductor into adsorbed oxygen leads to an electron-depleted space charge region, near the semiconductor surface. Almost any type of reducing gas is detected by this mechanism. A serious problem of solid state sensors (from point of view) is their strong dependence on temperature and humidity mainly (Yamazoe, 2006).

Materials and methods

The Figaro gas sensors TGS 826 (Figaro, 1978) were used for all the laboratory experiments with temperature and humidity changes. The porous material (wood sawing) with initial concentration of ammonia water was placed into laboratory apparatus where the humidity of material was changed (i.e. the changes of air with ammonia emissions that were measured). The experiments were repeated and were carried out for longer periods. One part of these experiments was focused on relation between resistance and temperature the sensor's sensing element when the humidity of air with ammonia emissions was the parameter. The temperature of the sensor's sensing element was changed by the power of the sensor's heater. More experiments were carried out with the increase of sensor's sensitivity. The first procedure was carried out with their connectivity into half the bridge where the second half of the bridge was replaced by the resistors. The second method was carried out with the aim to increase the sensitivity of gas sensor and was made by acting a X-ray radiation on sensor. The energy of this radiation was between 18-20 keV and 65-70 keV respectively. The radiation was applied at anytime of each start of sensor's ammonia concentration measurement. The experiments with X-ray radiation were carried out under standard conditions of the gas sensor.

Results and discussion

- The results obtained under standard working conditions of gas sensor (i.e. constant temperature and various humidity and ammonia concentration in air) are similar, as are described many authors (Ni, Heber, 2001).
- The temperature changes of the sensor sensing element vs. output resistance of the sensor 'appear' bring a small plateau (approximately in temperature range from 150°C to 250°C) for small ammonia concentration in air. The position of plateau depends on humidity of air too. The experiments were carried out in the small range only and similar situation presents (Yamazoe, 2006). The phenomena probably depends on the structure (size of particles, the thickness of the depletion layer etc.) and type of conductance of the sensing material (Russ, 2005).
- Two sensors TGS 826 connect in the half bridge (opposite two resistor) brings increase the sensor sensitivity but in narrow range of measured range of ammonia concentration, because the characteristic of the sensors are not the same (in their whole range).
- The application of soft X-ray radiation on sensor brings the similar course measuring of ammonia emission (for one value of air humidity only). It seems that in this case the 'starting point' of each measuring is the same and the courses of each measuring are similar. The sensitivity used sensor was not increased this way in many cases. This situation reminds using the command 'reset' before starting of system. More experiments must be done for reliable results and for explanation an influence of radiation on sensor.

Conclusions

- The gas sensors based on SnO_2 sensing material are fully and correct used in limited range (temperature, humidity and a concentration of gas).
- A part of sensor output characteristic is probably constant (from point of view of sensor temperature changes).
- Increasing the sensitivity of gas sensor by their connecting into the bridge is problem, because the characteristic of the sensors are not the same.
- The application of X-ray radiation on gas sensors may be brings better start conditions for measurement (similar start point).

Acknowledgments

This work was supported by grant NAZV, No QF 3140 from The Ministry of Agriculture Czech Republic.

References

Bendahan, M., P. Lauque, J.L. Seguin, C.H. Jacolin and K. Aguir, 2002. Development of an ammonia gas sensor. In.: Proceedings of the Int. Scientific Conf. 'EUROSENSOR XVI', Prague, September 15-18. Prague, Czech Republic, pp. 1049-1052.

FIGARO Gas Sensor. 1978. Specifications of the gas sensors type 824. Company material.

Ni, J.Q. and A.J. Heber, 2001. Sampling and Measurement of ammonia concentration at animal facilities-A review. ASAE Meating Paper No 01-4090. St Joseph, Mich.:ASAE.

Russ, S., A. Bunde, K.-D. Kohl and M. Ulrich, 2005. Percolation models for semiconductor gas sensors. In.: Book of Abstracts, 3rd Int. Symp. on Sensor Science, Jűlich, July 18-21.

Yamazoe, N., 2006. Prospects and Challenges for Gas Sensors. In: Proceedings of the Int. Conf. 'EUROSENSOR XX', September 18-20, Göteborg, Sweden.

Ammonia passive samplers: a helpful tool for monitoring measures to abate emissions

D. Matthaei, E. Seitler and L. Thöni
FUB, Research Group for Environmental Monitoring, Untere Bahnhofstrasse 30, P.O. Box 1645, 8640 Rapperswil, Switzerland

Introduction

In agriculture, there are high N losses in emissions of ammonia. Not to exceed critical loads, ammonia emissions in Switzerland should be abated for 42% in relation to the year 2000 (BUWAL, 2005). There are several approaches to reach this goal: (1) optimisation of construction and management of stables and free-cattle-ranges, (2) manure storing with reduced emission, (3) manure spreading with reduced emission e.g. with dragging hose system, (4) optimisation of nitrogen content in feed. Ammonia emissions and concentrations can be modelled to estimate how effective measures are to reduce ammonia losses. It is however required to calibrate these models with measurements. Ammonia passive samplers open a possibility for a simple and cost efficient measuring method. They are also important tools for monitoring measures that abate ammonia release to the atmosphere. 'FUB – Research Group for Environmental Monitoring' has been measuring ammonia for 13 years at differently levelled sites in Switzerland from background stations to high loaded places at farms. In this paper, measurements at farms with different manure storing and livestock husbandry systems are compared and illustrated. The aim of this study is to analyse how ammonia emissions from different farming practices are measured by passive samplers.

Methods

Concentration of atmospheric ammonia was measured using the Radiello® passive sampler, a radial designed diffusive sampler with an acid coated cartridge as absorbent. Three samplers per site and period were exposed in a box that prevented rain influence. The method is described in detail in Fondazione Salvatore Maugeri-IRCCS (2003). Three farms in the same region in Switzerland and consequently with a similar geographical and meteorological character were selected as sampling sites. At the fourth site, near a residential area in the same region, background ammonia concentration was measured. To prevent systematic errors due to the site location, the passive samplers were exposed at a similar distance from the ammonia source. Moreover it was attended that no influencing obstacles around the site could inhibit ammonia to diffuse into the passive sampler. All four sites were measured for one year (March 2004-March 2005). The samplers were changed fortnightly in summer (March-October) and monthly in winter (November-February).

Results and discussion

Ammonia concentration from a open manure silo without ventilation were measured. In summer (March-August), when the silo was full and manure was agitated from time to time, the average ammonia concentration 80 cm over the upper edge of the silo was about six times higher than in winter (September-March), when the silo was almost empty. Mean: 100 μg NH_3 m^{-3} in summer compared to 16 μg NH_3 m^{-3} in winter. In addition, two farms, both with 50 LU of cattle (LU means livestock unit, which is 500 kg of live weight) but different manure storing and livestock systems, were compared regarding ammonia emissions. Farm A had an open stable with a deep layer of mulch and the manure pit just under the stable. Farm C had a closed stable with an adjoining cow paddock, which was cleaned automatically twice a day, and a closed manure storing system including a biogas plant. At each of the two farms, the measurements have been taken just outside the stable on 2 m of altitude over ground. The sites were not changed during the study. Ammonia concentration at Farm A was about twice as high as at Farm C. Annual mean: 101 μg NH_3 m^{-3} at A compared with 49 μg NH_3 m^{-3} at C (see Figure 1). Background ammonia concentration was much lower: annual mean was 2 μg NH_3 m^{-3}. Passive sampler measurements also showed that ammonia concentrations were much lower when the cattle were on distant grazing land during two periods in summer (June and August 04, farm A).

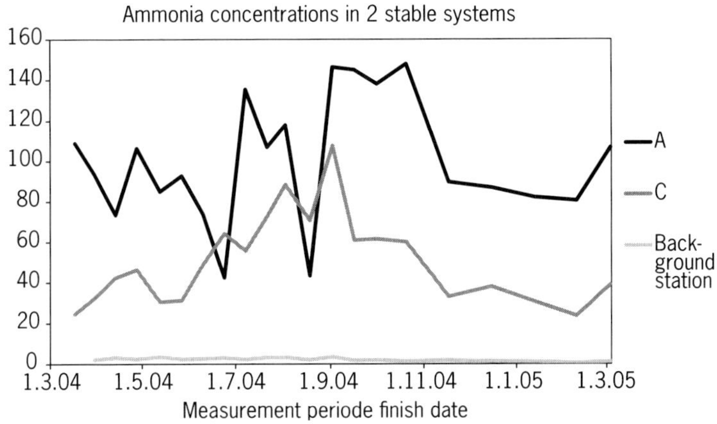

Figure 1. Ammonia concentrations in 2 stable systems. Both farms have 50 LU, the passive sampler sites had comparable conditions. Farm A: annual mean = 101 μg NH_3 m^{-3}, open stable, deep layer of mulch, manure pit just under the stable. Farm C: annual mean = 49 μg NH_3 m^{-3}, closed stable with adjoining cow paddock cleaned twice a day, closed manure storing system with biogas plant. Background station: annual mean = 2 μg NH_3 m^{-3}.

Conclusions

These measurements have shown that the type of farming and its effect on ammonia losses can be reflected in ammonia concentration in the air measured with passive samplers. Therefore, this simple and low cost method is highly appropriate to monitor emission reducing measures or to calibrate ammonia emission models, provided that the sites are well chosen and comparable.

References

Fondazione Salvatore Maugeri-IRCCS, 2003. Radiello®. Centro di Ricerche Ambientali, Padova, Italy. www.radiello.com.

Künzler, P., 2005. Weiterentwicklung des Luftreinhaltekonzepts – Stand, Handlungsbedarf, mögliche Massnahmen. Schriftenreihe Umwelt Nr. 379, Bundesamt für Umwelt, Wald und Landschaft, Bern.

Measurement methods - mechanically ventilated buildings and surface areas

Continuous monitoring method for ammonia emissions from poultry broiler houses in the United States

R.T. Burns[1], H. Xin[1], R.S. Gates[2], H. Li[1], L.B. Moody[1], D.G. Overhults[2], J. Earnest[2] and S. Hoff[1]
[1]Iowa State University, Agricultural and Biosystems Engineering Department, 3212 NSRIC, Ames, 50011 Iowa, USA
[2]University of Kentucky, Biosystems and Agricultural Engineering Department, 128 CE Barnhart Building, Lexington, 40546 Kentucky, USA

Introduction

To better quantify hourly and daily emissions of NH_3 from broiler production systems, accurate and responsive measurement systems are needed. The mass of NH_3 emitted from a facility is the product of the NH_3 concentration and volume of air exchanged through the facility. It is a challenge to reliably quantify NH_3 concentrations and airflow in broiler production housing on a continuous basis. The use of intermittent ventilation by cycling of the ventilation fans off and on, especially when the birds are young, makes it necessary to coincide in-house pollutant concentrations to periods of fan operation in order to calculate representative emissions. The NH_3 emission measurement system described in this paper uses a photoacoustic NH_3 analyser (Innova 1412, Innova AirTech Instruments A/S, Denmark) in conjunction with a multi-point sample acquisition system and calculation of exhausted air volumes based on curves developed using the Fan Assessment Numeration System (FANS) developed in the US. This NH_3 monitoring system has been used to continuously monitor ammonia (NH_3) emissions at two mechanically ventilated commercial broiler houses located in the South-eastern US since October 2005. The NH_3 emissions data will be used by the U.S. Environmental Protection Agency (US EPA) as representative of broiler facilities located in the South-eastern US. To date (15 months of continuous monitoring) the system has performed well. No failures have occurred with the air sample acquisition or airflow monitoring systems. A few failures in the *Innova* units are noted, otherwise they have proven to be accurate and stable and are an excellent choice for measuring ammonia concentrations in broiler housing systems.

Material and methods

Ammonia concentrations are measured with a photoacoustic NH_3 analyser (0-2000 ±0.2 ppm; Innova 1412). The Innova 1412 multi-gas analyser was configured for a 1s sample integration time and fixed flush time: 2s for the chamber and 3s for the tubing; the time required to complete one sampling cycle for NH_3, carbon dioxide and dew-point temperature measurements is approximately 22 sec. The response time of the analyser to step changes in gas concentrations was tested with both 22.8 to 60.8 ppm calibration gases. Four 22 second measurement cycles (88 seconds total) were required to reach the 98% response level for NH_3. Using this approach, the first three readings are discarded and only the fourth reading is used. The gas collection system uses individual sampling lines and pumps for each in-house and ambient sampling location. All air samples are continuously drawn from all sampling points within the house to minimise sample line purge time when cycling between sample locations. Each sample inlet point is equipped with two pleated paper dust filters used to keep large particulate matter from obstructing the sample line. Two filters provide redundancy in case of blockage in one of the filters. The filters are screened and connected using a 'T' connector to a single line and additionally filtered with a 20-μm PTFE-laminated polypropylene membrane filter. Another 5-μm PTFE filter is installed after the vacuum pump to provide additional protection. The gas sampling system is designed such that all solenoid valves, manifolds and associated connections are under positive pressure; positive pressure maintains integrity of the gas sample. Sampling lines are heated with heat trace or heat tape to prevent in-line condensation. House ventilation rates are calculated using measured house differential pressure and individual fan performance curves developed in situ using FANS (Gates *et al*, 2004). Since much more variability exists in the air volume measurements than the NH_3 concentration measurements, these curves are developed every 50 days to reduce measurement variability of the house ventilation rate. The runtime of each fan is monitored and recorded continuously by sensing ON/OFF state of current switches (CR9321, *CR Magnetics, Inc*, St. Louis, MO) attached to the fan power supply cord. This method has been successfully used in recent AFO air emission studies in the United States (Liang *et al*, 2005; Wheeler *et al*, 2006).

Conclusions

Two monitoring systems have been in continuous operation at two separate broiler facilities since October, 2005. To date the *Innova* 1412 gas analysers have proven to be accurate, responsive and to hold their calibration very well during 15 months of use. Both *Innova* units had mechanical chopper failures, one unit had a failure within a month of operation and the second unit has suffered two failures to date, one after 14 months, and a second at 15 months of operation. The chopper rotates and is used to pulse the infra-red light used by the unit. In both cases the chopper failure was caused by a worn drive unit. After 13 months of operation a sample air pump failed in one unit. In each failure, the US distributor, California Analytical Instruments, provided efficient repair service. Other than these component failures, the *Innova* 1412 units have proven to be an excellent choice for measuring ammonia concentrations in broiler houses. The gas collection system including the positive pressure sample distribution manifold and the magnetic induction current switches used to determine fan status have worked without issue during 15 months of continuous use.

References

Gates, R.S., K.D. Casey, H. Xin, E.F. Wheeler and J.D. Simmons, 2004. Fan assessment numeration system (FANS) design and calibration specifications. Transactions of the ASAE 47: 1709-1715.

Liang, Y., H. Xin, E.F. Wheeler, R.S. Gates, H. Li, J.S. Zajaczkowski, P.A. Topper, K.D. Casey, B.R. Behrends, D.J. Burnham and F.J. Zajaczkowski, 2005. Ammonia emissions from U.S. laying houses in Iowa and Pennsylvania. Transactions of the ASAE. 48: 1927-1941.

Wheeler, E.F., K.D. Casey, R.S. Gates, H. Xin, J.L. Zajaczkowski, P.A. Topper, Y. Liang and A.J. Pescatore, 2006. Ammonia emissions from twelve U.S.A. broiler houses. Transactions of the ASABE 49: 1495-1512.

Multi-element combined methods increase the reliability of emission factor measurement

M. Hassouna[1,2], P. Robin[1] and J.-M. Paillat[3]
[1]INRA, Agrocampus Rennes, UMR 1069, Sols, Agronomie, Spatialisation, 35000 Rennes, France
[2]INRA, UMR1079 Systèmes d'Elevage Nutrition Animale et Humaine, 35590 Saint Gilles, France
[3]CIRAD, INRA, Agrocampus Rennes, UMR 1069, 35000 Rennes, France

Introduction

Though many reviews were published, there is still a need to develop robust methods to increase the accuracy of emission factor estimates. For livestock buildings, the measurement period should take several months, it should be adapted to both short-term and mid-term variability of the climate and the farmer practices, and the method should allow simplifications for emission certification. The aim of this paper is to propose methods combination to perform the accuracy on emission factors.

Material and methods

In 2003, ammonia and nitrous oxide emissions were measured on a commercial pig farm (Robin *et al.*, 2004). The measuring equipment included devices suited to methods considered as reference ones (tracing gas), devices allowing the use of simplified methods (indirect ventilation estimates) or the cheapest method: a default mass balance. We used a meteorological station for continuous climate characterisation and thermo-hygrometers (sensor+logger) for minimum climate characterisation. A gas analyser coupled with a sampling and dosing multiplexer and a computer allowed measurements of H_2O, CO_2, CH_4, NH_3, N_2O at different points inside and outside the livestock building. A mixing system ensured the homogeneity of the gas concentrations and of the tracer injection. The effluent was sampled and weighed at the end of the experiment and the farmer gave information on the food, water, straw inputs, and animals.

Results and discussion

The kinetics of the gas concentrations revealed that several successive measurements on the same channel were necessary as well as sampling at least two sites outside and inside the building, to evaluate the measurement accuracy (Figure 1). The heat productions (total, sensible or latent), also used in indirect ventilation estimates, revealed unrealistic air flow rate estimates (Figure 2). The air flow rate estimates showed the sensitivity of the tracer methods to the given dose and the observed concentration variability. The mass balance of volatile and non-volatile elements provided a solution to control the quality of the emission factor estimate (Paillat *et al.*, 2005).

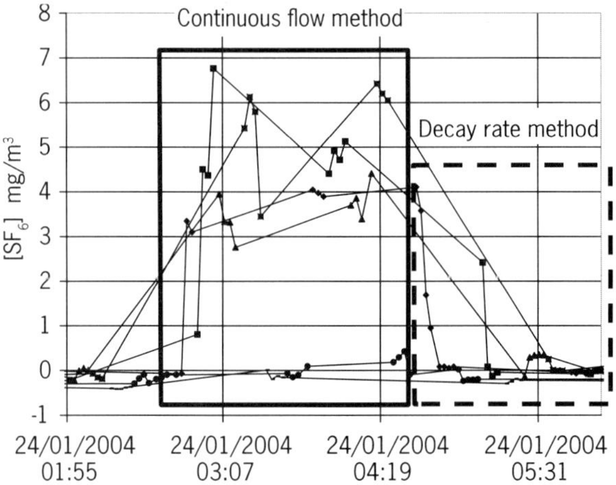

Figure 1. Kinetics of SF_6-tracer concentrations.

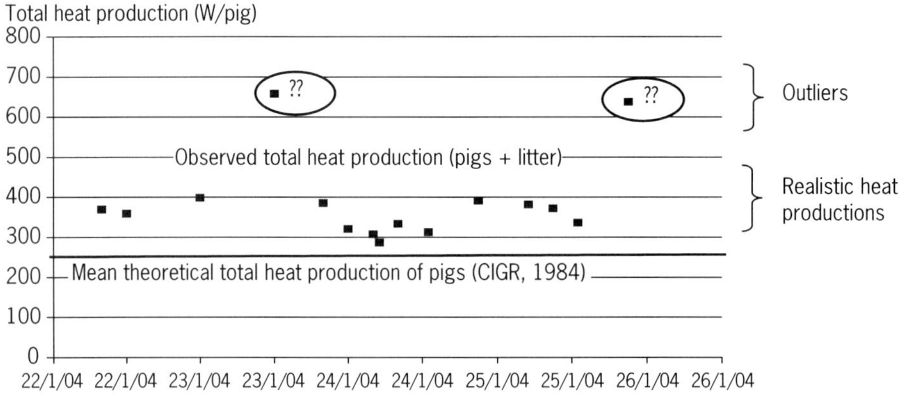

Figure 2. Observed and theoretical heat production.

Conclusions

The results showed that the multi-element approach (H_2O, C, N, P, K) allows the validation of the estimates and the simplification of the reference method, because the various elements have contrasted behaviours. Combining methods ('continuous flow' and 'decay rate' tracer methods; direct and indirect ventilation measurement; concentration ratios and mass balance of the effluent) make it easier to check the observations against realistic values of heat production within the building (instantaneous observations) or realistic values of gaseous losses (mass losses after some weeks or months).

References

Paillat, J.M., P. Robin, M. Hassouna, J. Callarec and P. Toularastel, 2005. Environmental assessment of composting pig slurry with wheat straw based on the Guernévez® process. International Workshop on Pork Production 'Porcherie Verte', Paris, France, pp. 25-27.

Robin, P., M. Hassouna and C. Texier, 2004. Emission d'ammoniac et de protoxyde d'azote des porcs engraissés sur litière de paille. Journées de la recherche porcine en France 36: 63-70.

Quantifying measurement uncertainty for the multi-state broiler ammonia emissions project

K.D. Casey[1], R.S. Gates[2], E.F. Wheeler[3], H. Xin[4] and Y. Liang[5]
[1]Texas A&M University, 6500 Amarillo Blvd West, Amarillo, Texas, USA
[2]University of Kentucky, Lexington, Kentucky, USA
[3]The Pennsylvania State University, University Park, Pennsylvania, USA
[4]Iowa State University, Ames, Iowa, USA
[5]USDA-ARS, Adams, Oregon, USA

Introduction

The body of scientific data necessary to establish baseline emission rates (ER) of modern U.S. concentrated animal feeding operation (CAFO) facilities has grown only recently. A multi-state integrated research and extension project that dealt exclusively with ammonia (NH_3) ER from poultry facilities (Liang *et al.*, 2005; Wheeler *et al.*, 2006) was recently completed. Building ER is proportional to the product of two primary measurements: gas concentration difference between discharge air and inlet (ambient) air, and the building ventilation rate. Considerable attention has been paid to accurate and robust methods to measure ammonia concentration and a number of different technologies exist. By contrast, building ventilation rate measurement has presented unique challenges and is believed to be a principal source of uncertainty for building emissions measurements (Gates *et al.*, 2005). The objective of this article is to present a component error analysis of broiler house ammonia ER for the instrumentation used in the multi-state ammonia emission monitoring project. This error analysis is useful for quantifying the uncertainty in ER calculation from measurements, for identifying the major sources of this uncertainty, and to provide insight into the quality of published ER by different groups using different measurement methods.

Methods

For this project, specialised Portable Monitoring Units (PMU) were developed and deployed for determining NH_3 and carbon dioxide concentrations, building temperature and static pressure difference (Gates *et al.*, 2005). The NH_3 measurements were provided from redundant electrochemical sensors (Dräger PAC IIIh) (±3 ppm). Building ventilation rate was obtained by adding together the contributory airflow rates computed from each operational fan, using calibrated fan performance curves and measured building static pressure difference. The fan performance curves were obtained for each fan using the Fan Assessment Numeration System (FANS) (Gates *et al.*, 2004). ER calculations for this project are fully described in Liang *et al.* (2005) and Wheeler *et al.* (2006). While these measurements are basically straight-forward, numerous systematic and random errors can enter the analysis, and create an unacceptably large error in the final ER value. ER was computed from measurements related to ventilation and concentration, as per Equation (1) below. The NH_3 concentration in the incoming air was neglected because firstly, in most cases this concentration will be near zero and secondly, the capability of the instrument used (Dräger PAC IIIh ammonia sensor) was limited by its lower detection limit of about 3 ppm.

$$ER = Q_T \times C_{NH_3} \times \frac{MW_{NH_3}}{MV} \quad (1)$$

where: ER = emission rate, Q_T = total exhaust ventilation rate, C_{NH_3} = concentration, MW_{NH_3} = molar mass of NH_3 and MV = molar volume at STP. ER uncertainty (ΔER) is introduced by uncertainties in measurements needed to obtain Q_T and C_{NH_3}. A component error analysis is useful for quantifying ΔER, and for identifying the major sources of this uncertainty. A measure of the overall uncertainty can be obtained by truncated Taylor Series expansion of Equation (1) (Doeblin, 1990):

$$\Delta ER = \frac{\partial ER}{\partial Q_T} \Delta Q_T + \frac{\partial ER}{\partial C_{NH_3}} \Delta C_{NH_3} \quad (2)$$

where ΔQ_T and ΔC_{NH_3} are the measurement uncertainties in Q_T and C_{NH_3}, respectively.

Results

For the instrumentation used, ER uncertainty ranged from 3% to 15% over a range of 10-100 ppm NH_3 and 1–8, 1220 mm ventilation fans operating. ER uncertainty was greatest at lower NH_3 concentration, with over 99% of this attributed to uncertainty in concentration measurement. At average project conditions of about 25 ppm NH_3, ER uncertainty was approximately 6.5%. As NH_3 concentration increased to 100 ppm, the proportion of uncertainty attributed to concentration measurement decreased to 85%, 96% and 98% for 1, 4 and 8 fans operating respectively. Increasing the Q_T uncertainty from ±1% for FANS derived fan characteristics to ±10% for manufacturer supplied fan characteristics, increased ER uncertainty to approximately 5% to 19% over the same ranges of NH_3 and Q_T, and ventilation contribution to ER uncertainty increased substantially at lower Q_T and higher concentrations. Upgrading the NH_3 sensor to an Innova Photoacoustic Multigas Monitor (±1%), reduced total uncertainty to about 2.5% for all concentrations and Q_T with the contribution from NH_3 measurement ranging 80, 94 and 97% for 1, 4 or 8 fans respectively. This analysis suggests that NH_3 concentration uncertainty was the primary contributor to ER uncertainty for the multi-state ammonia emission monitoring project; however, if lower accuracy methods are used for Q_T, then uncertainty in Q_T becomes the controlling factor in overall uncertainty.

References

Doeblin, E.O., 1990. Measurement systems application and design. 4th ed. New York, NY: McGraw-Hill Publishing Co. 960 pp.

Gates, R.S., H. Xin, K.D. Casey, Y. Liang and E.F. Wheeler, 2005. Method for measuring ammonia emissions from poultry houses. J. Appl. Poult. Res. 14: 622-634.

Gates, R.S., K.D. Casey, H. Xin and E.F. Wheeler, 2004. Fan Assessment Numeration System (FANS) design and calibration specifications. Transactions of the ASAE. 47: 1709-1715.

Liang, Y., H. Xin, E. F. Wheeler, R. S. Gates, J. S. Zajaczkowski, P. Topper, H. Li and K. D. Casey, 2005. Ammonia emissions from U.S. laying hen houses in Iowa and Pennsylvania. Transactions of the ASAE 48: 1927-1941.

Wheeler, E.F., K.D. Casey, R.S. Gates, H. Xin, J.L. Zajaczkowski, P.A. Topper, Y. Liang and A.J. Pescatore, 2006. Ammonia emissions from twelve USA broiler chicken houses. Transactions of ASABE 49:1495-1512.

Ammonia emissions in naturally ventilated cattle housing with an exercise yard: requirements and measuring concept using two tracer gases

S. Schrade[1], K. Zeyer[2], L. Emmenegger[2], M.K. Vollmer[2], M. Keck[1] and E. Hartung[3]
[1] Agroscope Reckenholz-Tänikon ART, Tänikon, CH-8356 Ettenhausen, Switzerland
[2]Empa, Swiss Federal Laboratories for Materials Testing and Research, Überlandstrasse 129, CH-8600 Dübendorf, Switzerland
[3]Christian-Albrechts-University Kiel, Institute of Agricultural Engineering, Max-Eyth-Str. 6, 24118 Kiel, Germany

Introduction

Cattle represent a large proportion of Swiss livestock and account for 74% of ammonia (NH_3) emissions of livestock farming. Whereas so far, tie stalls for dairy cattle have prevailed, the trend is now toward loose housing with natural ventilation and outdoor exercise yards. Improvements in animal welfare are offset by a significant increase in the soiled areas, leading to comparatively higher NH_3 emissions. There is thus a need for reliable emission inventories and implementable recommendations on emission reduction. The few published results vary considerably, are mostly based on short-term measurements and were mainly collected from housing with forced ventilation. The lack of emission data for natural ventilation and outdoor exercise yards is essentially due to difficulties in determining the air exchange rate.

Requirements and measuring concept using two tracer gases

To obtain reliable emission values for loose housing with natural ventilation and outdoor exercise yards, a measuring concept and a corresponding measuring arrangement are needed. They must be suitable for in situ monitoring on commercial farms and that meet the following requirements: To take account of the climatic variation during the year in housing influenced by the out-door climate, measurements must be taken in summer, winter and the in-between seasons. Due to variations in emissions throughout the day due to climate, use and management activities, individual measurements should always cover a minimum period of 24 hours. Daily means suffice in order to derive emission factors. A higher time resolution is desirable in order to record daily patterns, relevant impacting factors or the effects of short-duration events. The air exchange rate in open housing can react very dynamically to external factors. In order to quantify emissions in such housing representatively, the fraction of time recorded at each sampling location must be as high as possible. A high spatial resolution is necessary to take into account the large housing areas and volumes, and the large air exchange surfaces with multiple apertures. As individual housing areas vary in terms of their emission potential and affect one another, allocation of emissions to housing areas is necessary with a view to reduction measures. To classify the emission values, parameters such as herd size, feeding ration, management, surface, climate and nutrient content of the manure must be documented. The background concentration of the measurement parameters must also be determined. A comparison of published results of emission measurements for similar housing systems showed a considerable farm-to-farm variation. Taking measurements for a single housing system on several farms will at least reveal the farm effect. To meet these requirements, a tracer ratio method was selected and improved for emission quantification from housing with natural ventilation and an outdoor exercise yard. The use of two tracer gases permits not only validation of the tracers but also allocation of the emissions to individual housing areas. NH_3 emissions are determined from the known mass flow of added tracers and the concentration ratio of the tracer gases to ammonia. This concept requires that the dosing of the tracers mimics the emission sources and that the traces and NH_3 disperse equally. Gaseous tracers must have the following properties: (1) non-toxic to humans and animals, (2) chemically inert, (3) no surface adsorption, (4) low, constant background concentrations, (5) sensitive, precise analysis and (6) commercially available. In addition to the well established SF_6 tracer, trifluoromethyl sulphur pentafluoride (SF_5CF_3) was introduced as a second tracer. Atmospheric concentrations of SF5CF3 are very low (Sturges *et al.*, 2000), and its properties are similar to SF_6. A procedure was designed to produce SF_5CF_3 tracer gas and to determine its concentration by FTIR. An existing GC-ECD method for SF6 was improved to enable both tracers to be determined precisely, independently and sensitively in the same analytical process (Figure 1). The detection limit for both tracer gases is approximately 1 ppt. The tracers are continuously dosed directly to the emitting surface in the housing through critical orifices. Sampling is carried out quasi-continuously covering the spatial distribution of the stall apertures. Preliminary experiments were carried out in two dairy housing systems to test and improve the measurement concept and setup in situ. NH_3 was measured using a commercially available NH_3 trace gas analyser (TGA 310, Omnisens SA, Switzerland). This instrument operated reliably under shed conditions, and its time resolution, sensitivity and dynamic range was sufficient to follow the dynamics and daily patterns of NH_3. Based on these preliminary experiments, the described measuring concept using two tracer gases will be used in cubicle housing systems with an outdoor exercise yard to improve the data basis for NH_3 emissions.

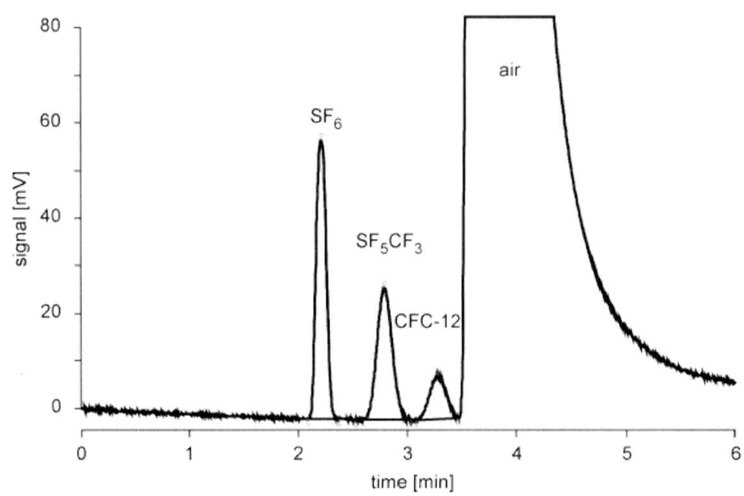

Figure 1. Chromatogram of SF_6 (approx. 108 ppt) and SF_5CF_3 (approx. 133 ppt); loop 2.4 ml.

References

Sturges, W.T., T-J. Wallington, M.D. Hurley, K.P. Shine, K. Shira, A. Engel, D.E. Oram, S.A. Penkett, R. Mulvaney and C.A.M. Brenninkmeijer, 2000. A Potent Greenhouse Gas Identified in the Athmosphere: SF_5CF_3. Science 289: 611-613.

Methodologies for measuring ammonia emissions from grazing livestock

T.H. Misselbrook[1], V. Camp[1], T. Demmers[2] and J. Williams[3]
[1]IGER North Wyke, Okehampton, Devon, EX20 2SB, United Kingdom
[2]Royal Veterinary College, Hawkshead Lane, North Mimms, Hatfield, AL9 7TA, United Kingdom
[3]ADAS Boxworth, Battlegate Road, Boxworth, Cambridge, CB23 4NN, United Kingdom

Introduction

Ammonia emission from grazing livestock represent approximately 10% of the total emission from UK agriculture (Misselbrook *et al.*, 2000). Current emission factors for grazing cattle and sheep rely on measurements made using the micrometeorological integrated horizontal flux (IHF) mass balance technique, generally from small paddocks receiving a relatively high fertiliser N input (>200 kg N ha^{-1}). In the UK, a large proportion of the beef herd and sheep flock graze on pastures receiving considerably less fertiliser N input. One criticism of the IHF mass balance technique is the assumption that the emission rate from the pasture directly upwind of the measurement point is representative of the emission rate from the whole pasture. This may not be valid for grazing livestock, where there may be an uneven distribution of urine patches and 'hot-spots' associated with e.g. drinking troughs or resting areas. The aims of this study were, therefore, to develop and compare suitable methodologies for measuring emissions from grazed pastures and to derive robust emission factors for grazing livestock on lower N input pastures.

Materials and methods

Ammonia emission measurements were made from beef cattle grazing on very low N input paddock (1.7 ha) at three different times of year (May, July and September) over 3-4 days on each occasion. Emission measurements were made using the IHF mass balance technique (Denmead, 1993) and a backward Lagrangian stochastic (bLS) modelling approach (Flesch *et al.*, 1995), both employing passive flux samplers (Leuning *et al.*, 1985). For the IHF technique, masts supporting samplers at heights of 0.25, 0.45, 0.85, 1.5, 2.8 and 5.0 m were located upwind and downwind of the grazed paddock. Samplers were changed every 24 h. Mean emission rate from the paddock was derived using two different calculation methods: firstly (IHF_1), using a step-wise integration of the horizontal flux between 0 and 5 m for each of the upwind and downwind masts to derive the overall net horizontal flux (and thereby vertical flux from the grazed paddock); secondly (IHF-2), by integrating the net horizontal flux (at each sampling height) between the limits z_o (the roughness length for the paddock) and z_p (the ceiling height at which net flux is zero). For the bLS approach, passive flux samplers were deployed in duplicate at four locations (one upwind and three downwind (spatially separated along the downwind edge), 1.5m above the ground. The upwind and one of the downwind sampling positions corresponded with the IHF mass balance mast positions. Samplers were changed every 24h. Ammonia concentration at each sampling point was derived by dividing the flux measurement by average wind speed at 1.5 m for the sampling period. Mean emission from the grazed plot was calculated using WindTrax software (Thunder Beach Scientific, Halifax, Canada), with plot dimensions, orientation and sampling positions, average period ammonia concentrations, wind speed and wind direction and plot roughness length as inputs. Calculations were made based on the two sampling points corresponding to the IHF sampling positions (bLS_1) and all sampling points (bLS_2). Atmospheric stability class was assumed to be neutral. In reality, there would have been changes in stability class over some of the 24 h sampling periods, but Sommer *et al.* (2005) showed stability class to have little influence if an appropriate sampling height (dependant on fetch length) was used.

Results and discussion

Table 1. Mean ammonia emission rates on a liveweight basis (g lu^{-1} d^{-1}, where lu = 500kg).

Campaign date	IHF_1	IHF_2	bLS_1		bLS_2	
	g lu^{-1} d^{-1}	g lu^{-1} d^{-1}	g lu^{-1} d^{-1}	% cover	g lu^{-1} d^{-1}	% cover
May	2.1	3.3	0.2	17	0.9	41
July	5.1	*	3.2	10	3.5	24
September	5.4	7.0	10.1	9	3.7	23

There was good correlation (r^2 0.89) between mean emission rates as measured using the bLS_1 and the IHF_1 approaches, although the slope of the line was 1.5. Mean emission rates were lower than those reported by Misselbrook *et al.* (2000), as would be expected for low N input grazing (Table 1). Emissions were greater using the IHF_2 approach, indicating that a 5 m mast was insufficient to sample the entire plume from the grazed paddock. The bLS_2 method gave consistently lower emissions than either IHF method, but sampled from a much greater proportion of the paddock (% cover, Table 1) than the bLS_1 (and by association IHF) approach. For various reasons (changeable wind directions, equipment failure) the data obtained to date are considered insufficient for a valid comparison of methodologies and further measurement campaigns will be conducted, including measurements with a tuneable diode laser, giving higher resolution ammonia concentration data.

Acknowledgement

This work was funded by the UK Department for the Environment Food and Rural Affairs.

References

Denmead, O.T., 1983. Micrometeorological methods for measuring gaseous losses of nitrogen in the field. In: J.R. Freney and J.R. Simpson (Eds.) Gaseous Losses of Nitrogen from Plant-Soil systems. Martinus Nijhoff/Dr W Junk, The Hague, pp. 133-158.

Flesch, T.K., J.D. Wilson and E. Yee, 1995. Backward-time Lagrangian stochastic dispersion models and their application to estimate gaseous emissions. J. Appl. Meteorol. 34: 1320-1332.

Misselbrook, T.H., T.J. van der Weerden, B.F. Pain, S.C. Jarvis, B.J. Chambers, K.A. Smith, V.R. Phillips and T.G. Demmers, 2000. Ammonia emission factors for UK agriculture. Atmos. Environ. 34: 871-880.

Leuning, R., J.R. Freney, O.T. Denmead and J.R. Simpson, 1985. A sampler for measuring atmospheric ammonia flux. Atmos. Environ. 19: 1117-1124.

Sommer, S.G., S.M. McGinn and T.K. Flesch, 2005. Simple use of the backwards Lagrangian stochastic dispersion technique for measuring ammonia emission from small field-plots. Eur. J. Agron. 23: 1-7.

Measurement methods - naturally ventilated buildings

An overview on ventilation rate measuring and modelling techniques through naturally ventilated buildings

S. Eren Ozcan, E. Vranken and D. Berckmans
Measure, Model and Manage Bioresponses (M3-BIORES), Catholic University of Leuven, Kasteelpark Arenberg 30, 3001 Heverlee, Belgium

Introduction

Measuring emissions from naturally ventilated buildings is an important issue and receives increasing attention while concerned measures are taken by governments and related bodies. To establish necessary standards, a robust measuring technique should be developed. Generally, emission rates are calculated from the product of pollutant concentration and ventilation rate. Pollutant concentrations can be measured by available instruments. However, there is a lack of reliable ventilation rate determination technique. Objective of this study is to review available techniques while putting emphasise on their accuracies. Additionally, advantages and disadvantages of each method will be discussed.

Materials and methods

Experimental comparison tests were performed at two laboratory facilities in which know ventilation rates were produced by a mechanical ventilation system. Firstly, a laboratory test chamber was used with dimensions of 3x2x1.5 m [LxHxW]. An orifice enabled an adequate measurement and control of the ventilation rate in a range 70-420 m^3/h and with an accuracy of ±6 m^3/h. Another set of experiments were conducted at the outlet section [0.6 x 0.6 m] of a standardised airflow rate calibration unit. Ventilation rate measurements were performed as a reference with an accuracy of 10 m^3/h in the range of 65 to 1500 m^3/h.

Results and discussion

Direct and indirect measuring techniques are available for determination of ventilation rate. Direct measuring methods include free running impeller, pressure transducers, propeller gauge, hot wire anemometer, particle image velocimetry, laser Doppler anemometer, and transit time sonic anemometer. Free running impeller and pressure transducers are able to measure total ventilation rate through the opening. However, these instruments cannot reproduce accurate results at low ventilation rates. Anemometers and other local measurements systems are accurate for velocity determination. Nevertheless, due to non-uniform and fluctuating flow behaviour through the inlets, measuring velocity at one position is not representative for average ventilation rate. Indirect measuring techniques are; heat balance, CO_2 balance, pressure difference, CFD analysis, tracer gas measurements, multizone modelling, and zonal models. These methods consider the whole system, and therefore, provide a useful tool for air flux through the building envelopes. Methods based on computer simulations (CFD, Multizone models) should be validated against experimental data. However, in most of the cases, those validations are lacking, or do not indicate the accuracy of the method. Tracer gas measurements are mostly used as reference method in validations. However, accuracy of this technique should also be studied. Therefore, in our current study, it is aimed to improve the accuracy of tracer gas measurements by using models of imperfect mixing within the building.

Table 1. Literature survey on measuring techniques for ventilation rate and air velocities.

Method	Advantage	Disadvantage	Inacc.
Direct Measuring Techniques			
Free Running Impeller[a]	– low cost – covers whole section	– need standard inlet/outlet – cannot detect low pressure variations	5-25%
Pressure Transducers	– easy to use – commercially available	– difficult to measure low pressures – local measurements	1–25%
Propeller Gauge	– depending on the uniformity, accurate measurements are possible	– local measurements – accuracy depend on pressure difference, temperature and density measurements	2–25%
Hot wire anemometer[a]	– accurate – can measure steady-state to transient temperature differences	– susceptibility to dust – local measurement – lack of physical robustness	0.2-25% [b]
Laser Doppler Anemometer	– accurate – non-intrusive – no need for calibration	– local measurements – accuracy is highly dependent on alignment of emitted and reflected beams – expensive	<2% [b]
Particle Image Velocimetry	– accurate – non-intrusive	– expensive – need expertness – vmeasuring volume is limited	<3% [b]
Transit Time Sonic Anemometer[a]	– covers whole section – no obstruction in the flow path – low maintenance cost	– high initial set up cost – susceptible to construction errors	0.1-4% [b]
Indirect Measuring Techniques			
Pressure Difference[a]	– cheap – local ventilation rates can be determined	– at low wind speed inaccuracy is bigger – high fluctuation in and around the buildings	20-35%
CO_2 balance	– used in naturally ventilated buildings	– imperfect mixing of CO_2 – CO_2 from other sources (respiration, manure, etc.)	8-40%
Heat Balance[a]	– only temperature measurements are enough	– difficulty in estimation heat gains/ loses	9-100%
Tracer Gas[a]	– can be used for total ventilation rate calculations	– mostly perfect mixing assumed	10-50%
CFD Analysis	– can be applied to any type of the building	– assumptions for boundary conditions – requires huge processing time	8-65%

Multizone Modelling	– mostly used in building design calculations	– boundary conditions should be well defined – inaccuracy due to turbulence	15-65%
Zonal Models[a]	– less zone is needed than CFD – less execution time	– more experimental validation needed	20-65%

[a]Techniques tested at laboratory test installation; [b] Accuracies in terms of velocity.

Conclusions

Among the tested methods, tracer gas technique is the best method for research purposes although inaccuracies can go up to 40%. Improvement of method was proven with increasing number of sampling positions (15%).

Study of the distribution pattern of the ammonia concentration inside a naturally ventilated dairy house

H. Cnockaert and B. Sonck
Institute for Agricultural and Fisheries Research, Technology & Food, Agricultural Engineering (T&V-AT), B. Van Gansberghelaan 115, 9820 Merelbeke, Belgium

Introduction

According to European demands the ammonia emissions in Flanders have to be reduced with 40% by 2010. A law of October 2003 prescribes that new pig and poultry houses must be built according to low ammonia emission building techniques. These techniques are described in a 'List of Building Techniques for the Reduction of Ammonia Emissions' and classified according to the different pig and poultry categories. However, at the moment only livestock buildings with a mechanical ventilation system are allowed on the list. Naturally ventilated animal houses are not accepted because there is no applicable method for accurate, reliable and online ammonia emission measurements. Therefore, the objective of this study is to develop an affordable method for the determination of ammonia emissions from a naturally ventilated dairy house. This research was carried out in cooperation with the M3Biores research group of K.U.Leuven who aims to develop a measurement tool for ventilation rate in naturally ventilated buildings. T&V-AT carried out the ammonia concentration measurements.

Materials and methods

T&V-AT has built out a mobile measuring team and developed two test devices. These exists of a photo-acoustic multigas monitor (Innova 1312 and 1314) combined with a 8-channel multisampler and according sampling tubes (FEB). Ammonia concentration measurements were performed in the experimental cubicle house (50 dairy cows) of Agrivet (University Ghent) during the winter period 2004-2005. Measurements were carried out with the eight sample points spread over a horizontal plane at 3.45 m height in the dairy house. During another experiment the sample points were situated at cubicle height throughout the cubicle house. Most of the time the sample points were positioned nearby the space-boarding. More specific in one line perpendicular to the centre of the space-boarding on the south-western side of the dairy cow house. The distance of the eight sample points varied during several experiments between zero up to 2.90 m from the space-boarding. The ammonia concentration was measured every 19 minutes at each sample point. So, the ammonia concentration was not measured simultaneously at all eight sample points. Therefore the dataset of one day was divided in twelve periods of two hours and the calculations were done with the two-hourly average of the different variables. A weather station of the Royal Meteorological Institute was 500 metres apart from the cubicle house. The following climate parameters were registered with a frequency of 10 minutes: wind speed at 10 metres height, wind direction, temperature, relative humidity, radiation and precipitation.

Results and discussion

The results showed that the outside wind speed had the greatest influence on the ammonia concentration at a specific point in the cubicle house. The wind direction was the second most influencing parameter. The higher the wind speed was the lower the ammonia concentration and vice versa. The average ammonia concentration measured with different wind directions at several places spread over the whole dairy cow house varied between 0.5 and 5 ppm. The wind direction determined to a large extent for which measuring points in the cubicle house the highest or the lowest ammonia concentration was measured. The (absolute) ammonia concentration gradient became smaller with increasing wind speed. It should be noticed that with higher wind speeds, little differences in ammonia concentration cause immediately (large) differences in ammonia emission because of the higher ventilation rates. During one month ammonia concentration measurements were recorded at eight different places (at cubicle frame height) throughout the cubicle house. The maximum difference in mean concentration between two measuring points was 35%. For one specific wind direction the difference increased to more than 100%. A large ammonia concentration gradient was found in the vicinity of the space-boarding. Table 1 shows the average ammonia concentration per wind direction in relation to different distances to the space-boarding on the south-western side of the dairy house. Only for the wind directions with more than twenty two-hourly measuring periods a statistical calculation (Duncan's multiple range test) was carried out. Remarkable is that both leeward and windward the ammonia concentration increased with increasing distance from the space-boarding. However, for the downward wind directions the equilibrium state of ammonia concentration was reached earlier (usually at 70 cm from the space-boarding) than windward (130 cm). Another conclusion was that with lower wind speeds the equilibrium state was reached earlier. However, the absolute concentration gradient was higher compared with the gradient measured with high wind speeds.

Table 1. Average ammonia concentration per wind direction in relation to the distance to the space-boarding on the south-western side of the dairy house, with mentioning of the average wind speed, temperature and relative humidity.

Wind direction	A	Ammonia concentration (ppm) Distance to space-boarding (cm)								Wind speed (m/s)	Temp. (°C)	RV (%)
		10	50	90	130	170	210	250	290			
N	5	2.68	3.22	3.41	4.36	4.46	4.56	4.54	4.62	3.96	-1.9	73.3
NE	20	2.99^{a}	3.80^{b}	4.12^{bc}	4.31^{bc}	4.29^{bc}	4.32^{bc}	4.34^{bc}	4.41^{c}	4.47	-1.0	80.4
SE	1	2.23	2.89	3.11	3.23	3.15	3.10	3.57	4.15	2.32	0.4	80.8
S	17	1.04	1.16	1.36	1.,85	2.20	2.48	2.63	2.79	6.03	-0.6	79.0
SW	4	2.72	3.11	3.39	4.04	4.17	4.24	4.19	4.26	3.76	-2.7	83.5
W	5	3.00	3.44	4.18	4.86	4.91	5.02	4.99	4.99	2.68	0.8	88.0
NW	28	3.46^{a}	4.08^{b}	4.79^{c}	5.48^{d}	5.52^{d}	5.61^{d}	5.59^{d}	$5.,64^{d}$	4.05	3.7	84.6
average	80	2.68^{a}	3.23^{b}	3.69^{c}	4.22^{d}	4.32^{d}	4.44^{d}	4.46^{d}	4.53^{d}	4.36	0.6	81.9

A: number of two-hourly measuring periods.

[a,b,c,d]Numbers in the same row with different superscripts are significantly different ($P<0.05$).

Conclusions

Ammonia concentration gradients were found when measuring on eight different places spread over the cubicle house. The maximum difference in mean concentration between two measuring points was 35%. For one specific wind direction the difference increased to more than 100%. Also in the vicinity of the space-boarding a large ammonia concentration gradient was recorded. Remarkable is that both leeward and windward the ammonia concentration increased with increasing distance from the space-boarding. However, for the downward wind directions the equilibrium state of ammonia concentration was reached earlier (usually at 70 cm from the space-boarding) than windward (130 cm). Further investigation is needed to gear the positioning of the ammonia concentration sample points to the positioning of the ventilation rate measuring points.

Methods to measure the ventilation rate from naturally ventilated livestock buildings

T.G.M. Demmers
The Royal Veterinary College, University of London, Hawkshead Lane, North Mymms, Hatfield, AL9 7TA, United Kingdom

Introduction

Continuous measurement of ventilation rate is routinely used to control the internal climate in mechanically ventilated livestock buildings, including pollutant concentrations, but to date no reliable simple measuring devices are available to implement this in naturally ventilated buildings. However, there is a need to quantify aerial pollutant emissions, as particularly ammonia has an important role in atmospheric chemistry and leads to subsequent damage of the environment (Apsimon *et al.*, 1987). Also, livestock production is the major source of ammonia emissions to the atmosphere (Asman, 1994). The challenge is to measure the ventilation rate of buildings fitted with multiple ventilation openings of varying size, where the ventilation is dependent on and varies with the wind direction, wind speed, ambient temperature and heat production by the livestock inside. A review by Phillips *et al.*, (2000) identified several techniques for measuring the emissions from livestock housing, some techniques requiring the measurement of ventilation rate others relying on the ratio between a tracer and the pollutant to estimate the emission directly.

Measurement of ventilation rate

A method to measure ventilation rate from naturally ventilated buildings was developed by Demmers (1997), based on the constant release of the tracer within the building. This method assumes perfect mixing of the air and hence the tracer inside the building, which in practice is rarely the case. Therefore, measurement of the tracer concentration at positions in openings at the perimeter was used as an alternative to measure inside the building. Airflow through the openings of the building alternates between inwards or outwards, and tracer concentration was used as a reliable indicator of the opening function, i.e. between air inlet or air outlet. The tracer itself needs to be stable and mix well with the air in the building, i.e. have a similar density to air, Carbon Monoxide, or be pre diluted with air, Sulphur Hexafluoride. This method using either location to measure the tracer concentration was validated in mechanically ventilated pig and broiler buildings. The instantaneous ventilation rate of a mechanically ventilated building was underestimated by 12% and 6% respectively, using the tracer method compared with a reference method using full size anemometers. The instantaneous estimate of ammonia emission derived from the tracer gas method underestimated the ammonia emission derived from the full size anemometer method by 14% and 16% for the piggery and broiler house respectively. For naturally ventilated buildings the accuracy of pollutant emission rate estimate using the constant release method to estimate the ventilation rate was validated in a full scale section of a livestock building. The presence of animals was simulated using heater banks and the ammonia source simulated using a distributed array of openings in lay flat tubing. The emission estimate was found to be overestimated by no more than 10% when the tracer and ammonia concentration were measured at the perimeter. Using an average sample from the centre of the cross section resulted in poor estimates of the emission rate varying from ±30% compared to the actual release rate.

Tracer ratio methods

Tracer ratio methods can be used to estimate the pollutant emission directly without the need to measure the ventilation rate (Phillips *et al.*, 2001)), although the ventilation rate can be estimated from the data. The ratio of the emission rate of the pollutant and the tracer are assumed to be equal to the ratio of the pollutant and tracer concentration measured at any location. For the tracer ratio method complete mixing of the air in the building is not a requirement. Merely, the dispersion of the tracer needs to be similar to the pollutant of interest. The location of the measurement can either be inside the building (internal tracer ratio) or down wind of the building (external tracer ratio). The internal tracer ratio method is less dependent on changes to the wind direction and is capable of measuring individual sources within a cluster (farm). Validation of the internal tracer ratio method was conducted using the same facility as described above and achieved close to 100% and 78% recovery of the artificial release rate of ammonia over 24 hrs using high and low release rates of ammonia, respectively (Scholtens *et al.*, 2004). There was little difference between any off three locations used for sampling (ridge and either side wall).

Conclusion

Robust measurement techniques are available for measuring ventilation rate from naturally ventilated buildings for the purpose of quantifying emission rates. However, these techniques are not suitable for use in feed back control systems for the internal climate.

References

Apsimon, H.M., M. Kruse and J.N.B. Bell, 1987. Ammonia Emissions and Their Role in Acid Deposition. Atmos. Env. 21: 1939-1946.

Asman, W.A.H., 1994. Emission and Deposition of Ammonia and Ammonium. Nova Acta_Leopoldina NF 70: 263-297.

Demmers, T.G.M., 1997. Ventilation of Livestock Buildings and Ammonia Emissions. Faculty of Agricultural and Food Sciences, University of Nottingham, Nottingham.

Phillips, V.R., D.S. Lee, R. Scholtens, J.A. Garland and R.W. Sneath, 2001. A Review of Methods for measuring Emission Rates of Ammonia from Livestock Buildings and Slurry or Manure Stores, Part 2: monitoring Flux Rates, Concentrations and Airflow Rates. J. Agric. Eng. Res. 78: 1-14.

Phillips, V.R., R. Scholtens, S.D. Lee, J.A. Garland and R.W. Sneath, 2000. A Review of Methods for Measuring Emission Rates of Ammonia from Livestock Buildings and Slurry or Manure Stores, Part 1: Assessment of Basic Approaches. J. Agric. Eng. Res. 77: 355-364.

Scholtens, R., C.J. Dore, E. Ayres, D.H. Wall, J. Aislabie, S. Bamforth, R. Bargagli, C. Cary, P. Cavacini, L. Connell, P. Convey, J.W. Fell, F. Frati, I.D. Hogg, K.K. Newsham, A. O'Donnell, N. Russell, R.D. Seppelt and M.I. Stevens, 2004. Measuring ammonia emission rates from livestock buildings and manure stores, Part 1: development and validation of external tracer ratio, internal tracer ratio and passive flux sampling methods. Atmos. Env. 38: 3003-3015.

Determination of ammonia emission from naturally ventilated animal houses

H.-J. Müller[1]*, B. Möller*[1]*, M. Gläser*[1] *and M. Cespiva*[2]
[1]*Leibniz-Institute for Agricultural Engineering e.V. (ATB), Max-Eyth-Allee 100, 14469 Potsdam, Germany*
[2]*Research Institute of Agricultural Engineering e.V. (VUZT), Drnovske 507, 16101 Praha 6, - Ruzyne, Czech republic*

Introduction

Keeping of animals in livestock buildings requires the ventilation of these buildings. On the one hand good climatic conditions for the animals in the livestock building have to be provided but on the other hand the emissions have to be kept at a low level. The air flow through the livestock building plays an important role for both opposing requirements. For targeted control of the climate in the livestock building, for the minimisation of emissions and for the application of dispersion models, knowledge about airflow and emission streams is necessary. One of the most important emitted components from animal houses is ammonia. To determine the ammonia emission, both ammonia concentration and ventilation rate must be measured. Especially for naturally ventilated buildings, the determination of ventilation rates leads to problems. In such cases the tracer gas technique can be used. The ATB uses CO_2, SF_6 (run measurements) and Krypton 85 (short term measurements) as tracer gases.

Materials and methods

The emission Flow rate is a product of both the concentration difference between emitted air and fresh air and the air flow rate:

$$\dot{m} = \Delta C \cdot \dot{V} \tag{1}$$

$\dot{m}$ g/s emission mass flow rate
ΔC g/m^3 concentration difference between emitted air and fresh air
$\dot{V}$ m^3/s air flow rate

To determine the emission mass flow rate we have to measure the concentration at the fresh air, the concentration at the exhaust air and the air flow rate. Both are difficult to measure the concentration and the air flow rate at naturally ventilated livestock buildings. Regarding the concentration it is to analyse which opening is the air inlet – it depends in particular on the wind direction. Another problem in naturally ventilated buildings is to measure the air flow rate. If we try to use velocity measurements than a lot of measuring points in the openings are necessary. The air velocity and the air flow direction must be measured. The experiences of the ATB show that a better way is to measure the air flow rate by using the tracer gas technique. During the last 30 years the ATB had used and had enhanced the tracer gas method. If CO_2 is used as tracer gas (CO_2 is emitted by the animals) than the activity of the animals is to consider (Müller *et al.*, 2006). The krypton method can be used to calibrate the carbon dioxide method. The krypton method is enhanced at ATB during the last years: reduction of measuring interval (1 second); extension of measuring points from 6 to 40; improvement of the interpretation (compartment method). A multitude of different naturally ventilated animal houses for cattle, pigs and poultries were investigated during the last years regarding ammonia emission.

Results and discussion

The air volume stream measured in a beef cattle house is shown in Figure 1. The black curve is calculated by carbon dioxide balance and the dots result from the krypton method (decay method). In principle an agreement between both methods exists, but in some cases bigger differences are to see. One reason is, that there is not considered the animal activity (constant CO_2-production) and the second one is, that the CO_2-method has a larger time constant. From these determined air volume streams together with the simultaneously measured ammonia concentration in the air outlets of the building results the ammonia emission flow.

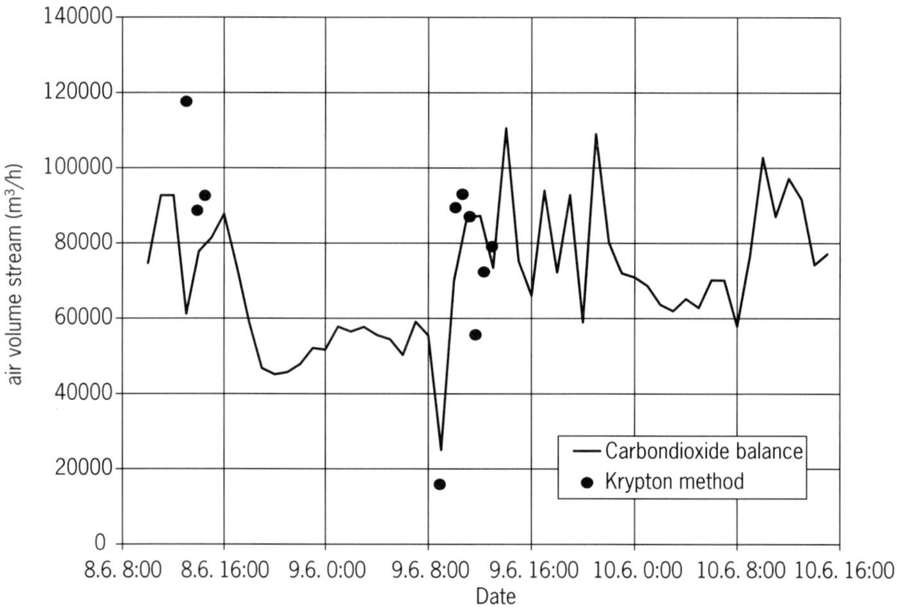

Figure 1. Air volume stream in a beef cattle house – determined by carbon dioxide balance and by krypton method (June 2004).

Conclusions

- The tracer gas technique is the best way to measure air volume streams in naturally ventilated livestock buildings.
- The ammonia emission mass flow increases together with the air volume stream and in the summer period a high emission results in spite of low ammonia concentrations.

References

Müller, H.-J., H.B. Rom and S. Pedersen, 2006. Comparison of methods of determination of ventilation flow in animal houses with different kinds of animals. CIGR World Congress Agricultural Engineering for a better World, 03–07.09. 2006 Bonn. Proceedings, pp. 485-486.

Measuring methods of determination of air exchange rates and ammonia emission mass flows in naturally ventilated livestock buildings

H.-J. Müller[1], B. Möller[1], U. Wanka[2], T. Heidenreich[2], S. Eren Ozcan[3], E. Vranken[3] and D. Berckmans[3]
[1]Leibniz-Institute for Agricultural Engineering e.V. (ATB), Max-Eyth-Allee 100, 14469 Potsdam, Germany
[2]Saxson State Institute for Agriculture, August-Böck-Stiegel-Str. 1, 01326 Dresden, Germany
[3]Measure, Model and Manage Bioresponses (M3-BIORES), Catholic University of Leuven, Kasteelpark Arenberg 30, 3001, Heverlee, Belgium

Introduction

To determine the ammonia emission mass flow, total air volume stream through the building and the concentration of ammonia in the air outlet openings should be measured. In naturally ventilated livestock buildings, no reference technique exists to estimate ventilation rate. Large building structures, large openings for natural ventilation and high air exchange rates limit the possibilities of existing measuring techniques. Tracer gas technique is the most popular technique for natural ventilation, since it gives an overall idea for net ventilation rate and disturbances from environment are minimal for an inert tracer gas. Different tracer gases such as CO_2, SF_6, N_2O, Freons and radioactive gases are applied. The ATB has developed a special method and technology using the radioactive gas Krypton 85 where simultaneous and fast measurements are possible up to 40 measuring points with a sampling frequency of one second. In natural ventilation, large building structures causes imperfect mixing of tracer gas. Therefore, depending on injection and sampling position; different results can be obtained (Van Buggenhout *et al.*, 2007). To overcome this problem, homogenous injection and multiple sampling strategies applied. Averaging results at different positions or modelling interaction between fictitious subzones are possible post-processing approaches for tracer gas calculations.

Materials and methods

Five different methods were compared that estimates ventilation rate through naturally ventilated buildings in a dairy cow house located at Kobschuetz, Germany. The decay method was applied with Krypton 85 as tracer gas (Müller and Möller, 1998). The tracer gas is injected suddenly into the stable space. At defined measuring points the decay of the homogenous starting concentration is recorded at the same time at 18 locations. The data was processed in two ways: firstly, average of 18 positions was taken; secondly, whole volume was divided into subzones, and refreshment rate at each zone was calculated to find total fresh air through the system. Studies with mechanically ventilated laboratory test installation showed that zonal modelling approach increased the accuracy of tracer gas measurements (Eren Ozcan *et al.*, 2006). In third method, CO_2 balance was set by measuring outside and inside CO_2 concentrations. CO_2 production rate per animal was assumed to constant at 475.3 g/h. Fourthly, a number of velocity measurement were performed at different openings, and multiplying with total area, total ventilation rate was calculated. Finally, assuming wind effect to be dominant on thermal buoyancy, an empirical relation was used to calculate ventilation rate as:

Total ventilation rate = Outside wind velocity x Area x 0.4

Results and discussion

Ventilation rate through the stable was calculated by different methods to show the variation between commonly used techniques. Figure 1 shows the results of above mentioned methods to estimate ventilation rate with time. Although similar trends were observed for inlet velocity and CO_2 balance methods; the average values were quite different. For all other methods, neither a linear shift nor a significant similarity between the mean values was observed. On the average a variation of 50,000 m^3/h was observed for an average ventilation rate of 190,000 m^3/h. This corresponds to 25% variation between the methods under investigation.

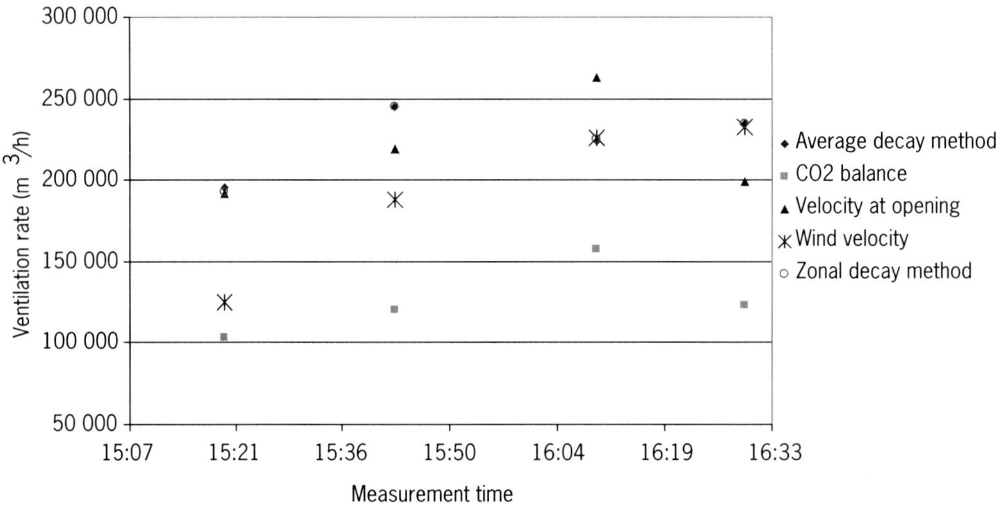

Figure 1. Ventilation rate calculated from different measuring and evaluation methods (20 April 2006).

Conclusions

There exists a huge variation between the results of different methods to calculate ventilation rate. Therefore, a deep understanding of current methods is needed at laboratory scale, controlled experiments to investigate best method to be used as reference technique. This technique will also improve current ammonia emission estimates.

References

Eren Özcan, S., E. Vranken, A. Van Brecht and D. Berckmans, 2006. Implementation of well-mixed zone model for naturally ventilated buildings. HB2006- Healthy Buildings, Creating a healthy indoor environment for people, June 4-8, 2006, Lisboa, Portugal.

Müller H.-J. and B. Möller, 1998. Determination of Air Change Rates in an experimental Cattle Housing using Tracer Gas Methods. ROOMVENT'98 Volume 2, pp. 511-516.

Van Buggenhout, S., A. Van Brecht, S. Eren Ozcan, E. Vranken, W. Van Malcot and D. Berckmans, 2007. Influence of Sampling Positions on Accuracy of Tracer Gas Measurements in Ventilated Spaces. Biosys. Eng. (in press)

Dispersion

Ammonia concentrations downwind of Idaho dairies and heifer facilities

M. Chahine[1] and R.E. Sheffield[2]
[1]Department of Animal and Veterinary Science, University of Idaho, Twin Falls R&E Center, P.O. Box 1827, Twin Falls 83303, USA
[2]Department of Biological and Agricultural Engineering, University of Idaho, Twin Falls R&E Center, P.O. Box 1827, Twin Falls 83303, USA

Introduction

The majority of odour and emission estimates on North American livestock facilities has been conducted in the past decade and has largely been focused on Midwestern swine production facilities. Information about odour concentrations and ammonia (NH_3) emissions from dairies, calf and heifer facilities, outdoor manure storage units, or open feedlots is available, but is incomplete. Confined animal feeding operations (CAFOs) produce large quantities of manure that may cause odours and gas emissions. An ever-increasing rural population is becoming involved in many conflicts with livestock and poultry producers. An accurate estimation of NH_3 and other emissions from CAFOs is needed to predict their potential adverse impacts and to facilitate the selection of the most effective control measures. The objective of this study was to quantify the odour and gas concentrations to understand the natural variability of odour and gas concentrations near various Idaho dairies and heifer feedlots using different manure management systems in order to develop a statewide odour management program.

Material and methods

Odour and gas samples were collected on and adjacent to 38 dairy and 15 heifer/calf raising operations in southern Idaho between August 2003 and April 2004. Dairy facilities were monitored during summer (August and September 2003), fall (October and November 2003) and spring (March and April 2004). The heifer facilities were monitored once during the spring of 2004. Each volunteer farm was assigned to one of the following groups for analysis: Openlot dairies with less than 1000 head (9 facilities), openlot dairies with more than 1000 head (10 facilities), freestall dairies with scraped/vacuum manure handling systems: 1200 to 4000 head (9 facilities), freestall dairies with recycled flush water manure handling systems: 1200 to 4000 head (10 facilities), heifer Facilities: 250 to 5000 head (15 facilities). Openlot dairies are comprised of sloped and compacted soil surfaces with concrete feed bunks and limited outdoor shelter where runoff is collected in an earthen storage basin. Prior to sampling, the most probable odour source on each of the volunteer study farms was identified by the investigators. The odour source on three of the nine openlot dairies less than 1000 head was determined to be the openlot itself, while the odour source on the remaining six farms were determined to be the wastewater storage basin. Similarly, the odour source on the heifer facilities was determined to be the openlot corrals. The odour source on the large openlot (>1000 head), freestall scrape, and freestall flush dairies were determined to be wastewater storage basins. During each sampling/application day, samples were taken at distances of 200 m, 50 m, and adjacent to each odour source, while NH_3 measurements were collected adjacent to the odour source and the odour panellist location. Odour assessments were conducted by a panel of trained assessors. On each sampling day, four panellists conducted odour assessments using the Nasal Ranger® field olfactometer and odour intensity (as compared to n-butanol). Odour intensity was measured using a 6-point scale from 500 to 16,000 ppm of n-butanol in solution of distilled water. During each evaluation, the panellists determined the detection threshold (DT) of the odour present at each site, application event, or manure management practice. DT is defined as the number of volumes of clean air required to make one volume of odorous air non-detectable by each panellist. The DT is equivalent to the concentration of odour, or the amount of odour, that is in the air. Gas concentrations for NH_3 in the field were estimated using Dräger diffusion (Dräger, Lübeck, Germany). Dräger hand pumps were leak tested according to manufacturer's protocols prior to each sampling event. During each odour sampling event, a single NH_3 sample was collected alongside the odour panellists as well as adjacent to the odour source. Data from this study were not found to meet the assumption of normality even when several transformations were tried. Differences between groups were tested for significance ($P < 0.05$) using the non-parametric Kruskal-Wallis test of the NPAR1WAY procedure of SAS (SAS 9.1). Kruskal-Wallis is the equivalent of ANOVA but is used in non-parametric analyses when the data distribution is not normal. A Pearson's correlation analysis was conducted between odour DT, odour intensity, and measurements of NH_3 for samples taken adjacent to, 50 meters, and 200 meters downwind of odour sources on farms.

Results and discussion

Average odour DTs from 38 dairies and 15 heifer lots in southern Idaho were found to be greatest at the odour source on scraped freestall dairies and lowest on open-lot dairies with less than 1000 head capacity. Similarly, during the summer season, NH_3 concentration was greater at the flushed freestall dairies than at the small open-lot dairies, large open-lot dairies and scraped freestall dairies (1.105±0.130 ppm, 0.464±0.132 ppm, 0.306±0.074 ppm, 0.431±0.110 ppm, respectively). Similar results were observed during spring with NH_3 concentration on flushed freestall dairies (0.615±0.103 ppm) being greater than the values observed at the heifer facilities (0.232±0.033 ppm), small open-lot dairies (0.244±0.046 ppm), large open-lot dairies (0.225±0.042), and scraped freestall dairies (0.236±0.040 ppm). This demonstrates that the type of manure handling facility has a greater effect on NH_3 concentrations than the number of animals. Flushed freestall dairies were found to have the highest average concentrations of NH_3. Lower concentrations from openlot dairies are likely due to the lower manure loading rates to wastewater storage basins compared to scrape and freestall dairies. Openlot basins in Idaho typically receive lot runoff, manure and wastewater from the milking centre and may or may not store collected manure from feed alleys. This equates to 45 -80% less manure being deposited in the storage basins compared to freestall facilities. Additionally, higher odour and NH_3 concentrations from freestall facilities may be due to the uncontrolled anaerobic breakdown and subsequent gas and odour production. No differences in NH_3 concentrations were observed during fall. NH_3 measurement was not seen to have a high predictive correlation to odour concentration ($NH_3 = 0.0062\ (DT) - 0.0142\ [R^2 = 0.10]$ where NH_3 = ppm and DT = dilutions to threshold measured by Nasal Ranger. Other studies have shown that NH_3, although an odorant, is not a strong single indicator of odour from livestock operations. Odour acceptability values were found to be lowest adjacent to odour sources. Panellist scores of odour acceptability at a party or event were found to be lower than the odour scores at rural residences, or at the property line of a livestock production farm. These findings show that citizens acknowledge and accept various levels of objectionable odours depending on land use conditions or how they are enjoying their property.

References

Sheffield, R.E., M. Chahine, B. Dye and M. Thompson, 2007. Odor Concentrations Downwind of Idaho Dairies and Heifer Feedlots. Transactions of the ASABE 50:

Ammonia concentrations around three pig farms in the central plateau of spain

M.J. Sanz[1], G. Montalvo[2], C. Monter[1], F. Sanz[1], A. Carpio[1], P. Illescas[2], C. Piñeiro[3] and M. Bigeriego[4]
[1]Fundación CEAM, Spain
[2]Tragsega, S.A., Grupo Tragsa, Spain
[3]PigCHAMP Pro-Europa, S.A., Grupo Proinserga, Spain
[4]Spanish Ministry of Agriculture, Fisheries and Food, Spain

Introduction

A large proportion of the ammonia (NH_3) emitted locally is, in contrast to other pollutants, deposited in the immediate neighbourhood of the source rather than transported over long distances. Quantitative information about the spatial location of emission sources, as well as estimations of the point source emissions is crucial for target-oriented abatement. A suitable spatial resolution and the acquisition of data in different climatic conditions is crucial for a realistic distribution of NH_3 sources and sinks, as well as for later appropriately modelling of atmospheric transport and deposition at local and regional scales. In general it is assumed that ammonia emission deposits for a larger part close to the source of emissions. The importance of the sector in Spain is being increasing in the last years, representing in 2005 the 15% of total EU-25 production (MAPA 2006). The aim of this paper is to describe the variability of NH_3 concentrations in the surroundings of several point sources (poultry farms) at field level which preliminary results were reported by Sanz *et al.* (MAPA, 2006). Few studies are undertaken that show the distribution of NH_3 emissions around point sources in the southern Europe. The aim of this paper is to describe the variability of NH_3 concentrations in the surroundings of several point sources (pig farms) at field level.

Material and methods

Three farms were selected as representative of most types of pig farms in the area: grower finisher pigs, sows with piglets, and one-site pig farm. Climatic conditions were similar in all trials, and concentrations and deposition were estimated over an area of 1 km in each of the farms (sources). Ammonia concentrations were determined by Ferm type passive samplers (Sanz *et al.*, 2006) located at 2 m above ground in four transects (N, S, E, W) for three farms (grower finisher pigs, sows with piglets, and one-site pig farm (586 sows with piglets to 100 kg) in 2005 and a systematic grid in 2005 for the sows with piglets farm. Emission rates were calculated based on the Gaussian dispersion equation and introduced in ISCST (EPA, 1995) model as input parameters, and the output of the model were compared with the experimental concentration measurements at surface level for Cantalejo and Aguilafuente farms. Modelling for other cases it is still ongoing.

Results and discussion

The highest concentrations of ammonia near the buildings for the sows with piglets and the growing finisher pig farms were similar (around 60 $\mu g/m^3$ NH_3), whereas around the pig farm of Valtiendas concentration reached 81 $\mu g/m^3$ NH_3. As could be expected due to the highest number of animals (two buildings of 2700 animals each one) in the latter with the first mentioned farm. The concentration predicted by the model ISCST (EPA, 1995) shows a good correlation ($r > 0.80$) with the passive samplers measurements, on two farms: grower finisher pigs and one site -pigs. Concentration fields decreased in all cases for both years to levels of 2 and 5 μg/m^3 NH_3 within distances of less than 1 km (600 m).

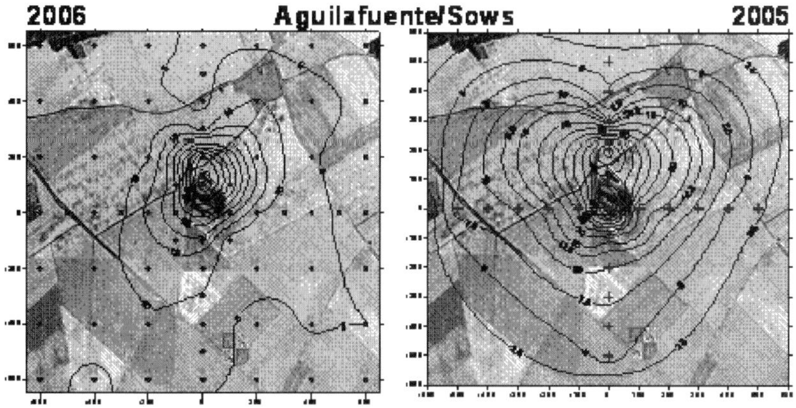

Figure 1. Interpolated air ammonia concentrations (μg/m^3) around a sows with piglets farm measured in summer of 2005 (b) and 2006 (a) for a period of 5 days.

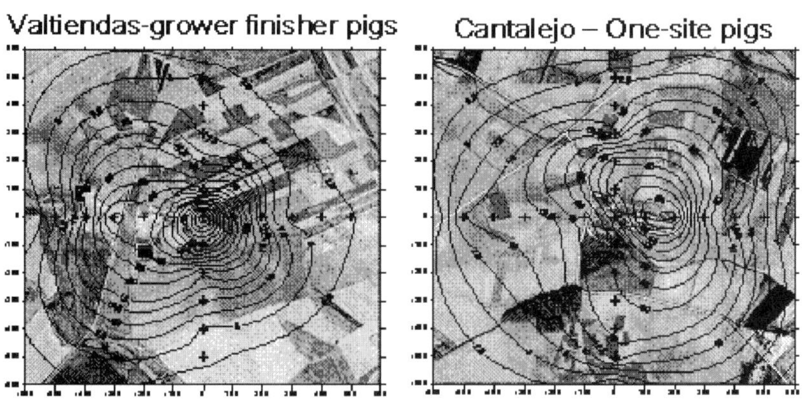

Figure 2. Interpolated air ammonia concentrations (μg/m^3) around a finishing pig farm and one – site pig farm measured in summer of 2005 for a period of 5 days.

References

EPA, 1995. EPA–454/B-95-003a. ISC3 User´s Guide for the Industrial Source Complex (ISC3) Dispersion Models – Volume I & II.

MAPA, 2006. Guía de Mejores Técnicas disponibles del sector porcino. http://www.mapa.es/ganaderia/pags /IPPC/ResumenGuiaMTDsSectorPorcino.pdf.

Sanz, M.J., C. Monter, M. Vázquez, F. Sanz, G. Montalvo, P. Illescas, C. Pineiro and M. Bigeriego, 2006. Ammonia concentrations around 5 farms in the Central Plateau of Spain. 12th Ramiran International Conference. Vol. I., pp. 191-193.

Ammonia concentration around two poultry farms in the central plateau of spain.

M.J. Sanz[1], G. Montalvo[2], C. Monter[1], F. Sanz[1], P. Illescas[2], C. Piñeiro[3] and M. Bigeriego[4]
[1]Fundación CEAM, Spain
[2]Tragsega, S.A., Grupo Tragsa, Spain
[3]PigCHAMP Pro-Europa, S.A., Grupo Proinserga, Spain
[4]Spanish Ministry of Agriculture, Fisheries and Food, Spain

Introduction

A large proportion of the ammonia (NH_3) emitted locally is deposited in the immediate neighbourhood of the source rather than transported over long distances. Quantitative information about the spatial location of emission sources, as well as estimations of the emissions, is crucial for target-oriented abatement. In addition a suitable spatial resolution and the acquisition of data in different climatic conditions is crucial for a realistic distribution of NH_3 sources and sinks. Few studies are undertaken that show the distribution of NH_3 emissions around point sources in the southern Europe, and no study is carried out in Spain. In general ammonia gas deposits close to the source of emission, studies in the surrounding of a poultry farm in UK reported deposition fingerprints of less than 1 km (Skiba *et al.*, 2006). The importance of the sector in Spain is being increasing in the last years, representing in 2005 the 13.4% of laying hen and 12% of broilers of total EU production (MAPA, 2006a, 2006b). The aim of this paper is to describe the variability of NH_3 concentrations in the surroundings of several point sources (poultry farms) at field level of which preliminary results were reported by Sanz *et al.* (2006).

Material and methods

Two farms were selected as representative of most types of poultry farms in the area (Segovia province) and most common types of exploitations for Spain: laying hen and broiler meat farms. A cup anemometer was installed before of beginning of measurements at each of the study sites (wind speed and wind direction with a 10 minutes average period were measured). Climatic conditions were similar in all trials, and concentrations and deposition were estimated over an area of 1 km in each of the farms (sources). Ammonia concentrations were determined by Ferm type passive samplers (Sanz *et al.*, 2006) located at 2 m above ground in four transects (N, S, E, W) for two farms (laying hen and broiler meat) in 2005 and a systematic grid in 2005 for the laying hen farm. Emission rates are being calculated based on the Gaussian dispersion equation and introduced in ISCST (EPA, 1995) as input parameters, and the output of the model will be compared with the experimental concentration measurements. Modelling for other cases it is still ongoing.

Results and discussion

Concentration fields decreased in the laying hen farm to levels of 3 - 5 $\mu g/m^3$ NH_3 (background levels for the area) within distances of less than 600 m, whereas in the case of broiler meat farm at 600 m concentrations were twice that. The measurements performed in 2005 in both farms showed that the concentrations around the buildings of the laying hen farm were much higher (around 50 $\mu g/m^3$ NH_3, even 70 $\mu g/m^3$ NH_3 between the two main buildings near the outlet of the ventilation system), whereas in the broiler farm concentrations near the buildings were around 25 μg/m^3 NH_3. However, concentrations attenuated faster in the laying hen farm with increasing distance than for the broiler meat farm. Same range of concentrations (25 μg/m^3 NH_3) was found within few meters (1000 m) downwind a poultry farm by Skiba *et al.* (2006). In summary, concentration in the fields decreased in the laying hen farm to levels of 3-5 μg/m^3 NH_3 (background levels for the area) within distances of less than 600 m, whereas in the case of broiler meat farm at 600 m distance downwind were twice higher in the case of the laying hen farm. Although near the farm buildings the concentrations were twice higher in the laying hen farms. That suggests that dispersion conditions were different.

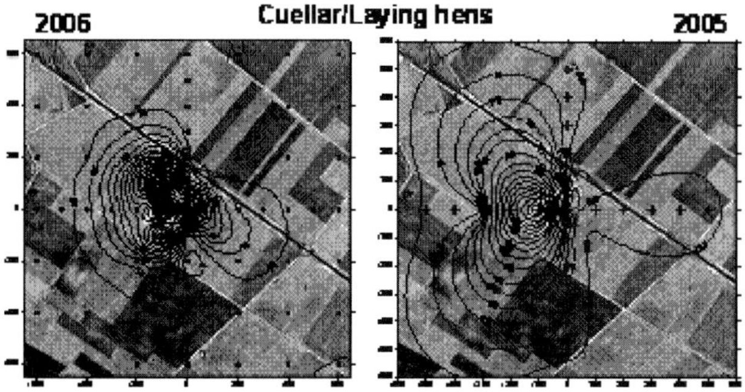

Figure 1. Interpolated air ammonia concentrations (μg/m^3) around a laying hen farm (75,000 animals) measured in summer of 2005 (b) and 2006 (a) for a period of 5 days.

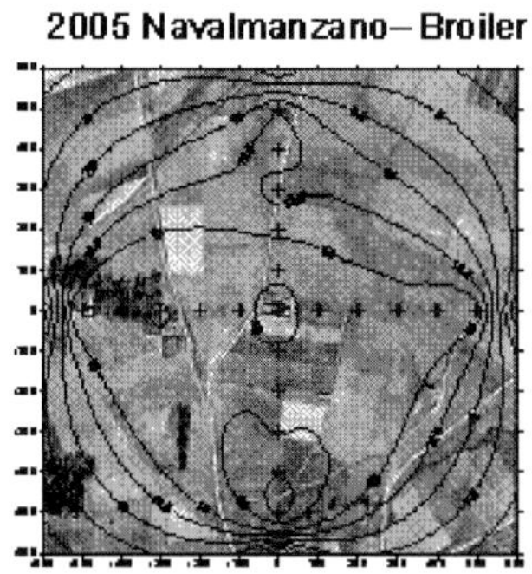

Figure 2. Interpolated air ammonia concentrations (μg/m^3) around a broiler meat farm (13,000 animals) measured in summer of 2005 for a period of 5 days.

References

MAPA, 2006a. Guía de Mejores Técnicas Disponibles del sector de la avicultura de puesta. http://www.mapa.es/ganaderia/pags/IPPC/ResumenGuiaMTDsSectorAviculturaPuesta.pdf.

MAPA, 2006b. Guía de Mejores Técnicas disponibles del sector de la avicultura de carne. http://www.mapa.es/ganaderia/pags/IPPC/ResumenGuiaMTDsSectorAviculturaCarne.pdf.

Sanz, M.J., C. Monter, M. Vázquez, F. Sanz, G. Montalvo, P. Illescas, C. Pineiro and M. Bigeriego, 2006. Ammonia concentrations around 5 farms in the Central Plateau of Spain. 12th Ramiran International Conference. Vol. I. pp. 191-193.

Skiba, U., J. Dick, R. Storeton-West, S. Lopez-Fernandez, C. Woods, S, Tang and N. Vandijk, 2006. The relationship between NH_3 emissions from a poultry farm and soil NO and N_2O fluxes from a downwind forest. Biogeosciences 3: 375-382.

Ammonia: air pollution and dispersion of agricultural and urban sources

Y. Buchleither, R. Lumpp, Ch. Lutz-Holzhauer and Z. Peranic
Karlsruhe GER, State Institute for Environment, Measurements and Nature Conservation Baden-Württemberg, LUBW, Griesbachstr. 1, 76185 Karlsruhe, Germany

Introduction

To protect sensible plants and sensible ecosystems from ammonia air pollution, a national ordinance (TA Luft, 2002) determines the minimal distance to ammonia emission sources such as livestock houses. Including the background offset the total ammonia air pollution is restricted to an annual average of 10 μg/m^3. In Addition ammonia emissions from non agricultural sources e.g. motor traffic are important (Frahm, 2006). Therefore, due to regional differences in climate and settlement as well as industrial or agricultural structures, specific ammonia background concentration has to be determined for different parts of Baden-Württemberg. The potential of ammonia air pollution has to be identified and quantified.

Material and methods

The ammonia air pollution was measured in different regions. In each region the ammonia concentrations were measured for a one year period with passive samplers from Radiello. Every two weeks the samplers were collected and analysed. The focus was at first on known, unknown or diffuse emissions in agricultural surroundings. The projects were started in 2003 (Sörgel, 2005) and finished in 2006 with measurements in urban surroundings. The results from the agricultural regions were validated with data from the Forest Research Institute Baden-Württemberg (Wilpert, 2006). For three sites in the vicinity of pig livestock houses the ammonia air pollution was additionally calculated with the specific dispersion modelling (AUSTAL2000). To have exact meteorological data for this dispersion model, for two sites the meteorology was measured on site also.

Results

Agricultural sources: Figure 1 shows the ammonia air pollution near the agricultural sources. In agricultural regions the ammonia concentration had a seasonal variation dependent on manure output. Therefore the average of the ammonia concentration was separated in 5 distance categories. The highest NH_3 concentration (>70 μg/m^3) was found next to the source. At a distance of 250 m the ammonia concentration was decreased to less then 10 μg/m^3.
Dispersion modelling: The exactness of the dispersion modelling for the ammonia concentration depended on the careful choice of the input variables especially for emission factors, meteorological characteristics, source configuration and topography.
Urban sources: Figure 2 shows the ammonia concentration for different sites of urban surroundings. During the whole measurement period the concentrations were lower and more homogeneous than in the agricultural regions. At traffic hot spots in cities ammonia concentration were 20 μg/m^3 and more. Near a highway in a suburban area the ammonia concentration was lower due to undisturbed air circulation flow. Another urban source for ammonia emission was the sewage plant with an average concentration of 3.4 μg/m^3.

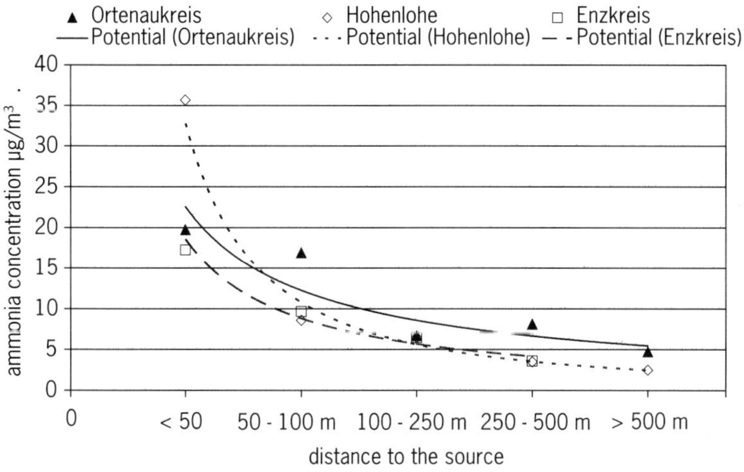

Figure 1. Ammonia air pollution in relation to the distance to the livestock house.

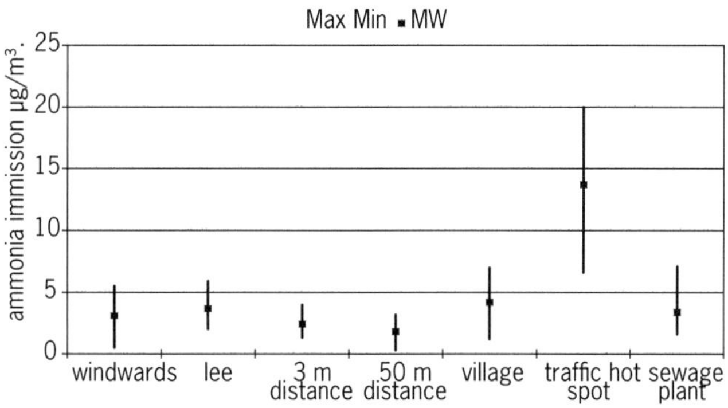

Figure 2. Ammonia air pollution in urban surroundings.

Conclusion

Differences in ammonia concentrations are such, that it is necessary to determine individual ammonia background concentrations for different regions. Two different categories of regions are identified. Ongoing long-term investigations will cover other categories of regions and will result in defined regional references for background air pollutions.

References

Frahm J.P., 2006. Der Einfluss von Ammoniak auf Stickstoff liebende Flechten in verkehrsbelasteten Gebieten. Immissionsschutz – Luftreinhaltung und Bauleitplanung. 4: 164-167.

Sörgel, C., 2005. Immissionen von Ammoniak im Umfeld zweier landwirtschaftlicher Betriebe. Karlsruher Schriften zur Geographie und Geoökologie Heft 18, 143 pp.

TA Luft, 2002. Technical Instructions on Air Quality Control (Germany). Erste Allgemeine Verwaltungsvorschrift zum BImSchG, 24.07.2002.

Wilpert, V.K., 2006. Stickstoff-Belastung von Waldökosystemen im Nahbereich von Massentierhaltungen. Forest Research Institute Baden-Württemberg Freiburg, Germany.

Dispersion of ammonia emissions in the surroundings of a big piggery

K.-H. Krause[1], T. Hinz[1], S. Linke[1], H.-J. Müller[2] and M. Mußlick[3]
[1]Federal Agricultural Research Centre (FAL), Bundesallee 50, D-38116 Braunschweig, Germany
[2]Leibniz-Institute for Agricultural Engineering e.V. (ATB), Max-Eyth-Allee 100, D-14459 Potsdam, Germany
[3]Thuringian Ministry of Agricultural, Environment Protection and Environment, Beethovenstraße 3, D-99096 Erfurt, Germany

Initial situation

The procedure of simulating the expansion of substances and odour is always the same. In addition to the determination of the mean value of the concentration of a substance the odour simulation demands a so-called fluctuation model to connect the mean odour value with the frequency of odour recognition. The fictive odour mean concentration cannot be measured, so it suggests itself to use ammonia as tracer gas in the surrounding of animal houses. The performance of a dispersion model can be proved by comparison of ammonia measurements with predicted emission values in refer to the mean concentration. The mean concentration is the basis for all dispersion models. To limit the costs of experiments ammonia was detected during six weeks by passive samplers at four locations, see Figure 1.

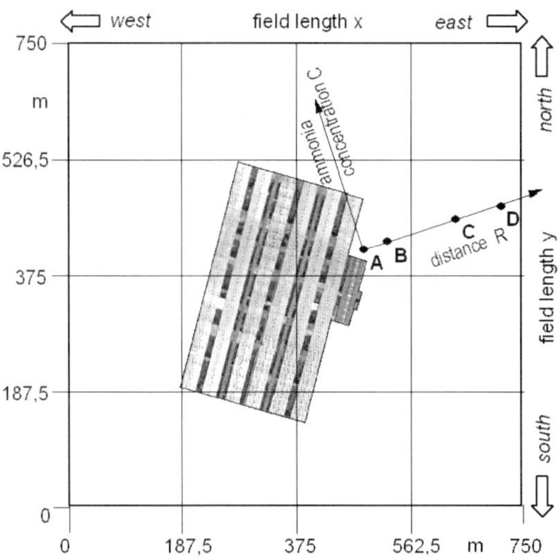

Figure 1. Location of the measuring points A, B, C and D of ammonia. 603 point sources characterise the emission situation of the piggery of investigation In order to reduce the waste of time for the input of such a great number of point sources they are substituted by one single area source or one single volume source. The area source includes 214 x 384 m = 82,176 m^2, the volume source by an additional vertical expansion of 3 m at a height of 6 m 82,176 m^2 x 3 m = 246,528 m^3.

The sampler had been constructed according to M. Ferm from the Swedisch Environmental Research Institute in Göteburg. This type of sampler is short, broad and therefore sensitive for relatively low concentration in ambient air of the environment. Thin porous membrane filter shall avoid turbulent diffusion inside the sampler. A filter covered with citric acid takes up the ammonia. The samplers were mounted on pales in a height of approximately 2.5 m above ground. After sampling the filter will be extracted and analysed. The result is an averaged concentration related to the sampling time.

Table 1. Data of ammonia measurements in 2005 by passive samplers.

Point (x, y) in m	Distance R in m	ammonia concentration in μg/m³ at week 38	39	40	41	42	43
A (486.4, 419.3)	0	349.1	254.6	373.9	141.9	108.7	202.3
B (527.2, 431.3)	42.5	274.6	201.4	233.5	123.4	83.0	144.3
C (639.8, 462.5)	159.4	163.0	178.7	98.9	113.0	58.0	69.2
D (709.3, 484.0)	232	39.3	109.2	56.1	76.3	37.5	43.6

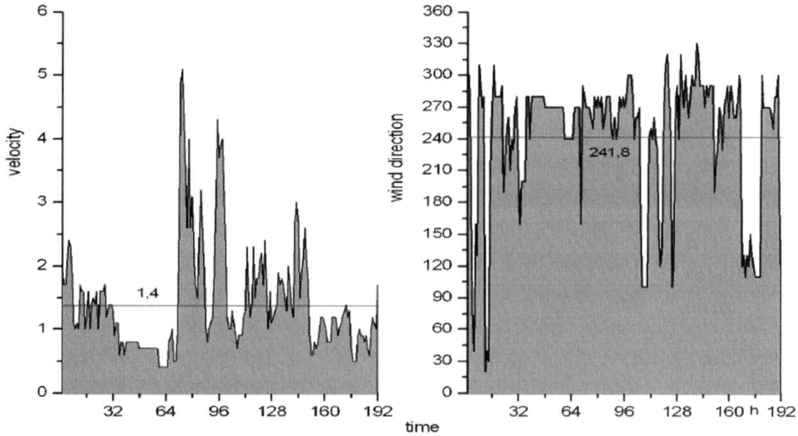

Figure 2. Time series of wind velocity and direction during ammonia sampling in the 38th week in 2005.

Application of the Lagrangian particle model AUSTAL2000

To calculate ammonia emission a Lagrangian particle model has to be applied in Germany according to the Clean Air Directive 2002. The promoted programme AUSTAL2000 involves several features to simplify the generation of emission. Because of the heterogeneous distribution of point sources in big animal plants it is not recommendable to make use of such a substitution. Moreover the experiments leads to the following results: the emission concentration is overestimated considerably at the far distance (Point D). The greatest deviations are produced by the approach of source point formulation, followed by the modelling of the emission by a

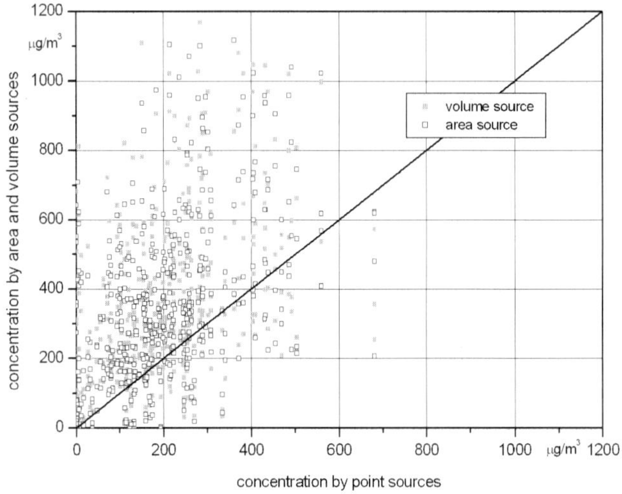

Figure 3. Comparison of the emission concentration of dispersion models with volume and area sources versus the superposition of the emission concentration of 603 point sources. Correspondence is expressed by location on the straight line. The results of the simulation with an area and q volume source differ from the point sources solution.

volume source and an area source. But at the moment of investigations one cannot rule out the possibility of great errors by an imperfect technology of measurement. Caution is advisable to calibrate an uncertain simulation model by uncertain measured values.

Open space workshops

Workshop animal feeding

Goal = to establish the most effective way of implementing animal feeding options to reduce environmental pollution – notably NH_3 - from animal husbandry

Topics presented in feeding and management:
- Altering dietary composition in terms of protein content, fermentable non-starch polysaccharides, acids or salts, and using additives show many possibilities to reduce ammonia emission.
- A lot of experiments are currently carried out with promising results.
- Reduction of ammonia emission in (dairy) cattle is more complicated but also feasible.

Topics presented in nutrition and management:
- Simultaneous measurement of several gasses recommended.
- Interesting results from USA.
- More insight in additive and/or interaction effects of dietary factors is required, also in relation to housing of animals.
- Using several additives to diets or faeces looks promising.
- Research from USA showed that using DDGS gave more ammonia and H2S emission.

Topics presented in free space: policy:
- Nutritional possibilities are not at all implemented in legislation.
- The feed industry is invited to take their responsibility for sustainable production.
- Legislators + feed manufacturers/farmers can cooperate to make progress in the whole chain there are good examples (e.g. environmental certification = Milieukeur varkensvlees).
- Communicate with consumers about added value they get.

Is there a gap between science and practice concerning the application of 'optimal' animal feeding?
- At the moment nutritional possibilities are not at all implemented in legislation.
- How to control and maintain?
  - Certification of feed.
  - Analyses of manure.
- Feeding strategy should also be investigated (e.g. feeding exactly according to production state of the animal).
- Dairy sector: a lot of home-grown feed; difficult to get a low protein diet on peat soils.
- Costs and benefits should be calculated.
- Producers are not well informed about benefits, or they do not believe the results of research.
- What is the responsibility of the feed industry? What responsibility should they take?
- When composing a diet not only the cost aspect should be considered, but also environmental aspects.
- A reduction in emissions is also improving the environment of the animal and of the worker.
- Several effective additives are not on the list of allowed compounds for feed.

- Feed formulators don't want to run any risk.
- Feed formulators don't have the benefits directly by themselves (e.g. lower CP).
- Legislators + feed manufacturers/farmers should work together.
- Farmer should not solely pay the costs for higher feed costs: also society should contribute (integral approach).
- Communicate with consumers about added value they get.

Do we have enough knowledge?
- No, because of lack of knowledge on:
  - Interaction between different feeding measures (protein, non-starch polysaccharides, (in)organic acids and salts).
  - Effect of different additives.
  - Environmental impact of production of additives and animal feed.
  - Effect of feeding measures in relation to housing.
- Can you develop a diet that will not give any waste?
  - Yes!!
  - Manure is not a waste!!

What instruments can be allocated to fully implement 'optimal' animal feeding?
- There should be incentives to implement feeding measures.
  - Permits, certification, include in the RAV.
- Random analysing of feedstuffs at feed companies by a controlling organisation (situation in Denmark).
- What other high protein sources are available?
  - Animal proteins.
  - Amino acids.
  - Ethanol by-products and rapeseed meal (however, poor quality of protein for monogastric animals; not for ruminants; also a high P-content).
- Question: how to implement in legislation?

Workshop ammonia abatement: heaven and hell

What were we aiming to achieve:
- Heaven
  - Fairly obvious.
  - The best outcome for ammonia abatement.
  - Try to identify the policies needed to achieve that outcome.
  - Identifying policies should indicate the most appropriate abatement techniques.
- Hell
  - By identifying the worst outcomes.
  - And the policies that would achieve them.
  - Should help in formulating policies to achieve desired goals.
  - by reducing the risk of the law of unintended consequences.
  - Also, hopefully, identify techniques that are not appropriate.

Heaven

1. Use the market:
- Consumers pay for environmentally-friendly produce.
- Make reduced-emission produce profitable.
- Combat homogenous offers in supermarket > traceability necessary and warrantable.
- Education of consumer.
- Options for farmers.
- Options to be provided by scientists.

2. Engage consumer:
- Increase the consumer awareness to consume less meat: improved health and environmental benefits.
- Reduce human protein intake to only what is needed.
- Make meat/milk without animals.

3. Welfare:
- Link abatement with animal welfare.

4. Technology:
- Look at chain-level solutions.
- Solutions to reduce ammonia should not result in emission shift and exchange.
- Feeding strategies.

5. Other:
- Improve spatial balance of animal production.
- Holistic solution.
- Combine abatement with energy production.

Hell

1. Excess regulation:
- Failing, over-complicated regulatory schemes.
- Big brother environmental audit.
- Consumer awareness will not change.
- Further growth of intensive, industrial size animal production.
- Loss of farmland due to driving farmers out of business.

2. Other:
- Many animals already in hell.
- Unchanged animal welfare.
- Segmented and Conflicting abatement options (shift and increase other emissions).

How do we use market to enable premium prices?
- Without government directive.
- Without driving production elsewhere.
- Does the market need incentives or taxes?
- Carbon tax to decouple crop and livestock production.
- Barrier.
- Not a level playing field; supermarkets only sell local food as a token.

Approach:
- Society needs to decide what it wants.
- Cheap protein or reduced emissions.
- Remove disincentives.
- Set standards – decide how to get there.

Policy maker's dilemma:
- There seems to be a consensus that the most effective approach will be to set goals and leave producers to meet them in the most cost-effective manner.
- However, producers currently not able to pass on extra costs to consumer.
- Policy
  - Market mechanisms.
  - To make work them: parameters need to be set.
  - Government should be the last resort.
  - Treatment and production decoupled.
  - Change the way the market works.
  - Approaches.
  - Performance re-evaluation of best available practices (IPPC techniques).
  - Work with nature of biological systems.
  - Nature paper: Develop a sustainable environmentally friendly technology that is a commodity.
  - NH_3 as fuel.

Conclusions - 1:
- Seems to be a consensus to use market forces rather than regulation.
- And abatement should not be at the expense of animal welfare.
- A range of options for individual enterprises.

Conclusions - 2:
- Doubt over how consumers can be persuaded to pay premium price.
- Education?
- Who pays for the investment > better milk education; how to arrange cash-flow.
- Solution: ring fence taxes for on-farm investment.

Workshop measuring and monitoring

- Errors in measured ventilation rate cause uncertainty in emission numbers.
- Applicable to all emissions: $NH_{3,}$PM, GHG plus important for energy use ...
- A reference technique needed to measure ventilation rate through naturally ventilated structures.
- Objective: come up with a technique with 20% inaccuracy for naturally ventilated structures.
- An international standard procedure/protocol is needed for:
- Sampling;
- data analysis;
- software;
- presentation of uncertainties.
- We submit a proposal for COST-Action incl. researchers, associations & companies.

Workshop inventories and modelling: the true story ?

Goal = to assess the potential of model (process models, farm models, national models) to represent the true world, and to provide guidance for improvement.

Scaling from micro (plot) to macro (national):
- Micro models can be used to improve macro models.
- Do sensitivity analysis on micro-models and only include the most important variables.

How accurate are the various models and how well do they represent the real world?
- Reliability of models depends on the quantity and quality of:
  - the data to parameterise the model;
  - the input data available.
- Accuracy poorly defined.

Model verification:
- Emission inventories difficult to verify above the plot/housing scale.
- Emission + dispersion models;
  - Continue work to understand 'Ammonia gap'

Which steps should be taken for improvement?
- Uncertainty analysis of models is very important.
- Uncertainty of models should also be communicated.
- Present inventories could be improved using uncertainty analyses.
- Problem that policy makers only want one value.
- How do we change this situation?

In conclusion:
- Models are powerful for policymakers.
- Good models/inventories mean.
  - model design for purpose:
  - good parameterisation:
  - adequate and reliable input data.
- Better dialogue between policymakers and scientist about uncertainty.

Workshop integrated measures: do we need them?

Goal = To assess the advantages and drawbacks of integrated N measures, versus more stringent NH_3 measures, and to advise the various stakeholders.

Questions to be addressed:
- Which measures, policies and instruments are present and/or needed to successfully abate losses of reactive nitrogen?
- What are the pro's and con's of integrated measures versus more stringent NH_3 measures?
- Is 'pollution swapping' a crass cliché, with no added value and inhibiting making hard decisions on reducing N pollution?
- Are there integrated measures/husbandry systems that are sustainable from the farmers' point of view?

Human food is inexpensive and high quality, but it does not take full account of environment:
- Market based solutions insufficient.
- A mix of tax, economic incentives and regulation is needed.
- Are consumers willing to pay?
- Can we learn from marketing of 'organic food' and 'fair trade'?

Integration is complex:
- Continuously changing priorities (i.e. need to change status quo):
  - Is ammonia loss more important than nitrate or N2O or phosphorous or microbial pathogens?
  - Farmer's role in preserving landscape.
- How to convey it to the farmer?

Do we have enough innovation?
- We need more low cost innovation.
- Involve farmers more; they are open for cooperation and dialog, but not so much to direct regulations.
- Can we learn from industry (e.g. refinery)?
- Good examples exist in some countries.

Solutions? *Technical*
- No '*one size fits all*' option.
- Optimised N-balance a good start supplemented further with technical measures targeting specific local issues.
- Good examples exist in some countries.

Solutions? *Regulatory and policy*
- Europe wide regulations necessary but not enough and should be a catalog which would be adjusted/tailored to local circumstances – more freedom and responsibility for local regulators?
- Education/outreach of the public/farmer.
- Emission/nitrogen trading?

Panel discussions

Panel discussion (Monday)

1. Ammonia reduction in Europe has been about 20% in the last years, which was mainly because of the reduction in the number of animals, spreading manure evenly, etc.; what will be other strategies in the future (manure spreading)?

M. Sponar:
Currently, reducing the number of animals as well as IPPC directive (with regard to large farms) – Best Available Techniques are part of the picture:
- Manure storage.
- Manure spreading (is reviewed).
- Implementation of Nitrogen Directive.
- Overfeeding: plants or animals (situation).
- Economic instrument at European level: using subsidise agricultural policy (make it economical/profitable on farm basis), cross compliance.
- Control is very complex.
- More focus on Nitrogen at farm level.
- Other sectors (such as transport) are pushing pressure because they have done their work.

2. Need emission control measures: NH_3, NO_2 – so that the farmer can choose.

J. Sliggers:
- Need something completely different.
- Make meat from animals is also old-fashioned.
- Reduce animal numbers?
- Too much meat in our diet?
- Future: make nutrients in a factory?
- Challenge: can we make proteins without animals?

A. Jongebreur:
- More complex: it is possible to produce meat in large units, in an industrial way, but the cost will rise in the future.
- Demand for meat is estimated to rise in the future (Asian countries).
- BAT: way of looking has been carried out because a lot of people had doubt about availability of technology to reduce emission.
- BAT must be done, otherwise we have no good idea what is achievable on farm level: BAT and monitoring are important.

S. Shaver:
- Proved compliance, certain technology important, validated baseline.

3. How comes, no one is looking on BAT? How can we make people use BAT?

S. Shaver:
- Operating permits necessary to trigger certain technology techniques with lowest achievable emission rate. In the US: only for particulate matter; other sources such as NH_3, odour are not controlled.

4. How can beef cattle feedlots become a part?

S. Shaver:
- Has to be based on non-fugitive emission.
- Cattle feedlot: will not be able to capture / control emissions easily.
- Feed lots and open are not controlled in the regulations.
- In some states are regulations, but no federal standard implemented.

5. On the emission maps some countries are decreasing (in emission). How sure are we that they are decreasing? How about uncertainties in measuring technique? Are there reliable techniques for prediction of emission from natural ventilated buildings?

K. van der Hoek:
- Inventories: calculation by multiplying number of animals and emission factor (except for horses).
- Emission factors (measuring protocol): e.g. pigs: 2 periods of 4 month each to establish an emission factor; included influence of season, first-middle-last part of growing period), certain ventilation system and capacity, protein level within feed should be in certain levels.
- Holds for mechanically ventilated animal houses (mainly in pig and poultry); dairy cattle: assumption that mechanical = natural ventilated houses.

M. Sponar:
- We have to live with uncertainties and ranges (emission factor could also change with climatic conditions).
- Need for improving emission factors.
- Different technologies for reducing emissions (discussed with experts).
- Safety net: measuring nitrogen deposition in different ecosystems/places.
- RAINS model, CAPRI, etc.: full nitrogen cycle (choose full nitrogen model in inventories).
- Uncertainties: consequences have to be discussed with EU policy.

6. In US there are certain regulatory advantages with reducing sources of emission. In European Framework, any experience with source reduction techniques?

M. Sponar:
- IPPC directive: BAT at farm levels (at large farms).
- Apart from that no direct regulation for the farms, member states are free in what they are doing with the emission ceilings; exact regulation is free.

7. Reduction of ammonia is going slowly. Political aim to produce Bioethanol, Biomass. How important is this ?

K. van der Hoek:

- One reason to have cattle, pigs and poultry: to get rid of wastes (e.g. from industry).
- Agricultural producing industry can not completely rely on these feed stuffs.
- Nitrogen contents of the by-products.

An abridged summary of panellist statements (Wednesday)

Dennis D. Schulte
University of Nebraska, USA

Statements of the participating panellists

Mark de Bode
Ministry of Agriculture, Nature and Food Quality, The Hague, The Netherlands

Mr. de Bode indicated that environmental solutions based on animal feeding and optimised nitrogen balances were of great interest to him. He emphasised that keeping cost-effectiveness in mind for technology development, and expression of uncertainty in presentation of scientific data was needed from the scientific community.

Petra Loeff
Ministry of Environment, The Hague, The Netherlands

Petra was especially drawn to abatement measures described in this conference, and was very glad to learn that the scientific community was working on such technologies and was sharing this knowledge with policy-makers.

Gerbrand Van 't Klooster
COPA

Mr. Van 't Klooster indicated that he and other farmers are the ones who have to pay the cost of ammonia abatement technology, and that it is very hard to pass on these costs on to consumers. He reminded the scientists and policy-makers that farmers must compete in global market too! He said that farmers prefer to have integrated options to choose from and like to see data on multiple impacts for use in deciding on which ammonia abatement technologies to use. He indicated that farmers want to take their fair share of responsibility and that they value and need more education. For example, he pointed out that feed companies are already well-received with their help on feeding strategies for ammonia abatement.

Michel Sponar
EU, DG Environment, Brussels, Belgium

Michael started by following up on Mr. Van 't Klooster's remarks and stated that animal feeding industry dialogue is indeed beginning, but that it needs to be EU- wide. He indicated that the recent IPPC directive was an example of a challenge to transform nitrogen balance strategies into action. Mr. Sponar disagreed with the thought that uncertainty is not accounted by policy-makers, but he acknowledged the need expert input to help with interpretation of uncertainty in scientific data. He closed by saying that there is a need of more integration of market mechanisms with regulations to achieve ammonia reduction goals.

Ulrich Dämmgen
Federal Agriculture Research Centre, Braunschweig, Germany

Dr. Dämmgen challenged the scientists in the audience by saying that there are a lot of data but only few can be used. He indicated that scientists need to perfect their way of going from trials to experiments and to improve the art of making measurements (i.e. metrology). Furthermore, Ulrich pointed out that, while inventories such as on ammonia emissions are communicated to policy makers, there exists a significant language difference which remains to be overcome. He used the debate on 'uncertainty' as an example of this language difference.

Sally Shaver
Environmental Protection Agency, Washington DC, USA

Sally Shaver began her comments by observing that farmers in U.S. are actually much like those in EU in that they do not like moving regulatory targets, and that they need more low-cost solutions for abatement technology. She stated that 'one-size fit-all types of regulations' and technical solutions that disregard practical and economical constraints are of limited use to USA and EU farmers. Mrs. Shaver also said that uncertainty in scientific data is hard to use, especially in direct regulatory enforcement actions, but that it should be accounted for in developing the regulations.

Questions and comments from the audience

Question: Is it really possible to have flexibility in solutions for farmers to comply with regulations, or is over-regulation a necessary evil?
Petra Loeff: Policy-makers are starting to introduce flexibility in technology choices to meet regulations, but compliance is hard to measure that way. Petra mentioned as an example the 3-year experimental integrated approach in the north of the NL for land application of manure with regard to ammonia emission.
Sally Shaver: Regulators do need to know if targets are being met, but the 'non prescriptive' approach is working in the USA where farmers can choose from a list of options to satisfy regulations.

Question: Are policy-makers investing too much in regulations versus investment in technology?
Michel Sponar: We need a combination of investments.
Petra Loeff: I think investments in regulations should definitely not discourage investments in innovation. Follow-up Comment from Audience: Regulations alone will not work due to market structure, there will always be some people who cheat, thus regulations will always be needed.

Statement from Dr. Smith (ADAS): An example of market structure versus regulations can be found in the lack of results from promotion of the value of manure, which has been going on for over 20 years. Farmers don't always accept our messages, thus we need to try harder to understand their reasoning. Another example is positive consumer response to environmental and animal welfare protection techniques. People will purchase those products, for example from the Fair Trade labelling program, indicating that educating the public does work.

Conference statement

Conference statement

On behalf of the organising committee, the scientific committee, and all conference participants, we wish to thank the sponsors of this conference. Special thanks are expressed to Christian-Albrechts University at Kiel in Germany and to Wageningen University and Research Centre in the Netherlands. Also, we sincerely thank EurAgEng, the CIGR, the Dutch Ministry of Agriculture, and the Dutch Association of Agricultural Engineers for their strong support of the conference.

We welcome the continuing participation of the international scientists and engineers in addressing technical research questions and we are looking forward to working with them and policy-makers through future mechanisms such as the UNECE. Solutions to problems caused by livestock ammonia emissions including relationships to particulate matter formation, loss of ecosystem diversity, and so forth, will require the combined efforts of scientists, engineers, and policy-makers on all continents.

Important conclusions were made at this conference. For example, concerning measurement and monitoring, an important conclusion was that there is a great need for standards. This will reduce uncertainties in measurements to acceptable and equal levels for all relevant emission sources, particularly area sources and naturally ventilated buildings. Financial support is needed for this to occur (Workshop measuring and monitoring).

On abatement of ammonia emissions, there is a long history of research and development. Important and promising options discussed in this conference were:
- animal feeding;
- technology, preferably low cost solutions;
- integrated measures.

When considering options for abatement of ammonia emissions, conference participants indicated that animal welfare must be taken into account as well. Also, this conference reminded us that we must be sure that pollution is not just shifted or swapped when considering abatement options (Workshops integrated measures, animal feeding and ammonia abatement)

A unique outcome of the conference is that, those who measure ammonia emissions need to be in contact with those groups that perform dispersion and deposition research. This is essential to close the 'ammonia gap'. As indicated by several speakers, ammonia is a part of the nitrogen cycle. Therefore an integrated approach is also needed with respect to political priorities. For example, farmers need to know what is expected of them. An important message for policy makers is also that over-regulation must be avoided. Furthermore, agricultural entrepreneurs need opportunities for local involvement with officials and flexibility in applying solutions (Workshop inventories and modelling and integrated measures).

Communication with and between producers and consumers on a practical level is also needed to increase the awareness of costs associated with preventing environmental pollution by ammonia. 'Fair trade' or 'organic products' are examples of how this communication can be successfully used to inform consumers (Workshop integrated measures).

In summary, the First International Ammonia Conference in Agriculture called for greater integration of research and policy-making with regard to ammonia and other volatile emissions and particulate matter. Stronger integration will improve the overall effectiveness of measurement, monitoring, abatement, inventory, legislative, and policy efforts of the future for producers and consumers.

Authors index